Economics of Forest Resources

Economics of Forest Resources

Gregory S. Amacher
Markku Ollikainen
Erkki Koskela

The MIT Press
Cambridge, Massachusetts
London, England

For information about special quantity discounts, please email special_sales@mitpress.mit.edu

This book was set in Palatino on 3B2 by Asco Typesetters, Hong Kong.
Printed and bound in the United States of America.

Library of Congress Cataloging-in-Publication Data

Amacher, Gregory S., 1962–
Economics of forest resources / Gregory S. Amacher, Markku Ollikainen, and Erkki Koskela.
    p.   cm.
Includes bibliographical references and index.
ISBN 978-0-262-01248-5 (hardcover : alk. paper) 1. Forests and forestry—Economic aspects.
I. Ollikainen, Markku, 1952– II. Koskela, Erkki, 1946– III. Title.
SD393.A46   2009
338.1′349—dc22   2008032762

10 9 8 7 6 5 4 3 2 1

To Our Families

Contents

Preface

Three things led us to undertake the level of effort needed to write this book. First, we hope there is pent-up demand for the book and that it will appeal to all resource economists. While many of us pored over Johansson and Löfgren's 1985 text (*The Economics of Forestry and Natural Resources*) to learn the theory of forest economics, that book is now more than 20 years old. The fact that it may still be consulted on a number of issues is a testament to its continuing value. However, given the growth of the field since that time, and the fact that their book has been long out of print, instructors currently do not have a book that can be used to teach forest economics to graduate students of any level. For the researcher, there is nothing in existence that remotely resembles a serious handbook, and even surveys of important new topics are absent from the published literature. Second, the field of forest economics has expanded rapidly during the past 20 years. None of this recent history has been chronicled or critically evaluated and compared. Any serious specialist requires such an evaluation before tackling the myriad problems falling within the field of forest resources. Finally, forestry problems themselves, particularly those involving deforestation, climate, and conservation of biodiversity, are increasingly dominating the attention of economists and policy makers. During the past 15 years, forestry problems have become an increasing part of what most people consider to be the economics of natural resources. Unfortunately, books dealing with the economics of natural resources aimed at the graduate level focus on fisheries or water resources in their renewable resource sections and relegate forestry to an obligatory chapter. Not surprisingly, only the most basic (and often outdated) material is ever in the hands of the student or reader.

The pages of this book provide a comprehensive and technical survey of forest resource economics, concentrating on developments within the past 25 years. The focus is especially directed toward understanding the economics behind forest policy problems and policy design, since the application of policy instruments is often a controversial area in which statements are sometimes not based upon rigorous study and reflection. However, the book is comprehensive enough to touch

on virtually all areas that form the theoretical foundations of forest resource economics. This book introduces the reader to the rigor needed to think through the
consequences of policy instruments targeting forests or the consequences of landowner actions and decisions. To provide context for the latest innovations, we
also discuss classical theories upon which modern forest economics is based. The
book is therefore a combination of a text for learning and for pondering. We hope
to stimulate new research, empirical and theoretical, with our presentation of
topics. We also hope to provide anyone with an interest in economics a gateway
to the rich and still wide-open field of forest resources.

Organization of the Book

This book is organized around a presentation of historical and classic core models that every serious student or researcher of forest economics must know. These models, along with numerous recent extensions of them, are discussed in part I (Basic Models). Part II (Policy Problems) is devoted to choice of policy instruments, deforestation, conservation of biodiversity, and age class-based forest modeling that builds upon the core models. Part III (Advanced Topics) surveys a rapidly expanding area of forestry problems, including uncertainty and stochasticity, and dynamic approaches to modeling forestry problems.

Most of the material in parts I and II of the text can be followed by someone with an understanding of multivariate calculus that beginning graduate students and advanced undergraduates in economics, agricultural and environmental economics, and forest and natural resource economics should have. Chapters 9–12 are the most advanced, but the material is generated from first principles to bring readers to the level needed to follow the main results. In addition, a mathematics review at the end of the book presents several concepts that will help in following proofs and interpretations in the text. A graduate course in microeconomics is not a necessary precursor to reading this book, although such training will obviously aid in understanding the more advanced concepts.

Because space is always finite when it comes to books, we have omitted some topics from consideration. We do not focus specifically on operations research (linear, goal, and some other programming-intensive) approaches to forestry problems, although many of these have been based on modifications of the models and policy problems we do discuss. Further, we do discuss in detail dynamic programming and both dynamic and static Lagrange and Kuhn-Tucker problems that are obviously related to the operations research literature. Comprehensive books devoted to operations research and numerical methods in forest economics on both intertemporal and spatial scales are Buongiorno, J., and K. Gilless, *Decision Methods for Forest Resource Management* (2003, Elsevier Science, London), and Hof, J., and M. Bevers, *Spatial Optimization in Ecological Applications* (2002, Columbia

University Press, New York). We also do not focus on econometric applications used to study forest markets or landowner behavior. This also means that the valuation of nonmarketed goods such as ecosystem services is not treated in any real detail. However, we do make extensive use of public goods values in our treatments of different forestry problems. A useful recent book that touches on many empirical areas in forest economics is E. Sills and K. Abt, editors, *Forests in a Market Economy* (2003, Kluwer, London).

The empirical relevance of the theories we survey in these pages should not be difficult for readers to see. It is our hope that the material in this book will open a range of new empirical applications that follow from more sophisticated modeling of forest landowner, market, and government decisions. It would certainly be impossible to cover every area of forest economics thought in one volume. Still, the chapters of this book cover nearly all of the economics-based models proposed within the past two decades to describe forest policy design, forest landowner behavior, or predict the effects of policy instruments on the bahavior of forest landowners.

Questions of policy and landowner behavior have occupied the minds of foresters and resource economists for more than 150 years, indeed even to the extent that one of the earliest contributions to modern forest economics, by the German, Martin Faustmann, in 1849, was conceived as a way of developing "fairer" taxation systems. We may never have another set of theoretical contributions to equal those that Faustmann's analysis has given us, but we are fairly sure that the value of any theory developed in the future will be worthless if it is not examined at some point using real data.

Part I   Basic Models

These chapters provide the basic and historically important models and problems that any student or researcher of forest economics should be familiar with. We briefly discuss the early history of forest economics in chapter 1. It is interesting to see the number of iterations concerning the same basic idea of choosing an optimal rotation age that were developed during the dawn of economic thinking. We will focus on three periods of debate: the prehistory of economic analysis concerning the optimal rotation age, the birth of the optimal rotation decision as a separate framework, and what is called the Faustmann revival period. Any serious student or researcher must have an appreciation of where the field originated if they want to eventually shape where the field will go.

In chapter 2, the focus is on the most basic rotation-based decision models. The rotation age problem was originally developed for a single stand as one of maximizing the timber volume produced on a given forest site over time. The pure focus on forest growth led economists to denounce this solution from the beginning. Economists have tended to side with the Faustmann solution to define an optimal rotation age, but this is quite different than the one historically adopted by foresters. Forests are a long-term economic asset, capable of jointly producing revenue and other benefits periodically. As we will see, the economically optimal rotation age strikes a balance between harvest revenue derived from timber yields and the opportunity costs forgone by delaying harvesting and effectively tying up capital in forestry (standing timber and land) instead of investing this money in other assets. Economic thinking essentially replaces the foresters' "biological" capital theory with financial capital theory. As we develop the Faustmann model, we will compare its solution with earlier solutions to illustrate of the debate and confusion that originally surrounded early thinking in the field.

Perhaps the most important part of chapter 2 concerns modifications to the basic Faustmann model. There have been literally several hundreds of articles written using this framework, and applications continue to show up in the literature even today. Some of the more interesting modifications we examine include a cost function approach that uses duality theory, and the decision landowners face concerning choices among competing land uses. A relatively new extension to Faustmann models, one that considers rotation age decisions simultaneously with consumption and savings decisions, is also studied. These have been developed in response to some evidence that the harvest decisions of landowners depend on owner-specific preferences and characteristics such as nonforest income and wealth. The idea is that landowners may have incentives to use their forest as a means of financing consumption expenditures, and this can be used to determine how binding financial constraints affect decisions about their forests.

Chapter 2 concentrates largely on rents from harvesting that forest users can capture through markets. However, it is well known that forests also provide flows of important public goods, known in the economics literature as "amenities" or "nontimber" goods. These include wildlife habitat, biodiversity, flood prevention, recreation, fishing and hunting opportunities, landscape aesthetics, and carbon sequestration, among others. Generally these are goods and services that may not generate income-based rent, although some may depend on harvesting. All are potentially important to the welfare of forest recreation users and nonusers, as well as landowners. Amenity services have two common features. First, in most cases we can think of them as public goods that are not priced in markets. Second, the time path of amenity production during the age of a stand depends on the amount of standing timber present through time. Hence, amenities are jointly produced with timber in the forest production technology. When a stand is harvested, the flow of amenity services changes in a corresponding and possibly complicated manner.

Chapter 3 builds upon chapter 2 by introducing amenities into the landowner decision problem. As in the previous chapter, we also return to the question of forest taxation and its importance in shaping landowner decisions when there are social costs present because landowners and society do not have the same preferences. Two theoretical generalizations developed within the past few years are then presented. First we consider a case where stands in a given area form a forest landscape and are potentially interdependent in producing amenity services over time and space, and second, we return to our examination of models that combine rotation age decisions with consumption and savings decisions from chapter 2, extending them to the case where amenities are present.

Chapter 4 addresses life-cycle models of forest management decisions. The central feature of these models is to maximize the net present value of forest rents over two periods, current and future. These theories follow from combinations of the classical fishery-relevant biomass model and the traditional Fisherian two-period model widely applied elsewhere in economics. The life-cycle model has proven useful in studying short-term timber supply questions and has simplified the analysis of uncertainty, capital market imperfections such as landowners facing borrowing constraints, and amenities that invalidate the traditional Fisherian (and Hartman) assumption of separability between the preferences of resource users and production decisions. The effects of forest policies and the development of timber supply within this framework are also considered, as are extensions to longer-run models of overlapping generations. These models accommodate the notion that timber and money transfers can be made across generations through either timber bequests or sales. It also allows an examination of various long-run forest steady states that have important implications for forest policy. Of special value in this chapter is a comparison of the results derived with those of rotation models because this comparison has often been confused within the literature.

Part II   Policy Problems

Chapter 5 turns to the problem of policy design from the perspective of a government or social planner, with an emphasis on identifying the socially best instruments for the forest sector. In examining this problem, the policy maker must always answer two basic questions: First, what types of tax instruments are "best," and second, what is the appropriate level for each instrument? Obviously, these choices depend on the target function maximized or minimized by the policy maker, as well as any constraints regarding the set of available instruments. The policy maker must also be able to anticipate the reaction of landowners to any choice of instruments. Often the known fact that private landowners and other agents may make decisions inconsistent with a social planner, or the fact that policy makers face several constraints on their actions, creates special scope for policy design.

In chapter 6 we consider the important global problem of deforestation, delving into property rights risks and illegal logging activities. Deforestation of temperate forests was very rapid in the industrial world up through the 1940s and since then, this activity has shifted largely to tropical forests in Africa, Asia, and Latin America. These forests play a special global role because they contain more than 80% of the Earth's biodiversity. Our focus in the chapter is on policies aimed at reducing incentives for forest users to harvest forests unsustainably. The policy choice environment here is quite different from the one studied in chapter 5. Both empirical evidence and the literature have made it clear that any policy design must take into account many institutional and economic factors prevailing in underdeveloped tropical countries. Imperfect credit and labor markets distort agents' decision making in tropical fringe areas. Insecure property rights often lead to illegal logging, and there is sometimes land clearing and suboptimal investments in plantation forestry. Poor and inefficient governments raise money through royalties applied to a concessions process, but there are nearly always improperly designed royalty systems and lack of monitoring. Corruption among public officials is also ever present. This, combined with migration pressures and other poorly designed government policies, imposes constant pressure on the world's remaining native forests.

Using two common frameworks, this chapter shows how these issues provide challenges for forest policy design. The first is a concession model in which we analyze how the size of concessions, royalty instruments, and enforcement should be designed jointly to raise royalties to finance government budgets but also to abate illegal logging and control deforestation. The modifications to policy instruments necessary in the presence of corruption among public officials is also examined. The second framework discussed is the land-use rent-based model. This type of model reveals how migration pressure, insecure property rights, and risks concerning illegal logging and expropriation affect both land use and deforestation.

Chapter 7 analyzes biodiversity conservation. Biological diversity is multifaceted and important to life in all possible forms on Earth. To make the concept operative, biologists typically distinguish among diversity of species, diversity of habitats, and genetic diversity with regard to number, composition, and relationships. Regardless of how it is defined, the provision of biodiversity by forests is a public good. The socially optimal level of conservation for forest land depends on the social benefits and costs of providing the biodiversity. While the loss of tropical forests is a major drain on these public goods, many challenges lie ahead for conserving biodiversity in boreal and temperate forests. Most important, the management decisions of individual forest landowners may not be consistent with practices designed to maintain or increase biodiversity. To wit, intensive forest management practices, such as planting single-species stands, improving timber stands, suppressing fires, and frequent harvesting have all replaced the natural disturbance dynamics that have driven forest renewal and helped to maintain the diversity of habitat for centuries.

This chapter focuses on forest species and habitat conservation in three parts. First, we review ideas from ecology concerning forest habitat networks based on modeling of site selection. Following the spirit of this book, we then examine two key policy questions in biodiversity conservation. First, we investigate how a green auction approach can be used to promote the voluntary participation of private landowners in building biodiversity conservation reserves on forest land. These approaches are becoming more common and are already in place in many countries. Second, we examine how forest policies can be used to promote biodiversity maintenance in commercial forests; our special focus here is on retention of green trees as a means of increasing the dead and decaying wood needed for threatened old-growth forest habitat. Finally, we discuss some other aspects of biodiversity and forests, including genetic diversity and invasive foreign species.

Chapter 8 turns to age class-based models in forest economics. These models allow the inclusion of multiple tree ages within a landowner decision model. An important question in modeling age classes has been an investigation of several long-run steady states that might be important to forest management. The efficiency of one steady state in particular, a normal forest, is also examined. The desirability of normal forests has long been a controversial subject within the forestry and economics literature. In this chapter we present recent research that sheds some light upon these debates. Second, we discuss policy design problems where age class-based frameworks should be applied, such as carbon sequestration and using forest structure to enhace forest amenities.

Part III   Advanced Topics

These chapters focus on the many types of uncertainties involved in forestry and their implications for policy. The long-term nature of forest production means that landowners may never know the value of all parameters when they make management decisions in current time periods. The future price of timber is probably the easiest type of uncertainty to envision, but there is uncertainty in real interest rates and the pattern of forest growth, all of which are important to forest management decisions. There is also the possibility that a landowner is uncertain about several parameters simultaneously when making decisions. Furthermore, there are many natural hazards that landowners face in any rotation, but for which there is imperfect information concerning their arrival over time. These "catastrophic" risks come in many forms, such as fires, ice and wind storms, and pest attacks.

Uncertainty in timber markets may change the ways forest taxation affects incentives for a landowner to provide timber and amenity services. As we discuss in many places within this book, governments can influence these incentives through policy choices. Thus we will ultimately consider in many of these cases how uncertainty should be taken into account when a government designs forest policies. The basic question here is whether forest policies can be tailored in ways that correct for the possible biases uncertainty causes in the decisions of private landowners. Answering this question requires understanding how uncertainty enters the objective function of the government, and whether the government should be regarded as risk averse or risk neutral when forming its policies.

In chapter 9 we explore how uncertainty has been studied within the two-period life-cycle model of chapter 4. Several types of uncertainties, including those associated with the forest stock and economic parameters, are shown to affect harvesting behavior, timber supply, and amenity production in the short run. We also examine how government policies can be used to correct the distortions that uncertainty induces in landowner decisions, if it is desired by the policy maker. We make extensive use of the economic theory of risk-bearing behavior, which dates back at least four decades to initial work in expected utility theory. However, we will take a slightly different and more recent approach that uses analytically simple and economically intuitive classes of models based on nonexpected utility theory. One important advantage of these models, aside from their ease of interpretation, is the fact that they can easily incorporate the risk attitudes of both forest landowners and policy makers.

Chapter 10 continues our presentation of risk and uncertainty for a different class of models. Here we examine catastrophic natural events such as fires, wind, ice, and pests. Such events often cause large, discrete jumps in the rents that forest owners can capture during a rotation. These events have always been part of the forest landscape, and in many ways they play an important ecological role. By destroying trees, natural catastrophes create open space to promote both regeneration and biological diversity. The chapter begins by reviewing the types of catastrophic risks that are normally present for forest land, and then begins a study of the models used to incorporate these risks into forest management decision making. An important piece of this modeling involves specifying a workable description of how catastrophic natural events arrive during a rotation. We consider both the case where amenities are not valued and the case where amenities are important, and we consider cases where the arrival of the natural hazard depends on the age of the forest and on costly protection efforts that the landowner may engage in. The chapter ends with a discussion of the large set of literature that has followed the first treatments of this problem in forest economics. Chapter 10 serves as an introduction to the more complex modeling of stochastic processes and forest management that arises from economic parameters, such as forest stand value, timber prices, and interest rates.

The incorporation of general economic risks as a real options problem is discussed in chapter 11. This chapter introduces a research area in forest economics active since the 1990s. This work reflects the fact that forest management involves other risks besides catastrophic single-loss events. Often, uncertainty in market and biological parameters evolves over time, so that changes in rents do not involve one-time jumps, and there is often volatility in certain parameters that takes on special forms. This is most typically the case for market prices and interest rates, where demand and supply tend to dampen fluctuations over time. The typical way of modeling this type of problem is to use stochastic processes to describe how unknown parameters change through time. In this chapter we examine a class of problems known as optimal stopping, where the landowner can either irreversibly harvest the stand and capture revenues, or continue to let it grow and retain the option of stopping to harvest in the future. A critical part of these models is specifying a workable description for both the trend and the volatility of economic parameters. Several types of stochastic processes are therefore considered, and the use of stochastic processes in single- and multiple-rotation problems is discussed. This chapter amounts to a rigorous survey of optimal stopping and real options models as they have been applied to understanding forest decision making under uncertainty.

The book culminates with a study of dynamic forest models in chapter 12. We try to cover in this chapter material that could comprise an entire book in and of itself. We discuss myriad applications of dynamic optimization to understanding problems of forest resource economics. There are two main purposes here: First, we show how the problems studied and the scope for results in dynamic forestry models compare with rotation-based models. Over the past 20 years, several important articles have been written using either optimal control or dynamic programming, covering such areas as stand management, land use, timber supply, strategic behavior among landowners, deforestation, forest preservation and old-growth features of forest stands, and carbon sequestration. Dynamic programming is increasingly being used to study complex questions involving uncertainty. The stochastic control and real options models discussed in chapter 11 present a way of modeling uncertainty in a stopping problem, which provides a close bridge to harvesting decisions in rotation models. However, these approaches restrict attention to uncertainty as a diffusion process that evolves over time. Our second objective is therefore to give the reader an appreciation for the link between dynamic models and practical questions involving the dynamics of forest stands and markets.

The primary dynamic optimization methods applied to forestry problems have been optimal control and dynamic programming. Dynamic programming has been used mostly for stand management problems and landowner's decision making under uncertainty, while optimal control models have been used for perhaps a wider variety of nonstochastic problems, including policy design problems. The main goal of this chapter is to introduce these approaches in a way that links them to other parts of the book, showing alternative ways that problems can be studied by abandoning the rotation model. The chapter begins with some preliminaries concerning dynamic optimization, using forestry problems as illustration. Second, optimal control is considered and four areas are studied: Faustmann and Hartman interpretations, old-growth forest and native forest interpretations, land-use interpretations, and other interpretations, such as carbon sequestration and biodiversity. Third, we take up dynamic programming approaches using both perfect foresight and stochastic interpretations.

An important component of this book is the summary at the end of each chapter. Taken together, these summaries not only point out where we have come as a scholarly field, but they also reveal certain weaknesses in current approaches or research. In the summaries we have tried to provide some hints as to where we think research is going or should go in these areas.

Suggestions for Use

The chapters and parts of this book can be tailored to reader preferences. Readers interested in an advanced undergraduate and beginning graduate survey of basic theories in general forest economics should concentrate on chapters 1–4. Those interested in the design of policy instruments should read chapters 1–8. Readers interested in a thorough survey of uncertainty and stochasticity should concentrate on chapters 9–11 and the latter parts of chapter 12, while readers seeking to understand the types of dynamic problems that have been studied in forest economics should focus on chapters 11 and 12.

We have used the material in this book for both master's and doctoral-level courses in natural resource and environmental economics, forest economics and management, risk and uncertainty, and dynamic optimization. The book could easily serve as a graduate textbook in forest economics, but our hope is that it will be used as a supplemental textbook for any natural resource or environmental economics course where the instructor intends to give a serious or current treatment of forestry problems.

Acknowledgments

It is probably impossible to thank everyone who helped us through the many years it took to write this book. Obviously our families deserve the most thanks for enduring our seemingly unending hours spent working on this project at their expense. Without encouragement from them it is hard to imagine we would have succeeded. We also involved countless graduate students at Virginia Tech and Helsinki Universities. Names that come to mind are Maria Bowman, Fanfan Weng (who checked many of our equations for us), Erivelthon Lima, Melinda Vokoun, Simone Bauch, Jani Laturi, Jenni Miettinen, Kimmo Ollikka, and Karoliina Pilli-Sihvola. We also are indebted to the all other students who waded through this book in courses where the chapters were tested. Several colleagues were also there throughout the process to hear our complaints and fears and to suggest important directions. Amacher especially thanks his good friends and colleagues Jay Sullivan and Arun Malik, who constantly dealt with questions concerning whether this book was worth writing, and who provided important suggestions throughout for content. Ollikainen thanks his collaborators Artti Juutinen, Mikko Puhakka and Timo Pukkala, as well as Jari Kuuluvainen, Ville Ovaskainen, Olli Tahvonen, Jussi Uusivuori and Lauri Valsta for valuable insights in various topics discussed in this book over the years. Amacher and Ollikainen jointly thank Chiara Lombardini-Riipinen, who kindly checked and further developed the empirical boxes present in several chapters, and Artti Juutinen and Jussi Uusivuori for comments on the material in chapters 7 and 8, respectively. Finally, we owe great gratitude to the tireless and highly professional staff at the MIT Press in Cambridge, Massachusetts. It's no wonder they have the reputation they do among economics authors. First, we thank John Covell whose encouragement from day one has been immeasurable and irreplaceable. Katie Helke has also been invaluable in helping us sort through the steps involved in publishing this book. Ruth Haas, the copy editor who read through our manuscript, was absolutely outstanding and led to a much better product than we originally submitted. Katherine Almeida also deserves thanks for fielding our numerous questions and moving the process along. This all said, any remaining errors are certainly our own.

I   Basic Models

1   A Brief History of Forest Economics Thought

Before we delve into the theory upon which the economics of forest resources is based, it is important to step back and review the early history of forest economics thought. This gives an appreciation of where the field originated and better allows one to understand the context of even the newest models advanced in this book.

The oldest traceable research in natural resource economics is arguably an analysis of a forest rotation, or the time at which a forest owner should harvest. If one were to describe the "history" of forest economics, certainly solutions to optimal rotations under various assumptions would demand considerable attention. Rotation analysis per se focuses on the optimal time to cut a single (homogeneous) stand of trees. For a forest area composed of multiple stands, analysis of the forest rotation is broader and seeks to determine how a given land area should be allocated to growing forest stands of different ages and, for each homogeneous stand, the optimal rotation age. While we take this up later, we should point out that the traditional forester's approach to managing multiple-aged stands is to strive for attainment of a *normal forest*. The idea of a normal forest is simple. If we denote by T^* a steady-state rotation age that repeats through time, then a normal forest refers to one in which there are $1/T^*$ forest age classes, each of which has trees of the same age contained within it (sometimes groups of these forest stocks are called "stands"). At each time period, the oldest stand is harvested and a new stand is planted to become the youngest age class. It is for this reason that some think of a normal forest as one with "even-flow" harvesting.

The classic approach for harvesting forests in order to eventually attain normality is called "forest regulation." This has rarely been embraced by economists, mainly because the optimality of such a forest was nearly always in question. As we will see later in this book, it is only recently that economists have developed conditions under which a normal forest might be an optimal and efficient steady-state solution. However, there remain many cases where normal forest structures are not desirable.

1.1   Prehistory of Economic Analysis of the Optimal Rotation Period

Early treatises on forest management and the optimal rotation decision date back to the Middle Ages.[1] Probably the first systematic discussion of the basic forest rotation question occurred in the seventeenth century, while the modeling of multiple forest stands and age classes began in earnest during the eighteenth century. It is interesting, and in a way confusing to the development of these theories, that foresters and economists have almost always participated in these discussions, often using different approaches and proposing completely different answers for what describes an "optimal" rotation age. The inevitable disagreements between the two groups over time have even provided comic relief, as Samuelson's famous paper can attest (Samuelson 1976). With this in mind, we now chronicle some of the important early history of rotation age solutions.

We can go back much further than the seventeenth century to find evidence of humankind's economic relationship with forests. The first discussions of economizing harvesting took place in the monasteries of Mauermunster, Germany, during the 1100s. Formal legislation targeting forests appeared in the 1300s in Germany and in the 1500s in Italy and France. Beginning in the 1700s, the term rotation age became a common word to describe age at harvesting for a particular stand of trees. The first treatment of forest management as a scientific discipline was probably the article "Sylvicultura Oeconomica," written in 1713 by Hans Carl von Carlowitz.

During the 1600s and 1700s, the concepts of rotation age, normal forest, and sustainable harvesting became guiding principles of forest management. These served as the basis for Prime Minister Colbert's forest law in France, which became a benchmark for subsequent lawmaking in other countries. In Prussia, King Fredrick the Great decreed that forestry should be conducted by applying the principle of achieving a normal forest, with rotation ages set at 70 years for each stand. The motivation behind this ruling was to prevent deforestation and ensure a steady supply of lumber for ships.

In general, society remained suspicious of economic arguments regarding the regulation of forests during this period. Economic-driven behavior was blamed for accelerated deforestation of the great mid-European forests.[2] During the 1700s, the first economic calculations of optimal rotation periods and an analysis of the optimal intensity for thinning forest stands during a rotation were success-

---

1. Section 1.1 utilizes findings by Viitala (2002, 2006).
2. It is interesting that economic motives were soon used to explain the deforestation that happened rapidly in North America during the 1800s. Today these same economic motives are still advanced as the main reason for continued deforestation of the world's tropical forests, a subject we study in detail in chapter 6.

fully executed. Fernow (1911) argues that the first understanding of thinning was due to the German foresters von Berlepsch, von Zanthier, and Oettelt.

Denmark and England were notable in the development of forest economics thought during the 1700s. The Dane, Count Reventlow, made intensive empirical studies on the growth of trees, eventually analyzing optimal forest management as a financial return (Helles and Linddahl 1997). In England, William Marshall computed an optimal economic rotation age for oaks. According to current belief, he was the first person to argue that at any given time during a rotation, the optimal rotation age should be determined by comparing the marginal benefit of continuing the rotation, explained by additional growth in tree value over the next period of time, with the opportunity costs of delaying harvesting (Scorgie and Kennedy 2000). The opportunity costs of delaying harvesting were correctly expressed by him in terms of first, the interest income forgone from not harvesting and investing the proceeds over the next period, and second, the cost of occupying the land with trees and therefore not beginning the next rotation sooner. The second cost is an important one that is now known to economists as the opportunity cost of land, or land rent.

1.2   The Birth of the Optimal Rotation Framework

Despite Marshall's early progress, the bulk of analytical work on an optimal rotation age was conducted by faculty at the German Forestry School in the early 1800s. It is thus not surprising that the world's first journal in forest science, *die Allgemeine Forst- und Jagt Zeitung*, was published in Germany in 1824. By 1850 this journal had become a central forum for the development of forest economics theories and applications. Notable works were published by Cotta and Hartig, and by the young German foresters Gottlieb König, Johan Hundeshagen, and Friedrich Pfeil. These all had a strong orientation toward economic analysis. In his book *Die Forstmatematik*, König tried to develop an economic theory of rotation age solutions. Even though he failed to adequately define the opportunity cost of land in his calculations, he did clarify the nature of forests as a long-term capital asset. He also suggested a method for analyzing the profitability of forestry using the concept of growth in value. Hundeshagen's work was complementary and also developed a basis for calculating investment in forest stands; he previously had devoted much of his life to developing forest statistics. Pfeil is generally regarded as the first scientist to argue that forest management should be based on ecological and site properties.

It was Edmund von Gehren who took the determination of land value and the importance of rotation age choices seriously in his 1849 article in *die Allgemeine Forst- und Jagt Zeitung*. Von Gehren asked how much one should compensate a forest landowner if his land was converted to agriculture. Von Gehren used a geometric average interest rate and proposed valuing a stand in current (undiscounted) terms, rather than in the correct present-value terms. Just two months after its publication, another German forester, Martin Faustmann, published a comment on this article in the same journal. Faustmann's main point was to criticize the weaknesses of von Gehren's approach, arguing that use of a geometric average interest rate would lead to land values that were too high. He suggested a new method for determining the value of bare land that was based upon the principles of discounting. He defined the value of land as the capitalized present value of returns from an infinite series of rotations, undertaken in a perpetual sequence of planting and harvesting. Using a spreadsheet-type analysis, Faustmann showed that the optimal rotation age of any stand could be determined as one that maximized the net present value of the land. Faustmann was initially heavily criticized by the scientists von Gehren and Oezel, but by 1850 his approach had been proven correct by the work of Max Pressler. Pressler was the first to solve the rotation problem using mathematics, publishing his theoretical model in *die Allgemeine Forst- und Jagt Zeitung*.

In retrospect, Faustmann correctly formulated the rotation problem and all the associated opportunity costs that follow from delaying harvesting in any period. However, he did so with an example and did not provide a general mathematical analysis of the problem. It was Pressler's mathematical framework that afforded the first qualitative representation of the optimal rotation age and of land rent in forestry problems. According to Pressler's work, Faustmann was not the only one thinking correctly. William Marshall had also correctly anticipated the theoretical rotation age solution. Later, but independent of Pressler, a young Swedish economist, Bertil Ohlin, also mathematically described the rotation age problem in 1921.[3] Faustmann, Presler, and Ohlin are collectively now thought to be the founders of rigorous forest economics thinking.[4] After these articles were published, economists began to call the determination of rotation age, through the maximization of the present-value returns to land, the Faustmann formula or model.[5]

Acceptance of the Faustmann formula among foresters and economists has not always been universal, as Samuelson (1976) so eloquently notes. Two dominant and competing schools of thought arose in the wake of Faustmann and Pressler.

3. Löfgren (1983) provides an entertaining history behind this contribution.
4. The seminal articles by Faustmann (1849), Pressler (1850), Ohlin (1921), and the influential article and survey by Samuelson (1976) were republished in the inaugural issue of the *Journal of Forest Economics* in 1995, (vol. 1, p. 1). Knut Wicksell also understood that the land value and the optimal rotation should be solved in a framework with an infinite series of rotations.
5. Löfgren (1999) provides on explanation concerning the equivalence of Wicksell's and Faustmann's approach. This equivalence is also discussed in Weitzman (2003, pp. 130–137).

The first was a group of scholars who advanced thinking about rotation age in a single-rotation model. The second were scholars of ecology who proposed using purely biological concepts to define the best rotation age. The second way of thinking is based on something now known as maximum sustained yield (MSY) models. These are discussed in chapter 2 when the basic theories of the field are presented.

The first school of thought developed in 1863, soon after Faustmann's contribution, when a famous German economist named von Thunen suggested that the optimal rotation age must be solved using a single-rotation interpretation (for details of von Thunen's thinking, see Manz 1986). In 1871 both Stanley Jevons and Irving Fisher also ended up with the same conclusion as von Thunen. Single-rotation models were used to show that the optimal rotation should be a period of time during the life of a forest stand in which the growth in value of the forest stock equals the market interest rate. This sounds plausible to the uninitiated because the investment opportunity lost from not harvesting in any period is the interest one could receive by harvesting and investing the proceeds at the market interest rate, while the benefit of delaying harvest must be the additional growth of the landowner's investment "on the stump." This of course ignored the important opportunity cost associated with occupying the land with trees and foregoing future use whenever harvesting is delayed.

The most serious early challenge to Faustmann came from the second school of thought. According to Samuelson (1976), the notion of sustained yield as an ecological principle first appeared in the 1788 Austrian cameral valuation method. Here the best rotation age was described as one that, in simple terms, maximizes timber volume produced per unit of land over time. The sustained-yield approach is frequently applied to determine something called an "allowable cut" for forest areas. According to this school of thought, harvesting of any stand should be regulated by how much the average tree age in a stand is above or below the age that maximizes steady-state lumber yield (and harvest revenues) per acre.

1.3   The Faustmann Revival

Since the publication of Paul Samuelson's seminal and quite critical article on the field of forest economics, there has been an explosion in the number of Faustmann-based rotation articles. Newman (2002) documents many hundreds of papers published on the subject in the past thirty years alone. In the current wave of interest in environmental and resource economics, Samuelson's article made the economic approach to forestry problems commonplace. In the same journal as Samuelson's paper, Richard Hartman (1976) published an important extension of rotation analysis to include the public goods provided by forests. Hartman formalized the idea that forests can provide a value in situ, i.e., as standing forest stocks. He showed that this value has important implications for determining the optimal rotation age. His work remains an important basis for forest policy decisions, and indeed it has become one of the most frequently cited articles in forest economics. The Hartman model is the subject of chapter 3.

Both Samuelson's and Hartman's contributions led to active research during the 1980s concerning the properties of Faustmann solutions. Common questions addressed have been the effects of catastrophic forest loss (e.g., such as from fire), uncertainty in future prices and growth, and the effects of various policy instruments on the optimal rotation ages. The interested reader is referred to the literature cited for this chapter, all of which was written after Samuelson (1976).[6] Some interesting works include Newman (1984, 2000), S. Chang (1984), Reed (1986), Wear and Parks (1994), Hyde and Newman (1991), and Montgomery and Adams (1996). Reviews related to timber supply aspects include Adams and Haynes (1980), Binkley (1987b, 1993), and Wear and Parks (1994).

1.4 Revival of Age Class Models

Faustmann models and their derivatives can be applied only to problems where the forest is a single even-aged stand of trees. A more general problem studied by forest economists in the early (but more modern) literature has been to understand the optimal harvesting program for a forest area composed of many different age classes. The oldest approach to solve this problem was presented intuitively and without any analytical considerations. Simple harvesting rules were derived under the explicit requirement that the forest stock approaches some long-run equilibrium characterized by an even flow of wood being harvested over time, i.e., the normal forest (Davis et al. 2001). The concept of a normal forest refers to an age class distribution where forest land area is evenly allocated over all existing age classes. The oldest age class is harvested in each period and replaced (planted) with the youngest age class. This harvesting–planting cycle is continued forever by harvesting and planting in each period.

Nearly all of the recent theoretical studies of normal forests date from Kemp and Moore (1979), who conjectured that a constant flow of timber might be a long-run equilibrium in a forest vintage model. However, they did not explicitly analyze the optimality of such a long-run state. Heaps and Neher (1979) developed a model with homogeneous forest land and assumed that a new forest is planted at the same time that harvesting is carried out. They also did not provide

---

6. For some interesting and useful surveys before Samuelson (1976), see e.g., Gaffney (1957), Bentley and Teeguarden (1965), and Pearse (1967).

an analytical solution. Later, Heaps (1984) developed the solution of the same model further and in doing so generalized the Faustmann problem using a continuous-time model and assuming that harvesting costs are a convex function of the harvest volume. Heaps demonstrated that if there exists a long-run equilibrium, then it must be a normal forest.

While Heaps (1984) used a continuous-time–continuous-space specification, a discrete-time–continuous-space model was first proposed by Lyon and Sedjo (1983) and Sedjo and Lyon (1990). In these models, an optimal control problem is formulated to study long-term timber supply as a steady-state outcome. An argument in favor of this kind of specification is that it is frequently used as a basis for forest policy analysis (see e.g., Sedjo and Lyon 1990). Moreover, to the extent that use of forest resources is characterized by some seasonality, this model can accommodate a solution where harvesting is concentrated in some specific period of time. An excellent review of other discrete-time models in forestry, particularly during the early period of forest economic thought, is presented in Getz and Haight (1989). (These authors also discuss examples of discrete-time models for wildlife management problems.)

Mitra and Wan (1985, 1986) have also used a discrete-time–continuous-space formulation to study normal forests. Under zero discounting and strictly concave utility functions, they showed that the normal forest is an asymptotically stable equilibrium, given any initial forest structure. However, if utility is discounted, Mitra and Wan (1985) showed numerically that equilibrium forest stocks can oscillate around the normal forest structure. Wan (1994) reduced the Mitra and Wan model by assuming two vintages (young trees and old trees), but he also found that cycles are possible. We will return to these issues in detail later.

2　The Faustmann Rotation Model

The stylized approach for solving the rotation age of a single forest stand was originally proposed by foresters as a problem of maximizing average timber volume on a given forest site over time. This approach and its solution have been called either the "maximum sustained yield solution" or the "culmination principle." According to this approach, a stand should be cut at a point in time when average forest growth (mean annual increment) is equal to marginal forest growth (current annual increment) (see e.g., Gregory 1987). Thus the optimal rotation age is determined solely in terms of forest growth without any regard for prices, costs, or real interest rates. Although economists have regularly denounced this principle, it is still used today in forest management and planning, especially in the determination of maximum allowable harvest strategies.[1]

The approach used by economists to define an optimal rotation age is quite different. Economists view forests as a long-term economic asset, capable of providing revenue periodically in the form of harvesting. The economically optimal rotation age strikes a balance between harvest revenue derived from timber yields and opportunity costs forgone by delaying harvesting and effectively tying up capital in forestry (standing timber and land) instead of investing it in other assets. Thus economists replace the foresters' "biological" capital theory with economic capital theory.

The economists' counterpart to the foresters' culmination principle is the so-called Faustmann formula, which first came into existence in 1849 when Faustmann showed that a stand should be cut when the marginal return of delaying harvesting for one unit of time is just equal to the forgone interest based on the value of the stand and the land.[2] As we will see, this forgone interest is a

---

1. This critique is outlined in several places, such as Gregory (1987, pp. 204–206); for the role of the MSY solution as the guide for U.S. Forest Service management of the national forests under the National Forest Management Act of 1976, see Cubbage et al. (1993).
2. There have been many mistaken economic solutions to the rotation problem before and after Faustmann's work. The most well-known attempts are the single rotation-period model by von Thunen and Jevons, and Kenneth Boulding's application of the criteria of the internal rate of return to the rotation analysis. Chapter 1 provides some discussion of these mistakes.

A Note to the Reader   By and large, most of the theories contained in these pages
are based upon calculus results. Throughout the book we will denote the derivatives
by primes for functions with one argument, while partial derivatives will be denoted
using subscripts for functions with many arguments. Hence, e.g., $f'(T) = \partial f(T)/\partial T$
for $f(T)$, while $A_x(x, y) = \partial A(x, y)/\partial x$ for $A(x, y)$, etc.

---

fundamental opportunity cost associated with delaying harvesting in any time period. It serves to distinguish economic models from rotation age solutions derived
in noneconomic models. A hundred and fifty years later, Faustmann's solution for
the optimal forest rotation is still regarded as correct by economists.

We start this chapter with a description of forest growth in section 2.1. Here we
also discuss the broader implicit economic assumptions behind the basic version
of the Faustmann model. In section 2.2 we analyze the determination of a Faustmann rotation age and show how it is related to timber supply. In section 2.3 we
discuss how forest taxation affects optimal rotation age and timber supply. In
section 2.4 we describe some modifications, including timber management effort;
duality and rotation age models; and competing land uses and timber supply.
We end by showing how to link the Faustmann model to a more dynamic life-
cycle model.

2.1   Forest Growth Technology

The early foundations of forest economics addressed even-aged forest management. The notion of even-aged forestry refers to a practice where a forest is composed of stands of trees of similar ages. In natural conditions, without human
interference, these kinds of forests regenerate through external shocks, such as
fire, ice storms, or damaging winds. To mimic natural disturbances and ensure regeneration of a new forest, even-aged forests are normally clear-cut (completely
harvested) at the rotation age, and then a new stand is established either naturally
or by planting. This type of management is typical in both boreal and temperate
forests. Sometimes people use the term stand management to refer to decision
making for even-aged forest stands.

An alternative is uneven-aged management. Many tropical and temperate
hardwood forests fall under this category, and it is gradually becoming more
common in boreal forests as biodiversity considerations are taken into account in
management decisions. As trees grow, they increase in both height and diameter.
Uneven-aged management entails growing trees of different diameter classes on
the same unit of land. This means that a given land unit has trees of all diameter
classes (and ages). Clear-cutting is not practiced, and instead trees in any diameter

class can be selectively harvested at any point in time, depending on the objectives for management. Therefore the forest plot remains covered in trees over time even if harvesting occurs every year.

Forest growth in uneven-aged forests is measured as an increase in the number of trees in each diameter class. The concepts of ingrowth and upgrowth are used to describe the transition of trees from one diameter class to another. Uneven-aged forests are typically harvested selectively and more frequently than even-aged stands. Also, sustainable use of forests is now given a new meaning. Sustainability in uneven-aged forests is defined by a harvesting frequency and intensity that keeps the distribution of diameter classes and the number of trees in each diameter class constant over time. Generally, optimal management leads to a distribution of the number of trees in the different age classes that resembles an inverted J (see Adams and Ek 1974, Haight 1985, and Haight and Getz 1987).

Chapters 2 and 3 deal with even-aged management. Chapter 4 discusses two-period forest economics models, which in many ways resemble uneven-aged management models. We will take up uneven-aged management questions in chapter 8.

2.1.1   Properties

In the theoretical description of even-aged management, all trees in a stand are assumed to be identical and of the same age. Thus, once one knows the growth function of a single tree and the number of trees in a given stand or land unit, one also knows the volume of the stand at any point in time. Forest biometricians have developed sophisticated methods to measure and describe the growth of trees. Their analyses demonstrate that the growth functions of any tree on any site have common features; namely, growth follows a sigmoid or S-shaped path when the volume of wood in the trees is graphed over time. Sigmoid growth means that volume first increases at an increasing rate and then eventually increases at a decreasing rate. Growth rates can then even become negative as the forest begins to decay and literally fall apart from old age.

An economically feasible and biologically justifiable description of sigmoid growth can be given as follows: Let $f(t)$ denote the volume of the forest in cubic meters (m^3) of wood at point in time t. Holding site factors constant, the volume of the stand increases by growth. The change in the volume of the stand over time is given by $f'(t)$, where $f'(t)$ is the time derivative of $f(t)$. Stand growth therefore equals $f'(t)$ and has the following mathematical properties:

Assumption 2.1   Forest growth function

(a) $f'(t) > 0$ for $t \leq t'$ but $f'(t) \leq 0$ for $t > t'$, where t' is the age of biological maturity of the stand

(b) $f''(t) > 0$ for $t < \bar{t}$ and $f''(t) < 0$, $t > \bar{t}$, where $\bar{t}$ is the inflection point of the growth function.

Assumption 2.1(a) implies that forest volume is an increasing function of time before the point of biological maturity, but after that, volume declines. Time periods when the stand is overmature are ultimately characterized by volume reductions, because the stand decays and begins to fall apart. Assumption 2.1(b) is consistent with the forest growth function having first a convex and then a concave shape over time. That is, volume increases at an increasing rate until a point in time where volume increases at a decreasing rate. In economics, $f(t)$ is interpreted simply as a classic production function with time as the primary variable input and volume of wood as the output.

It is common practice to describe the properties of the growth function with the help of three terms: (1) current annual increment (CAI), (2) mean annual increment (MAI), and (3) relative growth of the stand (see e.g., Pearse 1967 and Gregory 1972). Relative growth is sometimes called the "periodic annual increment" (PAI). Mathematically, CAI is simply the first derivative of the volume function evaluated at a particular point in time, $f'(t)$. CAI is therefore easily interpreted as the marginal physical product of the primary input (time). MAI is defined by the ratio of volume to time, $f(t)/t$. Thus MAI gives the average physical product of the time input. Finally, the PAI is defined by the ratio of the CAI to forest volume, $f'(t)/f(t)$. This is sometimes referred to as an instantaneous growth rate in the literature.

These concepts turn out to be helpful in providing interpretation for the mathematical formulas derived in the rotation framework. Box 2.1 shows the empirical realization of CAI, MAI, and PAI using two different estimated forest growth functions—one for Scandinavian boreal forests and another for pine forests in the southern United States.

Thinning is an important management regime for even-aged forests. It involves a landowner selecting some trees for harvesting during a rotation, thereby creating improved growing conditions for the remaining stand of trees. Thinning is often referred to as an intermediate treatment or stand improvement treatment, because it is undertaken sometime between establishment and the rotation age and is usually aimed at increasing the proportion of higher-quality trees at harvest time, along with obtaining possibly higher prices per cubic meter removed. The effect of thinning on a stand of trees is to decrease the volume in a discrete way; nevertheless, forest growth under thinning exhibits a sigmoid growth in a discontinuous fashion.

Box 2.1
Stand Growth and Its Properties

The measurement of stand growth is of vital importance for forest management. A number of growth models have been developed and applied over time. They can be grouped into two broad classes: they are either statically calibrated or process-based models. The former models estimate the parameters for the growth function specification using empirical growth data. Process-based models provide growth estimates using known causal biological processes and tree competition mechanisms (e.g., see Hyytiäinen et al. 2004). Both model types are useful and the choice between them in research mostly depends on the problem in question.

In this book we rely on statistical growth functions because our focus is on general policy aspects, and thus the role of empirical illustrations is more pedagogical and qualitative. To illustrate concepts in this and other chapters, we will use example growth functions for even-aged (Scandinavian) boreal forests and U.S. southern temperate forests. In this box we graphically describe volume (Q), MAI, and CAI.

Scandinavian boreal forests: $Q(t) = a * b * 1.64 * (1 - 6.36^{T/b})^{3.897}$.

where T is the rotation age. This Scandinavian growth function was developed by Fridth and Nilsson (1980). Parameters a and b refer to the average growth over the rotation and the age at which the mean growth peaks, respectively. For our graphs we use $a = 4$ and $b = 80$. Moreover, to make growth less rapid, we use as exponent a value of 3.897 instead of Fridth and Nilsson's 4.897 value. The following U.S. growth function refers to volume of southern (or loblolly) pine, typically planted throughout the U.S. South,

U.S. South temperate forests:

$$Q(t) = \exp\left(9.75 - \frac{3418.11}{d * T} - \frac{740.82}{T * 80} - \frac{34.01}{T^2} - \frac{1527.67}{80^2}\right).$$

The number 80 refers to the site index at age 100, and d refers to planting density, which is taken at 300 trees per acre ($d = 300$). This growth function has been used in several economic analyses, including first in S. Chang (1984). The function given here follows the parameterization in Amacher et al. (2005).

Figure 2.1 shows a graph of volume growth (y axis) in cubic meters over time (x axis) for both growth functions. This is a typical sigmoid nature of growth suggested by assumption 2.1. For the Scandinavian function, forest growth increases as an increasing rate until an inflection point (50 years here), after which growth increases at a decreasing rate. As figure 2.2 shows, the CAI peaks at about 50 years and MAI at about 100 years and cuts CAI at its peak.

The respective functions for the U.S. growth are given in figures 2.3 for volume growth and 2.4 for MAI and CAI values. The difference with the Scandinavian boreal forests is striking. Now the CAI peaks at 20 years and MAI at 30, indicating much more rapid growth.

Box 2.1
(continued)

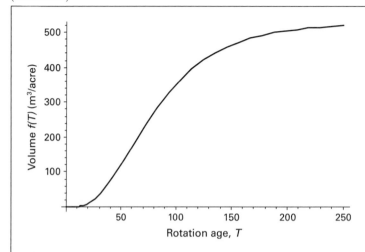

Figure 2.1
Scandinavian boreal forest growth function.

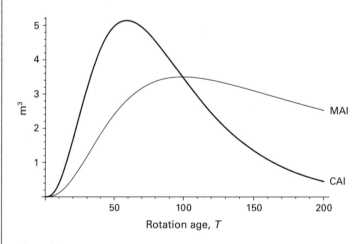

Figure 2.2
Scandinavian forest: mean annual growth (MAI) and current annual increment (CAI) (m^3).

Box 2.1
(continued)

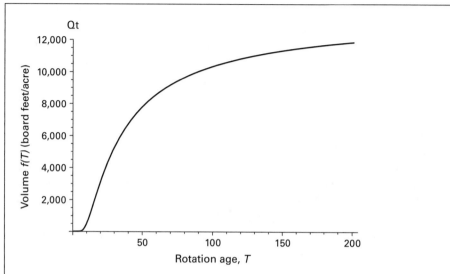

Figure 2.3
U.S. temperate southern pine forest growth function (board feet/acre).

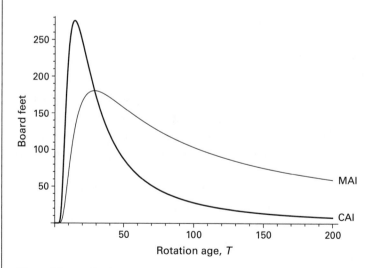

Figure 2.4
U.S. temperate forest: mean annual growth and current annual increment (CAI) (board feet).

2.1.2 Landowner Preferences and Assumptions

A key component in any economic model is the decision maker. The preferences of forest landowners are just as important as the biological possibilities of tree growth in defining the optimal rotation age and other choices.

Consider a landowner who has decided to manage a forest on a given land unit. Assume that forest growth follows the properties outlined in the previous section. What criteria should landowners follow when deciding to harvest? The answer in this chapter is simple. Irrespective of his preferences and consumption plans, the landowner is assumed to maximize the net present value of harvest revenue over an infinite cycle of rotations. A simple way of defining this net present value follows from several possibly restrictive assumptions of rotation analysis. These assumptions guarantee that the landowner's choice of rotation age is separable from consumption decisions. Separability here is a form of the Fisherian separation theorem (see e.g., Hirschleifer 1970, p. 63), which defines conditions such that production decisions do not depend on consumption.[3] Samuelson (1976) identifies these assumptions for rotation analysis and at the same time calls them heroic. We present them in assumption 2.2.

Assumption 2.2   Assumptions behind rotation analysis

1. Stumpage prices and regeneration costs are constant and known.

2. Future interest rates are constant and known.

3. The growth function of stands is known.

4. Forestland markets are perfect.

5. Financial capital markets are perfect.

The implications of certainty are evident. The landowner is able to predict all effects of his harvesting action on future net harvest revenue, ad infinitum. Perfect capital markets imply that the landowner can use these markets to finance consumption without distorting his forest management plans. A perfect land market implies that he can always sell forest land, at any period of time, at its capitalized value in the land market. This capitalized value is defined as the net present value of all future rents from establishing and harvesting successive rotations at the optimal rotation age forever. This guarantees that the landowner has incentives to follow efficient long-term stand management.

---

3. From his book Hirschleifer was clearly familiar with the rotation analysis and provides the solution for the Faustmann problem in his treatise; see Hirschleifer (1970, 82–92).

2.2 Computing the Optimal Rotation Period

2.2.1 Developing the Faustmann Formula

Consider a representative private landowner who begins with bare land and plants trees with a given fixed planting technology. The landowner has perfect foresight about the constant timber price and interest rate, and he chooses rotation age to maximize the net present value of harvest revenue over an infinite cycle of rotations. Denote the real interest rate by r and the regeneration cost by c. The stumpage price received for harvesting is p (we will use the terms stumpage price and timber price interchangeably). This reflects the practice in the literature of defining price as an in situ measure, i.e., applying it to uncut trees "on the stump." Thus the prices landowners receive for harvesting are assumed to be net of logging costs paid by the logger.

Assume also that stand management relies on a point-input–point-output production technology, meaning that there is a fixed forest establishment (planting) technology and no other variable production inputs other than time. The fixed factors of production are simply forest land with its natural productivity and the constant labor input required to establish new trees. Often this establishment input is thought to be an index capturing all of the inputs needed to begin a new rotation.

We denote the landowner's decision variable, rotation age, by T. Here, we will think of rotation age as the calendar time at which trees are harvested and the landowner begins the next rotation with bare land. Using assumption 2.2, the landowner's net present value from harvesting in the first rotation at time T is $pf(T)e^{-rT} - c$, where e^{-rT} is known as the discount factor and is a continuous time approximation for $1/(1+r)^T$. Here future harvest revenue has been discounted to the stand establishment date (time zero).[4] After harvesting, the bare land is replanted at cost c and a new rotation begins for the second T years. Because of assumption 2.2, this process continues ad infinitum with the same rotation age. Conveniently, we can write the net present value of an infinite sequence of identical rotations as follows:

$$pf(T)e^{-rT} - c + [pf(T)e^{-rT} - c]e^{-rT} + [pf(T)e^{-rT} - c]e^{-2rT} + [pf(T)e^{-rT} - c]e^{-3rT}$$

$$+ \cdots.$$

Factoring out the term $pf(T)e^{-rT} - c$ yields a geometric series for the remaining terms that is equal to $(1 + e^{-rT} + e^{-2rT} + e^{-3rT} + \cdots)$. By definition, this converges to $(1 - e^{-rT})^{-1}$.

---

4. We could conduct our analysis by compounding rather than discounting, as long as we are consistent for every rotation. The future value at time T of the first rotation in this case would be $pf(T) - ce^{rT}$.

We now arrive at the following equation describing the net present value of the first and all future rotations,

$$V = (1 - e^{-rT})^{-1}[pf(T)e^{-rT} - c]. \tag{2.1}$$

The numerator in (2.1) represents the present value of a single rotation, while the denominator is effectively a discount factor accommodating infinite future rotations. Equation (2.1) is sometimes called the Faustmann formula, bare land value (BLV), land expectation value (LEV), or soil expectation value (SEV). It represents the value of forest land defined as the present value of all future net rents that can be generated by an infinite sequence of rotations, where harvesting and establishing a new stand is assumed to continue forever. When land markets are perfect, this value also represents the market equilibrium price of bare forest land.

The landowner's economic problem is now to choose T to maximize (2.1). This task may seem complicated, but the fact that exogenous parameters are constant over time simplifies the analytics considerably. Recall that constancy also guarantees that the optimal rotation age is identical for all rotations. Differentiating (2.1) with respect to the rotation age T yields the following first-order condition:

$$V_T = pf'(T) - rpf(T) - rV = 0, \tag{2.2}$$

where V is defined by (2.1). The second-order sufficiency condition for the rotation age solution to indeed be a maximum requires that

$$V_{TT} = pf''(T) - rpf'(T) < 0 \tag{2.3}$$

using $rV_T = 0$ given the first-order condition. As one can see from (2.3), the sufficiency condition requires that the optimum lie within the concave part of the growth function, $f''(T) < 0$. It is common to express the first-order condition (2.2) in two equivalent ways:

$$pf'(T) = rpf(T) + rV, \tag{2.4a}$$

or

$$\frac{f'(T)}{f(T)} = r + \frac{rV}{pf(T)}. \tag{2.4b}$$

Equation (2.4a) shows that the optimal rotation is chosen so that the value of current annual increment, $pf'(T)$, captured by delaying the harvest for one period of time (left-hand side, or LHS) equals the opportunity cost of delaying harvest (right-hand side, or RHS). This opportunity cost equals rent the landowner can capture by harvesting and investing the proceeds at interest rate r for one period,

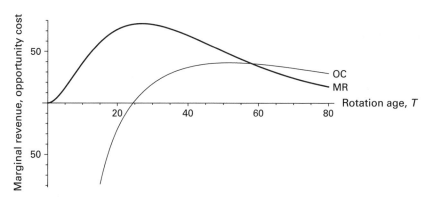

Figure 2.5
Determination of the Optimal Faustmann rotation age.

$rpf(T)$, plus the present value of investment returns lost by not harvesting now and beginning a new series of rotations on the land, given by the term rV. In this latter term, V is simply the bare land value for a series of rotations started in the current period defined by (2.1). This term also measures the rents accruing to the fixed production factor (land). As a result, rV is sometimes also called the "site rent" or "land rent" associated with the optimal rotation age solution. Equation (2.4b) expresses the same condition in a slightly different way. Here the optimal rotation age is defined by equality between the instantaneous growth rate (or PAI) given by the LHS, and RHS terms that include the real interest rate (first term) plus land rent weighted by the value of the trees at harvest time (second term).

The Faustmann harvesting rule can be stated more precisely as follows:

Proposition 2.1   Faustmann harvesting rule   The landowner should clear-cut an even-aged stand at an age at which the marginal return from delaying harvest is equal to the opportunity cost of delaying harvest, where the latter is given by forgone rents accruing to the future value of the stand and the land.

A graphical illustration of this result, based on equation (2.4a), is given in figure 2.5. This figure illustrates the baseline outcome reported in box 2.2. The graph shows the marginal return (MR) curves (LHS) and opportunity cost (OC) curves (RHS) of harvesting as a function of the optimal rotation age. Under an assumption that young forest trees are merchantable, the marginal return curve starts at the origin. This function first increases and then decreases once the maximum CAI is obtained. The opportunity cost curve first increases and then decreases. The MR intersects the opportunity cost line from above at roughly 58 years. This intersection point is the optimal Faustmann rotation age.

Box 2.2
Rotation Ages and Comparative Statics

Table 2.1
Comparison of Rotation Ages (Years) in the Faustmann, von Thunen-Jevons Single-Rotation, and MSY Models: Boreal Forests

	Faustmann	Single rotation	MSY model
Baseline	58.1	60.0	99.3
Sensitivity Analysis			
$p = €45$	57.0	no effect	no effect
$r = 4\%$	52.0	51.0	no effect
$c = 800$	58.7	no effect	no effect

For given growth functions, rotation ages are easy to determine once the values of timber price, real interest rate, and planting costs are known. Practically all common computer programs, such as Excel, Matlab, GAMS, or Mathematica, can be used to determine the optimal rotation ages. Here we illustrate the differences in rotation ages obtained in the Faustmann, von Thunen-Jevons single-rotation, and MSY models for the Scandinavian and U.S. growth functions presented in box 2.1.

Scandinavian Boreal Forests

We choose $p = €35$ per cubic meter as the stumpage price and let the planting costs be $c = €700$ per hectare (ha). The real interest rate is assumed to be 3%. This combination of these parameter values comprises our baseline. For a sensitivity analysis (i.e., comparative statics) we use the following values: stumpage price $p = €45$, regeneration costs $c = €800$, and a real interest rate of 4%. The models are compared in table 2.1.

Rotation ages in the baseline confirm the theoretical analysis in the text; the Faustmann model leads to the shortest rotation age, whereas the MSY model leads to the longest rotation age. The difference between the Faustmann model and the MSY model is a very large 41.3 years for the boreal forest example. Exogenous variables also have the effect on rotation age predicted by proposition 2.2. A higher price and interest rate shorten the rotation age, whereas a higher regeneration cost increases it. Note that in the von Thunen-Jevons single-rotation model, only the real interest rate affects the rotation age, as we discussed in the text.

U.S. Temperate Forests

For the U.S. temperate forest we use a $100 stumpage price per thousand board feet and a planting cost of $84 per acre; the real interest rate continues to be 0.03. Sensitivity analysis results are calculated for the same relative change as with the Scandinavian boreal forests.

Box 2.2
(continued)

Table 2.2
Comparison of Rotation Ages (Years) in the Faustmann, von Thunen-Jevons Single-Rotation, and MSY Models: Temperate Forests

	Faustmann	Single rotation	MSY model
Baseline	21.7	30.9	28.7
Sensitivity Analysis			
$p = \$100$	21.7	no effect	no effect
$r = 4\%$	20.3	26.87	no effect
$c = \$84$	21.7	no effect	no effect

The results in table 2.2 again confirm the theory. However, they demonstrate that in the Faustmann model, changes induced by planting costs and timber prices are so modest that using one (even two) decimals does not show any difference in the rotation ages. Only the interest rate makes a significant impact on rotation age decisions. However, the difference between the Faustmann rotation age and the MSY rotation age is considerable, 7 years.

We need to make two additional comments on the analysis of the first-order condition. First, the assumption that the landowner starts with bare land is not important. Suppose that the landowner's initial endowment is a growing stand of some age. If this stand is older than the Faustmann rotation age, it is financially overmature and the optimal action is to cut it immediately and begin a new series of rotations. If it is younger than the Faustmann age, the landowner simply delays the first harvest until the Faustmann rotation age and then harvests and begins a new sequence of rotations. The present value of an infinite sequence of rotations is revised accordingly. We leave it to the reader to produce this net present-value function. Second, consider what happens if the bare land value is negative. If forestry is the only potential use of the land in this case, then it is optimal for the landowner to either leave the land in an unmanaged (idle) state or to make an initial harvest without replanting the land. The issues of land values and other possible (competing) uses of land are taken up later in this chapter.

2.2.2 Comparison of Alternative Solutions

We are especially interested in comparing the Faustmann model with both the single-rotation period problem proposed originally by von Thunen and Jevons, and the foresters' MSY solution discussed in chapter 1.

Single Rotation-Period Model   Both von Thunen and Jevons assumed that it was sufficient to model the rotation decision as one where a landowner chooses rotation age to maximize the net present value of harvest revenue over a single rotation period. The net present value of a single rotation is simply given by

$$\max_T \ pf(T)e^{-rT} - c. \tag{2.5}$$

Partial differentiation with respect to T yields the following harvesting rule:

$$\frac{f'(T)}{f(T)} = r. \tag{2.6}$$

This implies that the stand should be harvested when its instantaneous growth rate, or PAI, is equal to the real market interest rate. Equation (2.6) includes as an opportunity cost only investment income lost from not harvesting at any point in time, $rf(T)$. Because future rotations are ignored, the single rotation solution does not incorporate the opportunity costs that are due to lost land rent. Given that land rent is not considered, the timber price paradoxically has no effect on rotation age. Therefore the rotation age solved from (2.6) cannot in general be equal to the Faustmann rotation age. The Faustmann solution for rotation age indeed incorporates a higher opportunity cost, and thus the Faustmann rotation age is unambiguously shorter than the von Thunen-Jevons rotation age.[5]

Maximum Sustained Yield: The Culmination Principle   The traditional foresters' approach seeks to maximize the volume of timber produced from a stand over time on a sustained yield basis. Given that trees in temperate and boreal forests become biologically mature only after a relatively long period of time, the sustained yield is defined slightly differently than in the case of a fishery; namely, as the average yield (growth) over the rotation period, i.e., as the mean annual increment. The problem is then to maximize MAI by choosing T, i.e.,

$$\max_T \ \frac{f(T)}{T}. \tag{2.7}$$

Differentiating with respect to T yields the following first-order condition:

$$f'(T) = \frac{f(T)}{T}. \tag{2.8}$$

---

5. The proof involves comparing (2.6) and (2.4b) and using the fact that $f''(.) < 0$ and $V > 0$ by assumption.

The LHS of equation (2.8) is the CAI and the RHS is the MAI (see figure 2.1 and box 2.1). According to this condition, a stand should be harvested when the current annual increment equals the mean annual increment. From the growth curves in figures 2.2 and 2.4 (see box 2.1), this point in time is just the culmination of MAI. As we argued in chapter 1, this solution is problematic because it neglects all opportunity costs and economic factors, such as prices, interest rates, and re-generation costs, all of which we know must affect the optimal rotation age if the landowner is concerned about financial returns.

How does the Faustmann rotation age relate to the maximum sustained yield age? Suppose first that the regeneration cost is zero in the Faustmann model, just as in the MSY formula (2.7). We can then express (2.2) as $pf'(T) = rpf(T) + rpf(T)e^{-rT}(1 - e^{-rT})^{-1}$, which, after some rearranging becomes

$$\frac{f'(T)}{f(T)} = \frac{r}{(1 - e^{-rT})}. \tag{2.9}$$

We can then re-express (2.8) as $f'(T)/f(T) = 1/T$. Thus for the RHS terms of (2.9) and (2.8) to be equal, it must now be true that $r/(1 - e^{-rT}) = 1/T$, which holds only for

$$\lim_{r \to 0} \frac{r}{(1 - e^{-rT})} = \frac{1}{T} \tag{2.10}$$

using L'Hopital's rule. From (2.10), the Faustmann and MSY rotation ages coincide only when the regeneration cost and real interest rate are zero. Moreover, given that for any positive value of real interest rate r we have a strict inequality between the two RHS terms in (2.9) and (2.10), it must be the case that the opportunity cost of harvesting in the Faustmann model is strictly greater than in the MSY model. This implies that the Faustmann rotation age is always shorter than the MSY rotation age.

Suppose next that regeneration costs are positive. The first-order condition for the Faustmann rotation age can be expressed as $f'(T)/f(T) = r\left(1 - \frac{c}{pf(T)}\right)/ (1 - e^{-rT})$ (see e.g., Binkley 1987a). Setting now $r\left(1 - \frac{c}{pf(T)}\right)/(1 - e^{-rT}) = 1/T$ and rearranging slightly, we obtain

$$\frac{c}{p} = f(T)\left[1 - \frac{1 - e^{-rT}}{rT}\right]. \tag{2.11}$$

If we now impose an assumption that land rents must be positive for ongoing forestry, i.e., $c/p \le f(T)e^{-rT}$, then we can obtain the following condition for the Faustmann age to be equal to or longer than the MSY age:

$$(1 - e^{-rT})(rT - 1) \leq 0. \tag{2.12}$$

Since $(1 - e^{-rT}) > 0$ for all $r > 0$ and $T > 0$, we must have from the term in the second parenthesis that $r \leq 1/T$ is a condition for having the Faustmann rotation age be longer than the MSY rotation age.

How realistic is it to suppose that the Faustmann model leads to a longer rotation age than the MSY? Binkley (1987a) notes that for slow-growing species like Douglas fir or white pine, the MSY is on the order of 100 years. Then, for the Faustmann rotation to exceed 100 years, the real discount rate must be less than the inverse of 100 years, or $r = 0.01$. This case is not very plausible. However, for fast-growing species, like tropical plantations, which reach culmination of MAI in much less than 20 years, there is a wide range of cost and price parameters, so that at a reasonable real interest rate of 5%, it may be true that the Faustmann rotation exceeds the MSY rotation age.

Finally, an interesting transform of the single-rotation model that yields the Faustmann outcome is due to Samuelson (1976). He was the first to notice that the single-rotation model can be written in a way that yields the Faustmann solution. One needs only to suppose that the optimization problem is augmented by the market rental value of bare land and that a zero-profit condition prevails in forestry. The market rental rate represents returns obtainable from a competing nonforest use of the land. Denoting the market rental rate of bare land by δ, the landowner chooses rotation age to maximize

$$V = pf(T)e^{-rT} - c - \delta \int_0^T e^{-rs} \, ds. \tag{2.13}$$

In this case, the market rental rate represents an explicit opportunity cost of not harvesting and occupying the site with trees in any period. The integral term reflects the observation that forgone rental rates are incurred each period that trees occupy the site. The first-order condition is $pf'(T) - rf(T) = \delta$. Under a zero-profit condition, $V = 0$, one then obtains from (2.13) that

$$\delta = \frac{r[pf(T)e^{-rT} - c]}{1 - e^{-rT}}. \tag{2.14}$$

Using (2.14) in the first-order condition of (2.13) establishes the equivalence with the Faustmann necessary condition.

2.2.3 Comparative Statics

The properties of the Faustmann model can be studied by developing its comparative statics. These results show how the rotation age solution depends on any exogenous parameters, holding all other parameters constant. Applying comparative statics first requires totally differentiating equation (2.2) with respect to the choice T and all parameters:[6]

$$0 = [f'(T) - rf(T) - rf(T)e^{-rT}(1 - e^{-rT})^{-1}] \, dp + r(1 - e^{-rT})^{-1} \, dc - \left(V + \frac{d}{dr}V\right) dr$$

$$+ [pf''(T) - rpf'(T)] \, dT \tag{2.15}$$

Denoting the Faustmann rotation age by T^F, it is straightforward to solve for the comparative statics results by applying the implicit function theorem to equation (2.15):

$$\frac{dT^F}{dp} = -\frac{[f'(T) - rf(T) - rf(T)e^{-rT}(1 - e^{-rT})^{-1}]}{[pf''(T) - rf'(T)]} < 0 \tag{2.16a}$$

$$\frac{dT^F}{dr} = -\frac{pf(T) + V + \frac{d}{dr}V}{[pf''(T) - rf'(T)]} < 0 \tag{2.16b}$$

$$\frac{dT^F}{dc} = -\frac{r(1 - e^{-rT})^{-1}}{[pf''(T) - rf'(T)]} > 0. \tag{2.16c}$$

Proof

Price effect Given that the denominator is negative and we have a minus sign in front of (2.16a), the sign of this expression is the same as the sign of the numerator. To determine the sign of the numerator of (2.16a), re-express the first-order condition (2.2) as $f'(T) - rf(T) - rf(T)(1 - e^{-rT})^{-1} = -p^{-1}[rc(1 - e^{-rT})^{-1}] < 0$.

Real interest rate effect By differentiation we can express the numerator of (2.16b) first as $-pf(T) - V - r(1 - e^{-rT})^{-1}Tpf(T)e^{-rT} - T[pf(T)e^{-rT} - c]e^{-rT}$. Rearranging the terms yields

$$-(pf(T) + V)\left(1 - \frac{rT}{e^{-rT} - 1}\right).$$

The first term in parentheses is negative. Applying L'Hopital's rule, one can see that

$$\left(1 - \frac{rT}{e^{-rT} - 1}\right) > 0,$$

---

6. The reader is referred to the general mathematics review at the end of this book for a refresher on the method of comparative statics. Simon and Blume (1994) also have an excellent discussion of these techniques.

so that the product of these terms is negative and the whole equation (2.16b) is negative as well. Q.E.D.

These results can be summarized as follows:

Proposition 2.2   The Faustmann rotation age solution depends positively on regeneration costs and negatively on the stumpage price and the real interest rate.

A higher stumpage price increases the marginal return of harvesting more than the opportunity cost of not harvesting, inducing the landowner to shorten the rotation age. A higher interest rate increases the opportunity cost of delaying harvesting and therefore shortens the rotation age. Finally, higher regeneration costs decrease land rent and thus decrease the opportunity cost of not harvesting. This induces the landowner to lengthen the rotation age. In box 2.2, we present solutions for rotation ages and comparative statics results using both the Scandinavian and U.S. volume functions from box 2.1.

2.2.4   From Optimal Rotation Period to Timber Supply

So far we have only discussed the rotation age of a single stand. Forests have commercial value because of the demand for timber. We can easily construct a timber supply function from the rotation model and then ask how the supply function depends on rotation age and exogenous parameters like stumpage price.

This requires us to first define the concept of timber supply for an individual land unit. It has become commonplace to distinguish between a short-run and long-run timber supply response (see e.g., J. Conrad 1999, pp. 68–70; Clark 1990). The short-run supply response describes the immediate change in the harvested volume that is due to a change in the rotation age. The long-run supply captures the change in the average harvested volume annualized over the entire rotation period. Annualized average harvest volume is defined by[7]

$$s^F \equiv \frac{f(T^F)}{T^F} = \frac{f[T^F(p,r,c)]}{T^F(p,r,c)}. \tag{2.17}$$

The short-run timber supply response can be derived directly from the comparative statics of how rotation age depends on exogenous parameters. When timber price or the real interest rate increases, the stand will be cut sooner, implying an increase in the short-run timber supply. The reverse happens when regeneration cost increases. Now the rotation age increases and short-run supply will decrease.

---

7. As J. Conrad (1999) points out, the aggregate long-run timber supply also depends on how land allocation between forestry and other (nonforest) uses changes as exogenous parameters change. When the profitability of forestry increases, new land enters into forestry production, and this tends to increase long-run supply. We provide a formal model of land allocation in section 2.5.

What happens to long-run timber supply when price, interest rates, and planting costs change is just the opposite of the short-term supply response. To see why, first denote exogenous variables by the vector $\psi = p, r, c$. Differentiating (2.17) with respect to ψ gives

$$\frac{ds^F}{d\psi} \equiv \frac{T_\psi^F}{T^F}\left[f'(T^F) - \frac{f(T^F)}{T^F}\right]. \tag{2.18}$$

From (2.18), the long-run supply response depends on two factors, the comparative static effect via changes in the rotation age, T_ψ^F, and the difference between the marginal and average growth of the stand. This last term is clearly positive if the original Faustmann rotation age is shorter than the MSY rotation age. In this (conventional) case, we find that the long-run supply response is negative for timber price and interest rate increases, but positive for planting cost increases; i.e., $s^F = s^F(\underset{-}{p}, \underset{-}{r}, \underset{+}{c})$. Naturally, if the Faustmann rotation age happens to be longer than the MSY rotation age, then the signs will be opposite because the expression in the brackets of (2.18) is negative. For an analysis of this case, the reader is referred to Binkley (1993).

Figure 2.6 shows the long-run timber supply curve as a function of stumpage price. The long-run supply curve is backward bending over the whole range at time periods longer than the MSY point, but it is increasing in price for time periods below the MSY point.

The unusual feature of a backward-bending long-run supply curve causes any economist to question it. Obviously, it reflects the limited ability of the fixed factor (land) to produce all possible levels of timber yield. A backward-bending

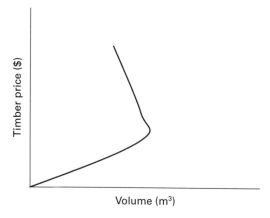

Figure 2.6
Long-run timber supply curve.

long-run supply curve would not necessarily be plausible for aggregate market supply functions in wood-producing countries, however. As Johansson and Löfgren (1985) argue, the problem with the supply function derived in the basic Faustmann model lies in the fact that not all factors are assumed to be variable even though a long-term analysis is being considered. If we make land devoted to forest production an endogenous choice variable for the landowner, or if we allow for forest management effort as a continuous choice, then we would arrive at a conventional positively sloped long-run supply function (see also Wear and Parks 1994 for a discussion of timber supply when land quality is variable, and S. Chang 1983 for timber supply under variable management effort).

2.3 Forest Taxation in the Faustmann Model

Forest taxation is an important policy instrument and has been the topic of probably more than a thousand papers in forest economics. Many types of tax instruments affect a landowner's choice of rotation age. The set of possible types of forest taxes that governments have at their disposal is large. How taxes affect rotation age is described by whether they are neutral or distortionary. A neutral tax is one that does not affect rotation age, while a distortionary tax does. This distinction will be especially important later when we study the socially optimal design of forest taxation in chapter 5. In this chapter we are more interested in examining how alternative forest taxes affect the landowner's rotation age choices at the margin; i.e., we are interested in the comparative statics of taxes.

The effects of forest taxation on the Faustmann rotation age are well known and derived, among others, in Amacher et al. (1991), Klemperer (1976), S. Chang (1982, 1983), and Johansson and Löfgren (1985). Amacher (1997) provides a comprehensive review of forest taxation effects. There are two broad classes of forest taxes in the literature and in practice: harvest taxes and property taxes. Harvest taxes include the yield tax, levied on harvest revenue, and the unit tax (also known as the severance tax) levied on the volume of harvested timber. Property taxes are levied on some assessment of land value. Within this class, two typically analyzed taxes are the lump-sum type of site productivity tax, paid annually and based on the yield potential of a given site irrespective of the actual harvests or standing timber. Another lump-sum property tax is a proportional tax on the land value, also called a site value tax. A property tax may also be levied on the value of trees, in which case it is often called a timber tax in the literature. In what follows, we will denote after-tax present value by $\hat{V}$ and analyze each tax individually. The proofs in this section are partly based on Koskela and Ollikainen (2001a).

2.3.1 Harvest Taxes

If the government levies a yield (τ) tax or unit tax (u) on harvesting, then the resulting net present value over infinite rotations is given by

$$\hat{V}(\tau, u) = [\hat{p}f(T)e^{-rT} - c](1 - e^{-rT})^{-1}, \tag{2.19a}$$

where $\hat{p} \equiv p(1 - \tau) - u$ denotes the after-tax stumpage price assuming the presence of yield and unit taxes. From the definition of $\hat{p}$, it is evident that harvest taxes work in the same way as a decrease in timber price. The comparative static effects of the yield and unit taxes are therefore as follows:[8]

$$\frac{\partial T^F}{\partial \tau} = -\frac{p}{(1 - \tau)} T_p^F > 0 \quad \text{and} \quad \frac{\partial T^F}{\partial u} = -T_p^F > 0. \tag{2.19b}$$

Hence both yield and unit taxes will lengthen the rotation age since they essentially represent a decrease in the net stumpage price and thus reduce the opportunity costs of continuing a rotation.

2.3.2 Property Taxes

We start with a site value tax, denoting the annual tax payment by b and defining its present value over time as

$$\int_0^\infty b e^{-rs} \, ds = \frac{b}{r}. \tag{2.20}$$

If the proportion of the value of forest land delivered in taxes is θ, then we have from equation (2.20) that $b = r\theta V$, so that the after-tax value of forest land is now $\hat{V}(\theta) = (1 - \theta)V$ (Johanssen and Löfgren 1985, p. 99). Maximization of this after-tax site value with respect to rotation age T gives the following first-order condition: $\hat{V}_T(\theta) = V_T = 0$, so that $\hat{V}_{T\theta} = 0$ and $T_\theta^F = 0$. Therefore the site value tax θ causes a lump-sum reduction in the net present value of forest rents, but it has no effect on rotation age.

The site productivity tax is administratively based on some site quality classification of forest land, which we denote by $a(i)$, where i refers to a measure of site quality (such as site index). The after-tax land value with this instrument is given by

$$\hat{V}[a(i)] = V - \frac{a(i)}{r}, \tag{2.21}$$

---

8. The reader can verify this by solving the comparative statics of taxes and comparing the explicit solutions with the price effect.

where $a(i)/r$ is the present value of the site productivity tax when levied ad infinitum (as in equation 2.20). Since $\hat{V}_T[a(i)] = V_T = 0$, we can conclude that the rotation age is not affected by a site productivity tax.

When a timber property tax, α, is levied annually on the stumpage value of growing timber volume, the objective function of the landowner becomes

$$\hat{V}(\alpha) = (1 - e^{-rT})^{-1}\left[V - \alpha\int_0^T pf(s)e^{-rs}\,ds\right]. \tag{2.22a}$$

The first-order condition with respect to rotation age T is given by

$$\hat{V}_T(\alpha) = pf'(T) - rf(T) - rV - \alpha[pf(T) - rU] = 0, \tag{2.22b}$$

where $U = (1 - e^{-rT})^{-1}\int_0^T pf(s)e^{-rs}\,ds$ denotes the present value of annual timber earnings, and V refers to the value of the forest land in the absence of the timber tax.

The effect of the tax rate α on the optimal rotation age turns out to depend on the sign of the expression $[pf(T) - rU]$. Under the assumption that $f'(T) > 0$ over the economically relevant range of time periods, we have the following:

Lemma 2.1 $pf(T) - rU > 0$, when $f'(T) > 0$.

Proof   Re-express $pf(T) - rU$ as

$$pf(T) - rU = \int_0^T pf(s)e^{-rs}\,ds\left(\frac{pf(T)}{\int_0^T pf(s)e^{-rs}\,ds} - \frac{r}{(1 - e^{-rT})}\right).$$

Given that $f'(T) > 0$ at the economically relevant rotation periods, one has

$$pf(T)\int_0^T e^{-rs}ds > \int_0^T pf(s)e^{-rs}\,ds \Leftrightarrow pf(T)\left(\frac{1 - e^{-rT}}{r}\right) > \int_0^T pf(s)e^{-rs}\,ds$$

$$\Leftrightarrow \frac{pf(T)}{\int_0^T pf(s)e^{-rs}\,ds} > \frac{r}{1 - e^{-rT}}.\ \text{Q.E.D.}$$

Lemma 2.1 states that when $f'(T) > 0$, the value of timber at harvest time $pf(T)$ is greater than the opportunity cost of harvest, rU. The first-order condition (2.22b) together with lemma 2.1 yields $\hat{V}_{T\alpha}(\alpha) = -[pf(T) - rU] < 0$, so that $T_\alpha^F < 0$. Hence a property tax on timber shortens the rotation age.

A summary of the effects of forest taxes on the private rotation age in the basic Faustmann model is given in the following proposition:

Proposition 2.3   Forest taxes in the Faustmann model

(a) As harvest taxes, the yield and the unit tax taxes will lengthen rotation age.

(b) As annual lump-sum taxes, the site value and site productivity taxes will have no effect on rotation age.

(c) A timber tax levied annually on the stumpage value of growing timber will shorten the private rotation age.

Harvest taxes reduce the net timber price, which in turn decreases both the marginal return of delaying harvest and the marginal opportunity cost of delaying harvest. Since the latter effect dominates, the rotation age increases. Lump-sum property taxes decrease land value but leave the relationship between the marginal benefit and the marginal opportunity cost of delaying harvest unchanged. Hence the rotation period remains unchanged as well. Finally, the timber tax decreases both the value of timber stock at harvest time and the opportunity cost of harvesting. The former effect dominates and the rotation age becomes shorter.

2.4 Modifications

Recalling assumption 2.2 and our discussion of a timber supply function, it is obvious that the basic version of the Faustmann model can be modified in many ways by changing the assumptions. Notable modifications in the literature include introducing uncertainty about future growth, prices, costs, and interest rates, allowing for imperfect markets, including management effort as a choice, and several others. Many of these issues will be dealt with later on. Here we review some modifications within the boundaries of certainty and perfect markets.

First, we should note some early modifications we do not have space to discuss but which are nonetheless quite noteworthy. McConnell et al. (1983) and Newman et al. (1985) explore a case where deterministic timber price and regeneration costs are allowed to change over time. They show that the effect on the optimal rotation age depends on how price changes affect the opportunity costs of continuing a rotation. Heaps (1984) and Heaps and Neher (1979) have studied the implications of various restrictions imposed on harvesting capacity. Kilkki and Väisänen (1969), and Näslund (1969) introduced thinning in the Faustmann model. In their model, the landowner has to choose the optimal timber volume and maintain it over time up to the final harvest age. Subsequent work includes applications of both optimal control theory and variational methods and will be mentioned in chapter 12.

Due to space constraints, we can discuss only a small selection of the many interesting modifications of the Faustmann model that are relevant for policy analysis. While many are mentioned in subsequent chapters, one recent study that is not is Hyytiäinen and Tahvonen (2001), who contrast Faustmann solutions with various legal requirements and recommendations in Finland. We start our modifications with the important management effort modification made by S. Chang (1983) and analyze how forest taxation affects both rotation age and management intensity. Building on the management effort model and drawing on Brazee and Amacher (2000), we then show how the Faustmann rotation age problem can be analyzed using duality theory. We also examine a simple model of land allocation by introducing land use as a choice variable for the landowner. This model is used to revisit timber supply analysis and build a foundation for our study of deforestation in chapter 6. Finally, we end by presenting a general life-cycle model of a landowner and demonstrating that the forest management problem is really separable under assumption 2.1.

2.4.1   Timber Management Effort

An interesting and much-applied modification to the Faustmann model was made by S. Chang (1982, 1983). He extended landowner choices to include decisions about silvicultural activities, which he labeled as management effort. The motivation for this extension is evident. The Faustmann model assumes a long-run time horizon. Therefore, keeping forest technology fixed over all future rotations is not very realistic. More often than not, landowners can undertake (costly) actions that improve the growth conditions of their stands; they may improve the quality of the site and the stand or they may use fertilizers to increase growth.

Chang introduced management effort (E) into the forest growth function by writing the forest growth function as $f(T, E)$ and assuming that $f_E > 0$ and $f_T > 0$, $f_{EE} < 0$, and $f_{TT} < 0$. The cross-derivative cannot be signed a priori, i.e., $f_{ET} \leq (>) 0$, because it depends on forest site characteristics and the type of management effort considered as Amacher et al. (1991) also point out. Let the unit cost of effort be denoted by w. The economic problem of the landowner is now to choose both rotation age and management effort according to the following problem:

$$\max_{T, E} \ V = [pf(T, E)e^{-rT} - wE](1 - e^{-rT})^{-1}. \tag{2.23}$$

The first-order conditions for (2.23) at an interior solution are

$$V_T = pf_T(\cdot) - rf(\cdot) - rV = 0, \tag{2.24a}$$

and

$$V_E = pf_E(\cdot)e^{-rT} - w = 0. \tag{2.24b}$$

Table 2.3
Cost Minimization and Effort-Based Faustmann Model: A Comparison

	Effort-based Faustmann[a]		Cost minimization[b]	
Parameter	E^*	T^*	E^C	T^C
Stumpage price	$+/-$	$+/-$	0	0
Effort cost	$-$	$+/-$	$-$	$+$
Interest rate	$+/-$	$+/-$	$+/-$	$+/-$
Yield tax	$+/-$	$+/-$	$-$	$+$
Harvest income tax	$+/-$	$+/-$	$-$	$+$
Forest increment tax	$+/-$	$+/-$	$+/-$	$+/-$

[a] Taken from S. Chang (1983) and Amacher et al. (1991).
[b] See Brazee and Amacher (2000).

Clearly, the determination of the optimal rotation age at the margin is qualitatively the same as in (2.2). Equation (2.24b) requires that management effort be employed to the point where the net present value of the marginal product of effort equals its unit cost.

As equations (2.24a) and (2.24b) reveal, the effort-based Faustmann model leads to a simultaneous two-equation system to be solved for rotation age and management effort. These choices depend on the exogenous parameters in the following ways:

$$T^F = T^F(\underset{+/-}{p}, \underset{+/-}{r}, \underset{+/-}{w}) \tag{2.25a}$$

$$E^F = E^F(\underset{+/-}{p}, \underset{+/-}{r}, \underset{-}{w}). \tag{2.25b}$$

With one exception, the signs are indeterminant and depend on whether time and effort are substitutes or complements in timber production. Therefore, management effort must be clearly specified. (Management may mean, for instance, choosing planting density, discing the soil, using weed control to clear understory vegetation, or fertilization.) Also, the effects of forest taxation on the optimal rotation age and management effort have been analyzed in Chang (1982) and Amacher et al. (1991). They show that the site productivity tax, as we have defined it here, is neutral both in terms of rotation age and management effort. We have indicated the effect of other forest taxes in table 2.3.

2.4.2 Duality
As is well known from the general economics literature, cost functions exist that are dual to any given profit functions. This issue was tackled verbally in the

history of forest economics by Hansen (see Helles and Linddal 1997); Löfgren (1984) also occasionally focuses on the issue.

Duality has recently been studied more rigorously by Brazee and Amacher (2000), and we rely on their work here. They consider a case where landowners make rotation age and management effort decisions to minimize the net present value of the landowner's costs. Thus we seek to derive a cost function that is dual to a forest rent–net present value function. Instead of using the Faustmann formula, Brazee and Amacher make use of Samuelson's formulation of a single-rotation model with exogenous land rents. The cost function is derived using the assumption that landowners would have to make choices to minimize costs subject to some output constraint, $\bar{F}$. This output constraint is not necessarily well defined in a forestry context. Assume it is realized as a constraint on timber volume, so that the landowner must achieve a given volume of timber (or revenue) in each rotation:

$$\bar{F} - f(E, T) = 0. \tag{2.26}$$

Constraint (2.26) parallels an output constraint in duality theory (see e.g., Varian 1992). Using (2.26) yields a cost function that is similar to the form where land rents arise through periodic rents the landowner could obtain in an alternative use. With this in mind, the cost minimization problem facing the landowner is

$$C(r, w, \bar{F}) \equiv \min_{E, T} \{wE + (1 - e^{-rT})A\}, \tag{2.27}$$

subject to (2.26), where $C(.)$ is the cost function and A is the (exogenous) sale price of land (implicitly, the individual landowner is a price taker in land markets). The first term in braces represents costs associated with forest establishment incurred at the beginning of the rotation, and the second term is the capitalized rents that accrue during each rotation.

The first-order conditions are obtained with respect to management effort, rotation age, and the Lagrangian multiplier associated with the constraint in (2.26):

$$L_E(E, T, \lambda) = w - \lambda f_E(E, T) = 0 \tag{2.28a}$$

$$L_T(E, T, \lambda) = rAe^{-rT} - \lambda f_T(E, T) = 0 \tag{2.28b}$$

$$L_\lambda(E, T, \lambda) = \bar{F} - f(E, T) = 0. \tag{2.28c}$$

These conditions imply, first, that the landowner equates the marginal cost of management effort to the marginal benefit of effort, second, that the landowner equates the marginal cost of delaying harvest to the marginal benefit of delaying harvest, and, finally, that the volume constraint is met for each rotation. The second-order conditions can be shown to hold (see Brazee and Amacher 2000).

Following duality theory (see e.g., Silberberg 1990, pp. 256–257), to solve for the appropriate cost function, one must identify conditions under which the cost-minimizing solutions from (2.28a)–(2.28c) equal the net present value-maximizing solutions yielded by the Faustmann model in (2.24a) and (2.24b). To do this, define the cost-minimizing solutions as E^C, T^C, and λ^C, and define the corresponding Faustmann values as E^*, T^*, and λ^*. Notice that if cost minimization is equivalent to profit maximization, then the constraint in (2.28a)–(2.28c) is simply $\bar{F} - f(E^*, T^*) = 0$. Thus one only needs to determine when the two conditions (2.28a) and (2.28b) are equivalent to the Faustmann solutions for effort and rotation age derived in equations (2.24a) and (2.24b).

The condition (2.28a) can be transformed by premultiplying through by $-(1 - e^{-rT^*})$. Substituting in the optimal output level and equating this transformed condition to the corresponding Faustmann first-order condition gives

$$w - e^{-rT^*} pf_E(E^*, T^*) = w - \lambda f_E(E^*, T^*), \tag{2.29}$$

which holds if the multiplier is defined by $\lambda = e^{-rT^*} p$. Substituting this into (2.28b) and rearranging (for details, see Brazee and Amacher 2000), one obtains a condition that A must satisfy:

$$A = \frac{pf(E^*, T^*) - wE^*}{(1 - e^{-rT^*})}. \tag{2.30}$$

Equation (2.30) requires that the sum of discounted net revenues from infinite future rotations, with an initial rotation being harvested now, must be capitalized into the price of land. A formal explanation for (2.30) follows from setting A equal to the price of land in the Samuelson version of the Faustmann model and adding revenues from the first rotation harvest prior to the sale of the land. Doing so, one obtains

$$A = \frac{e^{-rT^*} pf(E^*, T^*) - wE^*}{(1 - e^{-rT^*})} + pf(T^*, E^*) = \frac{pf(E^*, T^*) - wE^*}{(1 - e^{-rT^*})}. \tag{2.31}$$

Thus a sufficient condition for the cost function in (2.27) to be dual to the Faustmann net present value function is that the landowner must receive the value of both land and stumpage at the end of the first rotation when the land is sold. From, our previous discussion, we know that this is a result of perfect land markets.

Table 2.3 summarizes comparative statics derived in Brazee and Amacher (2000) and compares them with the effort-based Faustmann results. It is interesting that, with the exception of the forest increment tax (i.e., a site productivity tax), there is no longer ambiguity in the impact of the two types of taxes. Stumpage prices have no effect (as expected) on the optimal choice of rotation age and effort, but the interest rate effect remains ambiguous in a vein similar to that in the effort-based Faustmann model.

2.4.3   Competing Land Uses

So far we have assumed that forestry is the only use of land. In practice, however, forestry competes with other land uses, such as agriculture. Any change in the relative profitability of land uses will change the amount of land allocated to each use, and this will then change long-run timber supply, as discussed earlier. Land use is usually endogenized by varying either land quality or distance from land to the market. Both approaches lead to the development of economic margins between alternative land uses.

Heterogeneous land quality and distance models have been most often applied to analyze deforestation as a land-use problem, for instance, in Barbier and Burgess (1997), Hardie and Parks (1997), Parks et al. (1998), and Amacher et al. (2008). An interesting exception is Brazee and Amacher (2002), who examine various forest age classes on land units devoted solely to forest production. Our presentation here follows in part Amacher et al. (2008), but we simplify their model to two land-use forms only, agriculture and forestry.

Consider a single landowner holding a plot of land of varying quality. This land can be divided into a continuum of parcels for which land quality is uniform but varies across parcels. The landowner's economic problem is to allocate this land between a nonforest use (agriculture) and forestry in a way that maximizes the present value of returns from the land. Solving this problem involves choosing optimal production in each parcel for both uses and then allocating parcels to the use that has the higher net return.

In agricultural production, the landowner employs variable agricultural inputs each period, represented by l, at a constant marginal cost of w. Agricultural yield in each period is a concave function of the level of the variable input, i.e., $f(l; q)$, where $f_l(l; q) > 0$, $f_{ll}(l; q) < 0$. Higher land quality q increases the marginal product of the input, $f_{lq}(l; q) > 0$. Denoting the price of the agricultural crop as p_a, the present value of profits generated by agricultural production on any parcel can be written

$$\pi = \int_0^\infty [p_a f(l; q) - wl] e^{-rt} \, dt. \tag{2.32}$$

The landowner optimizes agricultural production by choosing the level of the agricultural input l according to the first-order condition, $p_a f_l(l; q) - w = 0$, which holds at each point in time. This implicitly determines the optimal level of l in each period, $\hat{l}^* = \hat{l}^*(p_a, w, q)$. Substituting this optimal level for l in (2.32) gives

the indirect net present value profit function per parcel, which has the following properties: $\pi^*(p_a, w, r, q)$.
$\qquad\qquad\qquad\qquad\qquad\quad\; {}^+ \; {}^- \; {}^- \; {}^+$

The net present value of returns on land units devoted to forestry is given by the Faustmann formula. Denote timber price by p_f. It becomes convenient to include land quality explicitly in the forest growth function by writing $F(T; q)$, with $F_q(T; q) > 0$ and $F_{Tq}(T; q) > 0$. The Faustmann formula is then written as $V = [p_f F(T; q) - c](1 - e^{-rT})^{-1}$ and it has the same properties that we derived earlier in the chapter. One can show that the indirect net present revenue function per parcel has the following properties in terms of exogenous parameters: $V^* = V^*(p_f, c, r, q)$.
$\qquad\qquad\qquad\qquad\qquad\;\; {}^- \; {}^+ \; {}^+$
$\qquad\qquad\qquad\qquad\;\;\; {}^+$

To formalize the land allocation decision, we now follow procedures described in Lichtenberg (1989) and rank land quality by the scalar measure q, with the scale chosen so that $0 \leq q \leq 1$. The cumulative distribution of q, $G(q)$ indicates the set of parcels having at most a quality level of q. Let $g(q)$ be the density function for $G(q)$, i.e, $g(q) = G'(q)$. Thus, the total amount of the land is given by

$$G = \int_0^1 g(q)\, dq. \tag{2.33}$$

The landowner allocates the area G to agriculture and forestry. The properties $\pi_q^* > 0$, and $V_q^* > 0$ indicate that both types of production are more profitable on land of better quality. To ensure that it is optimal to allocate the land to both uses, we make the following assumption:

Assumption 2.3 For land qualities and land uses, the following relationships hold:

(a) $V^* < \pi^*$ for $q = 1$, and
(b) $V^* > \pi^*$ for $q = 0$,

assumptions (2.3a) and (2.3b) define the relative profitability of forestry and agriculture. Together they imply that agricultural production is most profitable on higher-quality land while forestry performs better on lower-quality land. Given both assumptions, the land rent curves of agriculture and forestry must cross when graphed across the land quality continuum, ensuring that both agriculture and forestry are practiced in separate unique, compact land areas.

Denote the share of the land area devoted to forestry by h_f, so that the share devoted to agriculture is given by $h_f = 1 - h_f$. The landowner solves the following problem of choosing shares:

$$\max_{h_f} PV = \int_0^1 [V^* h_f + \pi^*(1 - h_f)] g(q)\, dq. \tag{2.34}$$

The necessary condition

$$\frac{\partial PV}{\partial h_f} = V^* - \pi^* = 0 \tag{2.35}$$

defines the critical land quality, q^c. This land quality divides all land into two compact segments of the continuum (L_f and L_a) devoted to agriculture and forestry as follows:

$$L_f = \int_0^{q^c} g(q)\,dq \quad \text{and} \quad L_a = \int_{q^c}^1 g(q)\,dq. \tag{2.36}$$

The comparative statics of land use now comes from defining first a vector of exogenous variables ψ; $\psi = (p_a, p_f, c, r, w)$. Differentiating (2.36) with respect to ψ yields

$$\frac{dL_f}{d\psi} = g(q^c)\frac{\partial q^c}{\partial \psi}, \quad \frac{dL_a}{d\psi} = -g(q^c)\frac{\partial q^c}{\partial \psi}. \tag{2.37}$$

As equation (2.37) reveals, an increase in one land use decreases the other because there are only two land-use forms and a fixed amount of land. Thus, once we know what happens to land allocated to forestry, we also know how agricultural land changes. This follows from examining how the critical quality of land use changes with parameter changes. To produce the necessary derivative for this result, $\partial q^c / \partial \psi$, we differentiate (2.35) totally to obtain

$$\frac{\partial q^c}{\partial p_f} = -\frac{V_{p_f}^*}{\Delta} > 0, \quad \frac{\partial q^c}{\partial r} = -\frac{V_r^* - \pi_r^*}{\Delta} \gtrless 0, \quad \frac{\partial q^c}{\partial c} = -\frac{V_c^*}{\Delta} < 0, \tag{2.38a}$$

$$\frac{\partial q^c}{\partial p_a} = \frac{\pi_{p_a}^*}{\Delta} < 0, \quad \frac{\partial q^c}{\partial w} = \frac{\pi_w^*}{\Delta} > 0, \tag{2.38b}$$

where $\Delta = (V_q^* - \pi_q^*) < 0$ owing to the second-order sufficiency condition for the optimal land area allocation. The signs indicate the direction of the shift in critical land quality as parameters change.

Combining (2.38a), (2.38b), and (2.37) reveals that a higher timber price (regeneration cost) increases (decreases) the amount of land allocated to forestry. It is interesting that the effect of the real interest rate remains ambiguous and depends on the relative importance of the interest rate to profits from forestry and agriculture. We also find that an increase in agricultural profitability via a change in price or input cost will decrease land allocated to forestry and vice versa. This suggests that "exogenous" forestry-independent shocks to changes in the long-run timber supply are possible. Except for this last case, it is clear that timber price and planting cost will have ambiguous effects on long-run timber supply. This is because the effects from land allocation run counter to the effects emerging from changes in rotation ages. Whether the long-run supply curve continues to exhibit a backward-bending property depends on the relative strength of the two effects, as Brazee and Amacher (2002) demonstrate in a different setup.

2.4.4 A Life-Cycle Interpretation

We have somewhat blindly followed assumption 2.2 throughout this chapter, applying the Faustmann model without ever referring to the landowner's other targets, such as consumption. Let us now consider a private nonindustrial landowner who uses both timber revenue and some exogenous revenue to finance his consumption. There is a large literature that makes these connections empirically (Amacher et al. 2002 review this area). In the theory sense, a model is needed in which the landowner chooses a consumption schedule and rotation age to finance consumption over time. A natural framework for examining this kind of decision is the life-cycle model, where agents optimize consumption over their life span. An important example of this type of model applied to forest landowners is found in Tahvonen and Salo (1999), who demonstrate how the Faustmann model emerges from a more general life-cycle dynamic optimization model, and we briefly present a simpler version of their life-cycle model here. We assume just as they did that this landowner lives forever, making the model similar to a dynasty model from macroeconomics, where all subsequent generations have the same preference. Thus the first generation chooses the optimal consumption schedule and rotation age.

Let the preferences of the landowner define a concave utility function over consumption, $u[c(t)]$, with $u'[c(t)] > 0$ but $u''[c(t)] < 0$. The rate of time preference of the landowner is ρ. The landowner chooses a consumption schedule $c(t)$ and a rotation age T to maximize this utility over on infinite time horizon:

$$\max_{c,T} U = \int_0^\infty u[c(t)]e^{-\rho t}\,dt. \tag{2.39}$$

The landowner's consumption choice is constrained by his exogenous income m and forest income V (defined by the Faustmann model). The intertemporal budget constraint over the infinite time horizon is

$$\int_0^\infty c(t)e^{-rt}\,dt = \int_0^\infty m(t)e^{-rt}\,dt + V. \tag{2.40}$$

The Lagrangian function for this problem is

$$L = \int_0^\infty u[c(t)]e^{-\rho t}\, dt + \lambda \left[\int_0^\infty m(t)e^{-rt}\, dt + V - \int_0^\infty c(t)e^{-rt}\, dt \right],$$

where λ is the budget constraint multiplier. Choosing consumption and rotation age produces the following first-order conditions:

$$\frac{\partial L}{\partial c(t)} = u'[c(t)]e^{-\rho t} - \lambda e^{-rt} = 0 \tag{2.41a}$$

$$\frac{\partial L}{\partial T} = \lambda V_T = 0 \Leftrightarrow \lambda[pf'(T) - rpf(T) - rV] = 0. \tag{2.41b}$$

Condition (2.41a) characterizes the optimal consumption path over time. Expressing it as $\lambda^{-1}u'[c(t)] = e^{-(r-\rho)t}$ reveals that the time path of consumption can be constant ($r = \rho$), decreasing ($\rho > r$), or increasing ($\rho < r$). Equation (2.41b) determines the optimal rotation age at the margin. It clearly shows that the preferences of the forest landowner given by utility do not affect the choice of the optimal rotation age because the optimal solution is independent of λ (the bracketed term must be zero in 2.41b). Thus the Faustmann solution is indeed the optimal solution for this life-cycle model and the Fisherian separation theorem holds. We will return to this model in the next chapter (section 3.6.2) to show that amenities defined from standing forests imply a breakdown of separability between consumption and forest harvesting.

2.5   Summary

This chapter presented the rudiments of the basic theory upon which most of modern forest economics is based, or at least derives from. We showed the assumptions that economists need concerning forest technology and examined the theories behind the basic solutions proposed for solving an optimal rotation age. We found that the Faustmann formulation, under the basic assumptions required, includes all the opportunity costs of growing trees over time, unlike other proposed formulations. In this model, a higher interest rate and timber price shortens the optimal rotation, while a higher regeneration cost lengthens it. These affect the short-run supply of timber, but the long-run effects turn out to be just the opposite. The determination of the optimal rotation age and its dependence on exogenous variables, including forest taxes, provides a basic building block and a useful yardstick for more sophisticated policy-oriented analyses, such as optimal taxation or the design of policies to promote stable climates and biodiversity, or to reduce exploitation of the world's remaining native forests, all of which are taken up in later chapters.

3 Hartman Models of Timber and Amenity Production

In chapter 2 the only returns obtainable from forestry were harvest revenues. It is well known that standing forest provides a steady flow of economically important amenities. These include ecosystem services, such as fish and wildlife habitat, biodiversity, flood prevention, recreation, fishing and hunting opportunities, landscape aesthetics, and carbon sequestration, among others. Generally these are goods and services that may not generate timber income, although some income may arise from harvesting. All are potentially important to the welfare of forest recreation users and nonusers as well as the landowners.

Amenity services have two common features. First, in most cases we can think of them as public goods that are not priced in markets. Second, the time path of amenity production during the age of a stand depends on the amount of standing timber present through time. Hence amenities are jointly produced with timber in the forest production technology. When a stand is harvested, or even as time goes by, the flow of amenity services changes in a corresponding and possibly complicated manner.

The dual features of joint production of timber revenues and amenities raise the question of how the analysis in chapter 2 should be modified to account for these types of goods. A solution was originally proposed by Richard Hartman, who developed an application of the Faustmann model by introducing amenity services (Hartman 1976). Hartman assumed that amenity services depend only on forest age (and therefore indirectly on the size of forest stock). He expressed the valuation of amenity services with something called "a felicity function" that described the monetary valuation of amenities as a function of a stand's age. The advantage of the felicity function was the accompanying assumption that a forest manager could determine rotation age by simply maximizing the present value of timber returns and amenity services (in economic terms, this made the landowner's target function quasi-linear). Subsequent work has largely adopted these assumptions as starting points, examining more closely the properties of rotation ages in the presence of amenities (for the immediate follow-up of Hartman's contribution, see e.g., Calish et al. 1978, Nguyen 1979, Strang 1983, and Bowes and Krutilla 1985).

In this chapter we explore Hartman's work and many subsequent contributions concerning joint production of timber and amenity services. We continue to take as given the description of forest production and the assumptions of chapter 2. We start by focusing on the types of amenities, their general description, and assumptions concerning their valuation. The basic properties of the Hartman model are then analyzed and the effects of exogenous parameters (prices, costs, and interest rate) on timber and amenity production are considered. We also examine how forest taxation affects rotation age and timber supply when amenities are accounted for.

Several recent theoretical generalizations of the Hartman model have been developed within the past few years. These include a case where stands in a given area form a forest landscape and are potentially interdependent. Interdependence could mean that the realization of amenity services varies over time and space. Biologists have long advocated that interdependent stands should be managed in concert, and recently economists have agreed. Bowes et al. (1984) and Bowes and Krutilla (1985, 1989) first raised the issue and undertook a preliminary analysis of stand interactions in forest-level decisions. Swallow and Wear (1993), Swallow et al. (1997), Snyder and Bhattacharyya (1990), Vincent and Binkley (1993), Koskela and Ollikainen (2001b), and Amacher et al. (2004) have since studied aspects of this problem. For our purposes, we will present newer concepts from this body of work needed to fully understand the importance of stand interdependence with regard to amenities.

We leave out some recent interesting literature, most of which focuses on specific empirical examples of amenity production. These include species conservation (see e.g., Csuti et al. 1997, Ando et al. 1998, Polasky et al. 2001) and ecosystem management in forestry (see e.g., Bevers et al. 1995, Albers 1996, Albers et al. 1996, Bevers and Hof 1999, Haight and Travis 1997, Montgomery 1995). These, together with the above-mentioned new analytical studies, confirm that the Hartman model and its theoretical extensions and empirical applications will continue to be part of forest economics thinking in the near future. As we will discuss in this chapter and later in chapters 7, 10, and 11, much interesting research remains to be conducted within the confines of this framework.

The rest of this chapter is organized as follows: In section 3.1 we present examples and classifications of how amenities can be defined. Section 3.2 presents a definition of landowner preferences concerning amenities. Sections 3.3 and 3.4 are devoted to deriving and discussing optimal rotation ages and to analyzing the effects of forest taxation. In section 3.5 we extend the Hartman model to include interdependent stands. Finally, in section 3.6 we present other modifications of the Hartman model, including discussions of the role of forests as a carbon sink, by analyzing how the land allocation and life-cycle models presented in chapter 2 can be adapted to accommodate amenity benefits.

3.1 Amenity Services

The classic examples of amenities are the ecosystem services produced by forests. In addition to those listed in the opening paragraph of this chapter, they also include wilderness, maintenance of water quality, and flood prevention. Foresters and forest economists have developed alternative classification schemes and descriptions for the technical dependence between stands and amenity services. The following typology is rather conventional and draws upon Calish et al. (1978), Bowes and Krutilla (1989), Swallow et al. (1990), and Hunter (1990), among others.

Recreation People have always attached value to forests for aesthetic reasons. The nature of recreational goods differs considerably across countries. While Nordic countries guarantee access to private and public forests with so-called "every man's rights" policies, many U.S. regions contain private forests that are not generally accessible to public users. Instead, access is open to users only on publicly owned forests. Harvesting by private landowners is usually seen to involve social costs by reducing the level of recreation below that which a social policy maker would choose, i.e., harvesting might be thought of as a negative externality.[1] In this case, there is a need for policy interventions aimed at modifying the behavior of landowners to achieve a socially optimal balance between recreation and timber production.

Fish and Wildlife Typical forms of recreation are hunting and fishing. Harvesting directly affects these activities, because forage and shelter conditions for species are changed by harvesting. Harvesting also indirectly affects water quality and growth possibilities of many fish stocks. Habitat management for wildlife, timing of harvests, and use of buffer areas around waterways are all potential management options when these types of amenity services are traded off against timber production.

Wilderness Preservation of ecologically and culturally important forested areas for future generations is the most extreme case of amenity and timber production because harvesting must be stopped for maintenance and production of these goods. While strict preservation holds as a rule for these areas, sometimes restricted management and managing to mimic natural disturbances are adopted. Thus amenity production here dominates landowner decisions, but small-scale timber production is sometimes possible.

---

1. For a definition of an externality, see Baumol and Oates (1988, chapter 3).

Water Quality Forests have long been known to play an important role in providing water quality and flood control. Maintaining forest cover in upstream areas reduces flooding and sedimentation downstream. Limiting or prohibiting harvesting close to waterways requires harvesting plans to be adjusted on a spatial scale.

Carbon Sinks The ability of forests to sequester carbon in standing timber is now thought to play an important role in reducing global warming. Solving for an optimal combination of harvesting and preservation of forest stocks as carbon sinks is an important issue in the multiple-use analysis of forests.

Biodiversity Since the United Nations Convention on Biological Diversity in 1992, the most urgent and comprehensive issue in forest management has become the conservation of biodiversity in forests. Biodiversity is typically divided into three dimensions: species diversity, habitat diversity, and genetic diversity; clearly, commercial forestry will affect all of these. It is interesting that when biodiversity conservation became part of the research agenda, the notion of ecosystem management, or landscape ecological management, also became important. Now forest economists and biologists refer less to the traditional terminology of multiple-use forestry, recognizing that a single stand-based approach can never be expected to capture the necessary complexities of biodiversity needed for policy analysis.

Aside from those services that are priced (such as hunting leases), determining amenity benefits over the life of a stand requires the application of economic valuation techniques. From a theoretical point of view, we are less interested in the absolute value of amenities than we are in the dependence of amenity flows on calendar time and rotation age. Some types of amenity values are likely highest when stands are young, while others are higher when forests are older. This suggests that a harvesting decision or policy used to promote only one type of amenity may weaken another type of amenity service. This is evident when the two amenities in question require different-aged forests to be produced. Tradeoffs are also possible when both amenities are highest in old stands. For instance, conservation of biodiversity habitats by promotion of old stands with plentiful understory vegetation may in fact decrease visibility, which is highly valued in certain types of recreational uses such as hiking.

3.2   Landowner Preferences over Amenity Services

We start with a simple theoretical description of amenity valuation. Let $F(s)$ describe an amenity valuation function at time s. Formally, this can be thought of as the present value of net monetary benefits from a flow of amenities at time s. Over an entire rotation period, the forest site provides a net present value amenity benefit equal to the sum of these values up to the rotation age, $\int_0^T F(s)e^{-rs}\,ds$. Differentiating this expression with respect to the rotation age T yields $F(T)e^{-rT}$ using Leibnitz's rule from calculus, which gives the (positive) marginal valuation of accruing amenity benefits over one rotation cycle, evaluated at rotation age T. The derivative of this expression with respect to rotation age then indicates how the marginal valuation of an amenity service behaves over potential choices of rotation age. Neglecting for a moment the discount factor, we can define several possible cases.

Definition 3.1   Marginal valuation of amenity services

$$F'(T)\begin{Bmatrix} > \\ = \\ < \end{Bmatrix} 0 \quad \text{as} \quad \begin{cases} \text{marginal valuation increases with stand age} \\ \text{marginal valuation is independent of stand age} \\ \text{marginal valuation decreases with stand age} \end{cases}$$

Definition 3.1 allows for three cases: a landowner who values an older stand $[F'(T) > 0]$, a landowner who values a younger stand $[F'(T) < 0]$, and a landowner whose amenity valuation is independent of stand age and instead might depend only on site-specific properties $[F'(T) = 0]$. General examples of increasing and decreasing marginal valuation are easy to find. Consider that the recreational value of a forest is typically higher when the forest is older. However, landowners who value deer, certain types of birds, or elk habitat, have a higher amenity valuation for younger stands. Generally, biodiversity benefits are thought to be higher in older forests, indicating that marginal valuation increases with the age of stand for this amenity.

Some caution is needed in interpreting definition 3.1. As Bowes and Krutilla (1989, pp. 107–108) point out, actually very little is known about the amenity values of a single stand. Usually these depend on the condition and management of neighboring forest areas in complex and not well-understood ways. Likewise, Knapp (1981) notes the problems of aggregating various amenities provided by a stand and cautions against trying to define the overall role of any stand in amenity valuation. Because of the almost infinite ways that amenity services arise over time, there is also no a priori reason to expect that the aggregate amenity valuation of a forest is always monotonically increasing or decreasing.

Using Calish et al. (1978), Swallow et al. (1990), and Steinkamp and Betters (1991), in table 3.1 we provide some parametric forms for amenities when they may be constant, decreasing, or increasing in time (rotation age). Box 3.1 illustrates some possible shapes of amenity valuation functions in terms of rotation age. Figures 3.1 and 3.2 represent cases where amenity benefits from old and young stands are valued, respectively, i.e., marginal amenity valuation is positive

Table 3.1
Types of Amenity Benefits and Specifications of Amenity Valuation Function

Amenity type: dependence on time	Specification of amenity benefit function	Empirical examples of applicable cases
Constant amenity benefits	$F(T) = F^0$	Existence value of an open space or a site-specific feature in forest
Decreasing amenity benefits	$F(T) = \beta T e^{-bT} + k$, or $F(T) = a - bT^2$	Wildlife species adopted to young forest, forage production
Increasing amenity benefits	$F(T) = k/(1 + e^{a-bT})$ $F(T) = b_0 T e^{-(1/b_1)T}$	Wildlife species adopted to old forest, all dimensions of diversity, recreation
All types of benefits over time	Combination of above	"Aggregate" benefits

or negative. Figure 3.3 illustrates a case where over some years amenity benefits increase and then decrease.

3.3 Determination of the Optimal Rotation Period

3.3.1 Optimal Hartman Rotation Period

Suppose a representative landowner values both net harvest revenue and amenity services from the forests.[2] Following Hartman, we assume that the landowner has the following quasi-linear objective function:

$$W = V + E, \tag{3.1}$$

where $V = (1 - e^{-rT})^{-1}[pf(T)e^{-rT} - c]$ and $E = (1 - e^{-rT})^{-1} \int_0^T F(s)e^{-rs}\,ds$. Here V is the net present value of harvesting over an infinite sequence of rotations, i.e., it is the Faustmann part of the model (equation 2.1 in chapter 2). The variable E is the present value of amenity services over an infinite sequence of rotations. Other symbols are as before [p is the stumpage price, $f(T)$ is the volume of the stand as a function of rotation age T, c is the regeneration cost, r is the real interest rate, and s is a variable of integration].

The economic problem of the landowner is to choose T to maximize (3.1). The first-order condition for an interior solution is in general, $W_T = V_T + E_T = 0$. We re-express the first-order condition in equation (3.2) and provide the second-order condition in (3.3):

---

2. Because the target function here is quasi-linear (i.e., separable in amenities and timber returns, but nonlinear in amenities), we continue by keeping the landowner type undefined. Thus the landowner in the Hartman model could be a private nonindustrial landowner, a forest industry owner, or a public landowner.

Box 3.1
Joint Production of Timber and Amenities

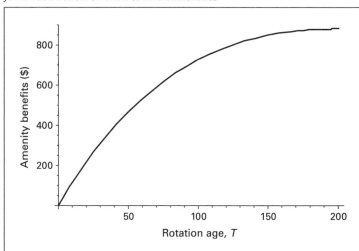

Figure 3.1
Amenity benefits increase in the rotation age.

Amenity benefits from forests may be increasing, decreasing, or constant in rotation age. Empirical simulations require workable specifications of the amenity benefit function for all these cases. The specifications should reflect the evolvement of amenity benefits over potential rotation ages, or calendar time, considering that the forest begins at time zero. Moreover, amenity valuation functions must be calibrated to match the socially relevant monetary estimates of amenity benefits at each point in time.

Table 3.1 in the text presents some often-applied specifications of amenity benefit functions in economic analysis. We illustrate the potential shapes of these functions by applying two amenity benefit functions from this table:

Amenity benefit function: $F(T) = b_0 * T * e^{-(1/b_1)T}$ and $F(T) = \alpha_0 - \alpha_1 * T^2$.

The first amenity benefit function (originating from Swallow et al. 1990) is widely applied because among other things it has a meaningful interpretation. In this formulation b_0 calibrates the model to reflect the monetary estimates of the amenity benefits. Parameter b_1 in turn defines the peak (maximum value) of amenity benefits. Setting b_1 high makes the benefit function increase in the rotation age, while low values of b_1 imply that amenity benefits increase in early years of the rotation but decrease in later years. The second (quadratic) amenity benefit function exhibits a decreasing shape over the entire rotation. Parameter α_0 defines the size of the benefits at the beginning of the rotation period and parameter α_1 defines how quickly these benefits decline during the rotation.

Figure 3.1 illustrates increasing amenity benefits using the first function. We produce this graph using parameter values equal to $b_0 = 0.235$, $b_1 = 250$. The peak in amenity values is obtained at 250 years. In boreal forests, amenity benefits may peak or they may not peak at all because the older the stand, the higher its biodiversity value. Figure 3.2 illustrates decreasing amenity benefits. The graph is produced by using the following parameter values and the second function: $\alpha_0 = 126$, $\alpha_1 = 0.07$.

Box 3.1
(continued)

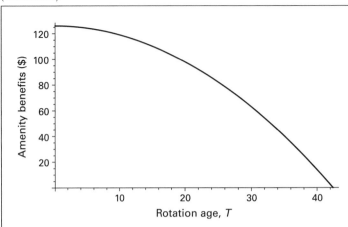

Figure 3.2
Amenity benefits decrease in the rotation age.

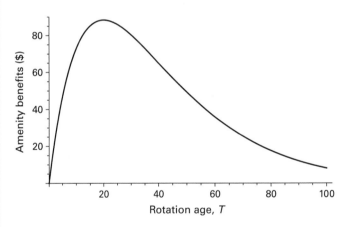

Figure 3.3
Amenity benefits first increase, then decrease in the rotation age.

In Figure 3.2 amenity benefits decrease monotonically for the first 40 years, reaching zero at 42 years. Perhaps a more relevant case than figure 3.2 is the case where the benefits peak in early years of rotation and decrease thereafter, as figure 3.3 reveals. This would be the case for wildlife that requires young stands to thrive, such as many species of birds. In figure 3.3 the path of amenity benefits by setting $b_0 = 12$ and $b_1 = 20$ in the first function.

One can expect that the amenity benefits in figure 3.3 result in a longer rotation period than the monotonically decreasing benefits in figure 3.2. This is confirmed in box 3.2.

$$W_T = pf'(T) - rpf(T) - rV + F(T) - rE = 0. \tag{3.2}$$

$$W_{TT} = pf''(T) - rpf'(T) + F'(T). \tag{3.3}$$

The analysis of equations (3.2) and (3.3) contains many interpretations because the second-order condition (3.3) may fail to hold. There are three possible cases: the second-order condition holds globally, only locally, or does not hold at all. We initially assume that the second-order condition holds globally, so that an interior solution is obtained from (3.2). The second-order condition always holds (globally and locally) for $F'(T) \leq 0$, but when $F'(T) > 0$, we need to also have that $pf''(T) - rpf'(T) < -F'(T)$ over all rotation ages.

Expressing (3.2) as Hartman (1976) originally did gives

$$pf'(T) + F(T) = rpf(T) + r(V + E). \tag{3.4}$$

The LHS of equation (3.4) is the marginal return (also called a marginal benefit or marginal revenue term for this case) from delaying harvesting written as the sum of harvest revenue and amenity benefits. The RHS is the opportunity cost of delaying harvesting in terms of harvest and amenity production. Thus condition (3.4) provides an interpretation similar to the Faustmann model; delaying harvesting for one unit of time allows the landowner to capture future amenity and income benefits, and this is taken into account. Both opportunity costs discussed in chapter 2 are also explicit in the RHS of (3.4). Here the first term represents investment opportunities lost by not harvesting, while the second term represents lost opportunities in terms of harvesting and amenities by not beginning the next stand sooner (i.e., this reflects what was called "land rent" in chapter 2). With these interpretations we have the following proposition:

Proposition 3.1 Hartman harvesting rule The landowner should harvest a stand of trees when the sum of the marginal harvest revenue and amenity benefit from delaying harvest for one period equals the opportunity cost of delaying harvest, where opportunity cost is defined as rent on the value of the stand (timber plus amenities) plus rent on the value of the land.

Define the Hartman rotation age solution to (3.4) by T^H. How does this solution relate to the Faustmann rotation age T^F? Comparing equations (3.2) and (2.2) reveals that the difference depends on the sign of the two new amenity valuation terms, $F(T) - rE$; that is, on the relative size of the amenity benefits at harvest time and the opportunity cost in terms of additional amenities that accrue from delaying harvest. It is useful to see how this difference behaves under alternative assumptions for the marginal amenity valuation function. Lemma 3.1, demonstrated originally by Koskela and Ollikainen (2001), provides the result.

Lemma 3.1 $F(T) - rE \begin{Bmatrix} > \\ = \\ < \end{Bmatrix} 0$ as $F'(T) \begin{Bmatrix} > \\ = \\ < \end{Bmatrix} 0.$

Proof Re-express $F(T) - rE = F(T) - r(1 - e^{-rT})^{-1} \int_0^T F(s)e^{-rs}\, ds$ as follows:

$$\int_0^T F(s)e^{-rs}\, ds \left[\frac{F(T)}{\int_0^T F(s)e^{-rs}\, ds} > (\le) \frac{r}{(1 - e^{-rT})} \right].$$

If $F'(T) > (\le) 0$, then $\int_0^T F(T)e^{-rs}\, ds > (\le) \int_0^T F(s)e^{-rs}\, ds$

$$\Leftrightarrow \frac{F(T)}{r}(1 - e^{-rT}) > (\le) \int_0^T F(s)e^{-rs}\, ds \Leftrightarrow \frac{F(T)}{\int_0^T F(s)e^{-rs}\, ds} > (\le) \frac{r}{(1 - e^{-rT})}.$$

Hence we have

$$F(T) - rE = \int_0^T F(s)e^{-rs}\, ds \left[\frac{F(T)}{\int_0^T F(s)e^{-rs}\, ds} > (\le) \frac{r}{(1 - e^{-rT})} \right] > (\le) 0,$$

so that

$$F(T) - rE > (\le) 0 \quad \text{as } F'(T) > (\le) 0. \text{ Q.E.D.}$$

According to lemma 3.1, if marginal amenity valuation does not change with the age of the stand $[F'(T) = 0]$, then $F(T) = rE$ and we can conclude that the Faustmann and Hartman rotation ages coincide. If the marginal amenity valuation increases with the rotation age of the stand, then the timber production part, V_T, of equation (3.2) will be negative, and the Hartman rotation age will be longer than the Faustmann rotation age. The opposite holds when the marginal amenity valuation decreases with rotation age; in this case the Hartman rotation age will be shorter than Faustmann rotation age. A more detailed proposition concerning rotation age results is as follows:

Proposition 3.2 The Hartman rotation age is longer (or shorter) than the Faustmann rotation age if the landowner's marginal valuation increases (or decreases) with rotation age. Hartman and Faustmann rotation ages coincide if the marginal valuation of amenities is constant in rotation age.

Drawing on previous descriptions of typical amenity services, possible rotation age outcomes are shown for the cases of increasing and decreasing amenity benefit valuations in figures 3.4 and 3.5 assuming that the second-order condition (3.3)

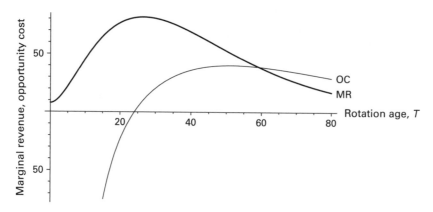

Figure 3.4
Determination of the Hartman rotation age: old stands valued.

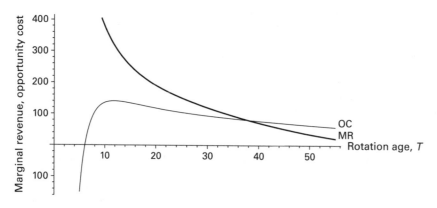

Figure 3.5
Determination of the Hartman rotation age: young stands valued.

holds. These figures are graphs of the marginal revenue (MR) and opportunity cost (OC) of delaying harvest as functions of rotation age. They reflect the baseline solution for the Hartman rotation age that is reported in box 3.2.

Recall that in both cases of amenity valuation the optimum requires that the marginal revenue curves cut the opportunity cost curve from above. Also, in both figures the decreasing MR curve cuts the OC curve from above just once, and therefore we can define unique optimal rotation ages. These graphs are clearly consistent with the theoretical results. The Hartman rotation age is close to 60 years when amenity benefits from old stands are valued, but only about 38 years when amenity benefits from young stands are valued.

Box 3.2
Optimal Rotation Age in the Hartman Model

Table 3.2
Hartman Rotation Ages (Years) under Increasing and Decreasing Amenity Benefits

	Faustmann	Hartman Benefits increasing	Hartman Benefits decreasing	Hartman Benefits first increasing, then decreasing
Baseline	58.1	59.3	38.0	53.8
Sensitivity Analysis				
$p = €45$	57.0	57.9	39.8	53.6
$r = 4\%$	52.0	53.0	38.7	49.5
$c = 800$	58.7	60.0	38.6	54.4

In this box we combine the forest growth function presented in box 2.1 and the chosen economic parameter values in the baseline case with the parametric specifications of the three amenity benefit valuation functions presented in box 3.1. We solve and report the optimal Hartman rotation ages and compare them with each other and with the Faustmann rotation age.

The optimal Hartman rotation age is determined using the same parameter values for the amenity benefit functions as in figures 3.1, 3.2, and 3.3 of box 3.1. The results, together with the previously solved Faustmann rotation age, are reported in table 3.2.

Table 3.2 confirms the theoretical conjectures in the text. When the amenity benefits from old stands are valued (figure 3.1, box 3.1), the Hartman rotation age is clearly longer than the Faustmann age, although in our example this is true only slightly, by 1.2 years. The Hartman age is shorter than the Faustmann age under the quadratic amenity function where young stands are valued. Depending on the peak and the slope of the decreasing part of the amenity benefit function, the Hartman age is either shorter or longer than the Faustmann age when the amenity valuation function is first increasing and then decreasing (figure 3.3) (our specification here leads to a shorter rotation).

The sensitivity analysis, for decreasing amenity benefits, verifies the possibility, shown in the theory, that a higher stumpage price may lead to a longer rotation age in the Hartman model by changing the relative profitability of the timber income and amenity valuation parts of the landowner's objective function. Other parameters have conventional effects on the optimal Hartman rotation age.

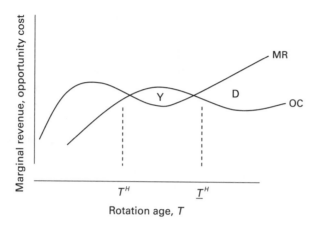

Figure 3.6
Global versus local analysis of the Hartman rotation age.

In some cases the second-order condition (3.3) does not hold globally. This was the subject of early discussions surrounding the Hartman rotation age in the literature, primarily started by Strang (1983) and Swallow et al. (1990). Strang (1983) was first to demonstrate that for $F'(T) > 0$, there are two cases where the optimal rotation age is infinite and it is never profitable to harvest. These are likely when amenity services increase steeply with the age of a stand. The first case holds when $pf''(T) - rpf'(T) < -F'(T)$ over all rotation ages; that is, when the amenity valuation increases more rapidly than the harvest revenue part decreases. In this case the opportunity cost curve always lies below the marginal revenue curve. This leads to an infinite rotation age and the stand is never harvested because it provides higher benefits when it is devoted to amenity production only.

Strang's (1983) main point was to demonstrate that in some instances the second-order condition may hold locally but not globally. To show this, we reproduce Strang's graphical argument in figure 3.6. The MR and OC curves correspond to the case where $F'(T) > 0$. We have drawn these so that the MR curve crosses the OC curve first from above and then from below.

In figure 3.6 the two intersection points define the locally optimal Hartman rotation age T^H, and a local minimum, given by $\underline{T}^H$. Suppose, however, that the landowner lets the stand grow beyond these ages. Then the marginal benefit of delaying harvest would always be higher than the opportunity cost of delaying harvest, and it would be never optimal to harvest the stand. Strang's contribution was to suggest that the value of the objective function must be checked at the end points $T = 0$ and $\lim T \to \infty$. Since the landowner starts with bare land, from (3.1) one has $W(0) = 0$ when $T = 0$. For the limit when T goes to infinity, one must have that $W(T^H) > \lim_{T \to \infty} W(T)$ for the local optimum to be a global optimum. If this does not hold, then the stand should never be harvested. In terms of figure 3.6, this means that area D is greater than area Y.

Strang was also the first to notice that rotation age choices become more complicated in the Hartman model when the landowner has an initial stand of age A. Recall from the initial stand case in chapter 2 that the rotation age of all future stands after the first harvest is determined by equations (3.1)–(3.4). To determine the optimal harvest for the initial stand, recall that after the first harvest the landowner continues with bare land, so that the future stream of value after the first harvest is simply W defined by equation (3.1). Let T now define the time of the first harvest, so that $t = T - A \geq 0$ denotes the length of time until the first harvest. The landowner's economic problem becomes one of maximizing,

$$W^0 = pe^{-r(T-A)}f(T - A) + \int_0^T F(x - A)e^{-r(x-A)}\,dx + e^{-r(T-A)}W, \qquad (3.5)$$

where the first two terms describe the present value of harvesting the initial stand (since regeneration costs are sunk, they are not present in 3.5), and $e^{-rT}W$ denotes the present value of all future rotations. The first-order condition for an interior maximum simplifies to

$$pf'(T - A) + F(T - A) = r[pf(T - A) + W]. \qquad (3.6)$$

The interpretation of equation (3.6) is similar to that of equation (3.4). If A turns out to be less than T, then the time to the first optimal harvest is $t > 0$. If A exceeds T and an interior solution (finite rotation age) is optimal, then the stand is harvested at once ($t = 0$) and a new series of rotations is begun. But if the stand is "old enough," it may be profitable to never harvest it in order to produce amenity services, depending on the nature of these services over time. In terms of figure 3.3, this would be a case where we are beyond the crossing point of MR and OC curves, and the MR curve would be intersecting from above (defining T^H). In this case, the marginal benefits from increasing marginal valuation of amenities always lie above the opportunity cost curve.

3.3.2   Comparative Statics of the Hartman Model

To examine the comparative statics of this model, an interior solution will now be assumed to exist. Surprisingly, the comparative statics results have only recently been derived by Koskela and Ollikainen (2001a), although Bowes and Krutilla (1989) did provide some earlier findings under specific assumptions. We first write the optimal Hartman rotation age as a function of exogenous variables, $T^H = T^H(p, r, c)$. Substituting $T^H = T^H(p, r, c)$, for T in $W_T = 0$ (equation 3.2), and totally differentiating with respect to exogenous parameters gives the following results:

$$\frac{dT^H}{dp} = -\frac{f'(T) - rf(T) - rf(T)e^{-rT}(1 - e^{-rT})^{-1}}{W_{TT}} < (\geq) 0 \tag{3.7a}$$

$$\frac{dT^H}{dr} = -\frac{V + r\frac{d}{dr}V + E + r\frac{d}{dr}E}{W_{TT}} < 0 \tag{3.7b}$$

$$\frac{dT^H}{dc} = -\frac{r(1 - e^{-rT})^{-1}}{W_{TT}} > 0. \tag{3.7c}$$

The sign of the last equation (3.7c) is obvious, and we can conclude that regeneration costs affect the Hartman rotation age in a way similar to that in the Faustmann model. For the remaining results we have the following:

Proposition 3.3   The Hartman rotation age is increasing in regeneration costs and decreasing in the interest rate, while the effect of timber price is ambiguous.

Proofs

Price effect To determine the sign of the numerator of (3.7a) write the first-order condition (3.2) as $f'(T) - rf(T) - rf(T)e^{-rT}(1 - e^{-rT})^{-1} = -p^{-1}[rc(1 - e^{-rT})^{-1} + F(T) - rE]$. Using lemma 3.1 we have two cases to consider. First, if $F' \geq 0$, then $F(T, \tau) - rE \geq 0$, and the RHS is negative. Second, if $F' < 0$, then $F(T) - rE < 0$, and the sign of RHS depends on whether $rc(1 - e^{-rT})^{-1} + F(T) - rE \geq (<) 0$. Q.E.D.

Real interest rate effect The numerator of (3.7b) has the term $V + r(d/dr)V$, i.e., the "Faustmann part" of the derivative, which was shown to be negative in the previous chapter. Therefore, we have to study the sign of $E + r(d/dr)E$, i.e., the "Hartman part" of the derivative. Notice that $(d/dr)E = -Te^{-rT}(1 - e^{-rT})^{-1}E - (1 - e^{-rT})\int_0^T sF(s)e^{-rs}\,ds$. Integrating the last term in $(d/dr)E$ yields $(1 - e^{-rT})\int_0^T sF(s)e^{-rs}\,ds = TE - E$, so that $(d/dr)E = -Te^{-rT}(1 - e^{-rT})^{-1}E - TE + E = Ed$, where $d = [1 + r - rT - rT(e^{rT} - 1)^{-1}] = (1 + r)[(e^{rT} - 1) - rTe^{rT}](e^{rT} - 1)^{-1}$. Next apply L'Hopital's rule for d. Differentiating the numerator and the denominator of d with respect to rT yields $\hat{d} = (re^{rT} - rTe^{rT})/e^{rT}$. Taking the limit gives

$$\lim_{\underbrace{rT \to 0}} \hat{d} = \frac{(1 + r) - 1}{1} = r > 0.$$

Hence we have $(E + rE) = E[1 + r - rT - rT(e^{rT} - 1)^{-1}] < 0$. Now the overall term is

$$-\left(1 - \frac{rT}{e^{rT} - 1}\right)[pf(T) + V] - \left(1 + r - rT - \frac{rT}{e^{rT} - 1}\right)E < 0,$$

so that $T_r^H < 0$. Q.E.D.

Proposition 3.3 is intuitive. If the timber price increases, then the relative profitability of timber and amenity production changes in favor of timber production, irrespective of the landowner's amenity valuation function. The most interesting message of proposition 3.3 is that this higher profitability of timber production may show up, not only in the possibility of a shortened rotation age, but also an unchanged or even longer rotation age. If the landowner values old stands, then the rotation age unambiguously shortens. If, however, the landowner values young stands and timber production becomes more profitable, then the landowner increases rotation age toward the Faustmann age from below, or the rotation age remains unchanged in the special case where regeneration costs and amenity valuation terms cancel each other out. A higher real interest rate affects both timber and amenity production in similar ways, decreasing the opportunity costs of harvesting in terms of both timber and amenity production. Thus, the rotation age unambiguously shortens. Finally, higher planting costs imply that the profitability of amenity production increases relative to timber production, so that the optimal rotation age becomes longer. Box 3.2 illustrates the price effect.

3.3.3 Timber Supply in the Hartman Model

The derivation of a timber supply function in the Hartman model depends on amenity valuation. Recall the distinction between short-run and long-run timber supply response presented in chapter 2. The former is captured directly by the rotation age effect and the latter by the change in the average annual harvested volume over the rotation period. In the Hartman case, the long-run supply, or average annualized harvested volume, can be expressed as

$$s^H \equiv \frac{f(T^H)}{T^H} = \frac{f[T^H(p,r,c)]}{T^H(p,r,c)}.$$

The Hartman volume coincides with the Faustmann model only when both rotation ages are identical (i.e., when amenity valuation is invariant in time). However, it is smaller when amenity benefits from younger stands are valued and higher when amenity benefits from older stands are valued; the latter holds if the Hartman rotation age is shorter than the MSY rotation age.

The way in which short-run timber supply responds to a higher timber price also depends on the nature of the amenity valuation function. If older stands are valued, then an increase in timber price effectively makes the stand overmature.

It will be immediately cut, implying a higher short-run timber supply. However, if a young stand is valued for amenities, then a price increase may increase or decrease short-run timber supply. A higher interest rate increases short-run supply, but an increase in planting cost decreases short-run timber supply because the stand is grown longer, just as with the Faustmann model.

The long-run supply effects can be obtained by differentiating the average annual harvest with respect to parameter ψ (recall $\psi = p, r, c$):

$$\frac{ds^H}{d\psi} \equiv \frac{T_\psi^H}{T^H}\left[f'(T^H) - \frac{f(T^H)}{T^H}\right]. \tag{3.8}$$

The long-run supply response depends, again, on the rotation age effect, T_ψ^H, and on the difference between marginal and average growth. The sign of the last (bracketed) term will vary in the Hartman model. If young stands are valued, or if amenity valuation is independent of the rotation age, then the Hartman rotation age is shorter than or equal to the Faustmann rotation age. Plausibly, it is also shorter than the MSY rotation age, so that the second term in brackets in (3.8) is positive. But if older stands are valued, the Hartman rotation age can exceed the MSY age and the second term would then be negative.

Suppose now that timber price increases. If the landowner values older stands for amenities, we know that the Hartman rotation age shortens and if it approaches the MSY age from above, timber supply increases. However, if the Hartman rotation age is originally shorter than the MSY age, then timber supply decreases because now the second term in (3.8) is negative. When the landowner values younger forests, the timber supply response is ambiguous because the effect of rotation age is ambiguous.

The effects of the interest rate and regeneration costs on rotation age are negative and positive, respectively, so that the effect of these on long-term timber supply depends on whether the Hartman rotation age exceeds or falls short of the MSY age. In the former case, a higher interest rate increases timber supply and in the latter case, it decreases it. The regeneration cost works in an opposite manner, remaining ambiguous in its effect. Thus the long-term timber supply function of the Hartman model contains some ambiguity, $s^H[T^H(p, r, c)] = s^H(\underset{-/+}{p}, \underset{-/+}{r}, \underset{+/-}{c})$.

3.4 Effects of Forest Taxation

We now focus on the question of how various types of forest taxes affect the relative profitability of harvesting versus amenity production. This question is of considerable policy relevance. If there is any divergence in the preferences of landowners and society concerning the flow of amenity services provided by

forest stands, then a government could use tax instruments to change private rotation ages, thereby moving both timber and amenity production toward socially
optimal outcomes.

Chapter 5 considers these policy design questions in detail. In this section we
will not yet define what we mean by the socially optimal set of policies or the socially optimal design of taxation. However, the comparative statics we derive here
will be important once we turn to these social optimality issues in chapter 5. The
comparative statics of the same forest taxes we studied in chapter 2 are presented;
namely, harvest taxes (yield and unit tax) and property taxes (site productivity
and site value tax).

3.4.1 Harvest Taxes

If a government levies a yield tax (τ) or a unit tax (u) on harvesting, then after-tax
net harvest revenue is equivalent to the Faustmann case of equation (2.6a), while
the amenity part, E, of the Hartman objective function remains unchanged. We
then have

$$\hat{W}(\tau, u) = (1 - e^{-rT})^{-1}[\hat{p}f(T)e^{-rT} - c] + (1 - e^{-rT})^{-1} \int_0^T F(s)e^{-rs}\, ds, \qquad (3.9)$$

where $\hat{p} \equiv p(1 - \tau) - u$ denotes after-tax timber price assuming the presence of
yield and unit taxes. As in chapter 2, it is again evident that harvest taxes work
just like a decrease in timber price. Recalling the comparative static effect of timber price, we can simply re-express it for the yield and unit tax as follows:[3]

$$\frac{\partial T^H}{\partial \tau} = -\frac{p}{(1-\tau)} T_p^F \quad \text{and} \quad \frac{\partial T^H}{\partial u} = -T_p^F. \qquad (3.10)$$

Therefore we obtain the following result:

$$T_\tau^H, T_u^H \begin{Bmatrix} > \\ = \\ < \end{Bmatrix} 0 \quad \text{as } rc(1 - e^{-rT})^{-1} + F(T) - rE \begin{Bmatrix} > \\ = \\ < \end{Bmatrix} 0. \qquad (3.11)$$

The effects of harvest taxes on the Hartman rotation age depend on both the nature of amenity valuation and the size of regeneration costs. Harvest taxes have
no effect on the amenity part of the objective function. Therefore, if $F'(T) > 0$,
then harvest taxes will increase the profitability of amenity production relative to
timber production, and the landowner will lengthen the rotation age as in the

---

3. The reader can see this by solving the comparative statics of taxes and comparing the explicit solutions with the price effect.

basic Faustmann model. On the other hand, if $F'(T) = 0$, then harvest taxes will have the same qualitative and quantitative effects on rotation age as the Faustmann model does. Finally, in the case of $F'(T) < 0$, harvest taxes make timber production less profitable relative to amenity production (recall that now the Hartman rotation age is shorter than the Faustmann age). Provided that regeneration costs are "small enough," the landowner shifts toward amenity production by shortening the rotation age.

3.4.2 Property Taxes

Recalling equations (2.7) and (2.8) from chapter 2, we can express the landowner's objective functions for the site value tax and the site productivity tax, respectively, as

$$\hat{W}(\theta) = (1 - \theta)V + E, \tag{3.12}$$

and

$$\hat{W}[a(i)] = V - \frac{a(i)}{r} + E. \tag{3.13}$$

For the site value tax θ, we have: $\hat{V}(\theta) = (1 - \theta)V$, and the first-order condition for the maximization of (3.12) is given by

$$\hat{W}_T(\theta) = (1 - \theta)[pf'(T) - rpf(T) - rV] + F(T) - rE = 0. \tag{3.14}$$

Differentiating (3.14) with respect to θ and using lemma 3.1 yields

$$\hat{W}_{T\theta}(\theta) = -[pf'(T) - rpf(T) - rV] \geq (<) \, 0 \quad \text{as } F'(T) \geq (<) \, 0. \tag{3.15}$$

Using equation (3.15) and the fact that sign $(T_\theta^H) = \text{sign}(W_{T\theta})$, we arrive at the following results:

$$T_\theta^H \begin{Bmatrix} > \\ = \\ < \end{Bmatrix} 0 \quad \text{as } F'(T) \begin{Bmatrix} > \\ = \\ < \end{Bmatrix} 0. \tag{3.16}$$

According to equation (3.16), the site value tax does not affect rotation age (it is neutral) if amenity valuation is site-specific, i.e., when $F'(T) = 0$. If, however, $F'(T) > (<) \, 0$, then a rise in the site value tax makes amenity production relatively more (less) profitable than timber production. Consequently, the landowner lengthens (shortens) the rotation age.

For the site productivity tax, the first-order condition can be shown to equal $\hat{W}_T[a(i)] = V_T + F(T) - rE = 0$. The site productivity tax is neutral with respect to

the landowner's amenity valuation because it does not affect the relative profitability of timber and amenity production.

For a property tax on stumpage value, α, the objective function of the landowner is given by

$$\hat{W}(\alpha) = V - \alpha(1 - e^{-rT})^{-1} \int_0^T pf(s)e^{-rs}\, ds + E. \tag{3.17}$$

The first-order condition is

$$\hat{W}_T(\alpha) = pf'(T) - rpf(T) - rV - \alpha[pf(T) - rU] + F(T) - rE = 0. \tag{3.18}$$

Differentiating (3.18) with respect to α yields $W_{T\alpha}(\alpha) = -[pf(T) - rU] < 0$ (to
see that this partial derivative is negative, recall lemma 2.1 in chapter 2). Hence
the timber tax shortens the rotation age irrespective of the sign of the term
$F(T) - rE$ because the sign of the timber production part in (3.18) does not determine the sign of $\hat{W}_{T\alpha}(\alpha)$. This holds for any timber growth function with
$f'(T) > 0$. The interpretation is the same as with the Faustmann model.

Summarizing, we have obtained the following results for forest taxes in the
Hartman model:

Proposition 3.4   The harvest and site value taxes affect the Hartman rotation age
a priori ambiguously, with their effects depending on how the amenity valuation
function behaves in terms of stand age. While the site productivity tax has no effect on rotation age, the timber tax will shorten it.

It is interesting that while harvest and timber taxes continue to be distortionary,
the site value tax also becomes distortionary when it is moved from the Faustmann model to the Hartman model. The site productivity tax is the only tax that
remains neutral. The harvest and site value taxes either lengthen or shorten the rotation age, depending on when amenity valuation increases or decreases with the
age of the stand, respectively. They all decrease the timber production component
of land value, but they do not affect the amenity valuation part, making amenity
production more profitable relative to producing timber income. This shift toward
amenity production implies a longer (or shorter) rotation age if amenity valuation
increases (or decreases) with age of the stand. The timber tax unambiguously
shortens the rotation age independent of amenity valuation.

3.5   Amenities from Interdependent Stands

As should be obvious, forest ecosystems consist of complex site-specific interactions among trees, other plants, and animal species. In many instances, the man-

agement of each stand in a given region should not be undertaken independently of other stands. Bowes and Krutilla (1989) pointed out that harvesting even one stand may sometimes pose a threat to the maintenance of the entire ecosystem, and it would certainly affect the pattern of amenity services produced in the forest area.

We now develop a single-stand extension of the Hartman framework that can be used in a preliminary analysis of the choice of rotation age at a forest landscape level. Bowes and Krutilla (1985, 1989) initially proposed a linear programming approach to maximize the rents associated with multiple stands under a single (government) owner. Swallow and Wear (1993) and Swallow et al. (1997) were the first to formulate explicit spatial interactions for nontimber amenity benefits using the idea of a "focal" stand (the one of interest) and an exogenous "adjacent" stand, but they mainly concentrated on numerical analysis of the problem. Drawing on their work, Koskela and Ollikainen (2001b) presented an economic description of stand interdependence, examining the rotation age decision for a single landowner making decisions for a focal stand under the assumption of the landowner taking the adjacent stand as exogenous. Important other contributions have also been made by Snyder and Bhattacharyya (1990) and Vincent and Binkley (1993). The former analyzed a case where maintenance costs are associated with nontimber values that prevent the values from vanishing through a process of decay and will be discussed in chapter 12. The latter used production possibilities frontiers to show that specialization of land use may sometimes be efficient in parts of a forest ecosystem.

Here we develop a description of interdependence by providing two definitions of spatial dependence and its evolution over time, and then showing how to solve for the optimal rotation age of a focal stand. Our discussion here relies mainly on Swallow and Wear (1993) and Koskela and Ollikainen (2001b).

3.5.1 Spatial and Temporal Interdependence

Consider two adjacent stands, denoted by 'a' and 'b', assumed to be owned by two different landowners, A and B. Assume further that the stands are interdependent in terms of amenity production, but independent with regard to timber production. Problems have been examined for the management of one stand, called the focal stand. Denote this as stand a. The assumption is also made that the rotation age of the adjacent stand b is taken to be exogenously given by the landowner of stand a. Assume that landowner A maximizes utility from net harvest revenue and amenity services by explicitly accounting for interdependence between the focal and the adjacent stand.

The important question to consider then becomes: How do we model stand interdependence in the landowner's objective function? Swallow and Wear (1993) proposed incorporating the rotation age of the adjacent stand into the focal land-owner's amenity benefit function, denoted by $v(T, \tau)$, as follows:

$$v(T, \tau) = \int_0^T F(s, \tau) e^{-rs}\, ds, \tag{3.19}$$

where $F(s, \tau)$ is the flow of amenities from the focal stand of age s when it is po-tentially affected by an adjacent exogenous stand of age τ. As we did earlier, we can obtain the discounted marginal valuation function at a given rotation age of the focal stand by differentiation: $v_T(T, \tau) = F(T, \tau) e^{-rT}$. Definition 3.1 then holds for the properties of this amenity valuation function with respect to the rotation age of the focal stand, T.

The sign of the cross-derivative, $v_{T\tau}(T, \tau)$, indicates how the discounted mar-ginal valuation of the focal stand depends on the age of the exogenous stand. To explore this interdependence more precisely, we follow Koskela and Ollikainen (2001b) and define the "static" concept of Auspitz-Lieben-Edgeworth-Pareto (ALEP) complementarity or substitutability between forest stands using the fol-lowing definition:

Definition 3.2 Spatial interdependence Two stands of trees are substitutes, independents, or complements in the ALEP sense when an increase in the age of the exogenous stand decreases, leaves unchanged, or increases the marginal valu-ation of amenities for the focal stand, respectively, i.e.,[4]

$$F_\tau(T, \tau) \begin{Bmatrix} < \\ = \\ > \end{Bmatrix} 0 \quad \text{if stands are} \begin{cases} \text{substitutes with respect to amenities} \\ \text{independents with respect to amenities} \\ \text{complements with respect to amenities} \end{cases}$$

Thus, if the stands are spatial substitutes, then the marginal amenity valuation of each stand decreases with the age of the adjacent stand. If the stands are spatial complements, then the opposite is true, i.e., the marginal amenities of each stand increase with the age of the adjacent stand.

It is not difficult to think of empirical examples that illustrate the various cases in definition 3.2. Swallow and Wear (1993) imagined a landowner who values for-age production consistently with big game production, where big game requires both forage and cover. The focal stand and the exogenous adjacent stand then function as substitutes in their production if both stands can provide forage and cover simultaneously. The stands will be complements if the focal stand provides forage while the adjacent stand provides cover. If both forage and cover can be

---

4. For a definition of ALEP, see Samuelson (1974) and further discussions in Chipman (1977), Kannai (1980), and Weber (2000).

obtained in the focal stand and the adjacent stand does not contribute at all to game production, then the stands are spatially independent according to the definition.

Because we are interested in forest stands over all possible rotation (calendar) ages, it is also important to know how substitutability or complementarity is affected by the choice of rotation age for each stand. This is obtained by differentiating the amenity valuation function in definition 3.2 with respect to the own (focal) stand rotation age. The resulting second derivatives of the marginal amenity valuation function define how spatial dependence between stands evolves with the focal rotation age. This is called "temporal interdependence" and is given by the following definition:

Definition 3.3 Temporal interdependence Temporal interdependence between two stands is constant, increases, or decreases when substitutability or complementarity between the stands remains unchanged, increases, or decreases with a higher focal stand rotation age, i.e.,

$$
F_{\tau T}(T, \tau) \begin{Bmatrix} < \\ = \\ > \end{Bmatrix} 0 \quad \text{if} \quad \begin{cases} \text{spatial dependence decreases with stand age} \\ \text{spatial dependence is unchanged} \\ \text{spatial dependence increases with stand age} \end{cases}
$$

From definition 3.3, temporal interdependence between two stands may be constant, increasing, or decreasing, depending on how the spatial substitutability or complementarity between the stands changes with increases in the rotation age of the focal stand. *Constant temporal dependence* holds when substitutability or complementarity is not associated with time. This would be the case if substitutability or complementarity is merely associated with site-specific properties that remain constant as the rotation age of the focal stand changes (an example would be the slope or aspect). Note that if the amenity valuation is site-specific, then $F_\tau = 0$ and the stand is also temporally independent. The reverse does not necessarily hold (the reader should think about why this is true). *Increasing temporal dependence* between stands means that for complements, the complementarity between stands increases with the focal rotation age. However, for substitutes, substitutability between stands decreases with the focal rotation age. *Decreasing temporal dependence* implies just the opposite: complementarity weakens while substitutability becomes stronger for increases in the focal rotation age.

Providing empirical examples for increasing or decreasing temporal dependence is more complicated than the examples associated with definition 3.2. Consider the previous example of spatial interdependence. Suppose that the focal and adjacent stands are originally spatial substitutes and the rotation age of the adjacent stand is increased. If this decreases (or increases) forage and cover within the focal stand, perhaps owing to changes in understory vegetation, then we have a case of decreasing (or increasing) temporal dependence between stands. One therefore expects that temporal dependence is important for conserving biodiversity. Typical old-growth wildlife species require significant stands of older trees for colonization. While some old-growth species are highly specialized in terms of habitat needs, others are specialized to a lesser extent. Therefore, if increasing the rotation age of old-growth stands makes the forest more (less) suitable to more (less) specialized species, this would indicate stronger (weaker) interdependence between stands.

3.5.2 Optimal Rotation Age and Stand Interdependence

Using the definitions 3.2 and 3.3, we now turn to the landowner's choice of rotation age. The landowner chooses T to maximize the present value from future harvest revenue and the benefits of amenity services over an infinite cycle of rotations:

$$\max W = V + E, \tag{3.20}$$

where $V = [pf(T)e^{-rT} - c](1 - e^{-rT})^{-1}$ and $E = (1 - e^{-rT})^{-1} \int_0^T F(s, \tau)e^{-rs}\, ds$.

The first-order condition, $W_T = V_T + E_T = 0$, can be expressed as

$$W_T = pf'(T) - rpf(T) - rV + F(T, \tau) - rE = 0. \tag{3.21}$$

The second-order condition is

$$W_{TT} = pf''(T) - rpf'(T) + F_T(T, \tau) < 0, \tag{3.22}$$

which we assume to hold. The interpretation of equation (3.21) should already be familiar to the reader. The landowner chooses the roation age to equate the marginal benefit of delaying the harvest at age T, defined by $pf'(T) + F(T, \tau)$, to the marginal opportunity cost of delaying harvest at time T. The opportunity cost is defined by interest lost on the timber and land, $rpf(T) + r(V + E)$. We have a different type of externality present here, because the flow of amenity benefits for the landowner of the focal stand depends on the rotation age chosen by the landowner of the adjacent stand.

Now we ask: What happens to the optimal focal rotation age when the rotation age of the exogenous adjacent stand changes? Using a procedure similar to the one described earlier, we obtain $T_\tau^H = -(W_{TT})^{-1}W_{T\tau}$, where

$$W_{T\tau} = F_\tau(T, \tau) - r(1 - e^{-rT})^{-1} \int_0^T F_\tau(s, \tau)e^{-rs}\, ds. \tag{3.23}$$

Applying integration by parts to (3.23) yields

$$\int_0^T F_\tau(s,\tau)e^{-rs}\,ds = \frac{1}{r}\left[F_\tau(0,\tau) - F_\tau(T,\tau)e^{-rT} + \int_0^T F_{\tau T}(s,\tau)e^{-rs}\,ds\right],$$

so that equation (3.23) can be re-expressed as

$$W_{T\tau} = F_\tau(T,\tau) - (1 - e^{-rT})^{-1}\left[F_\tau(0,\tau) - F_\tau(T,\tau)e^{-rT} + \int_0^T F_{\tau T}(s,\tau)e^{-rs}\,ds\right]. \tag{3.24}$$

The terms in equation (3.24) have a natural interpretation. The first term describes the effect of the adjacent stand on the amenity valuation of the focal stand at the time of the first harvest. The first and the second RHS bracket terms give the present-value effect over all rotations of a change in the adjacent rotation age τ on the marginal amenity valuation of the focal stand at the time of harvest. Finally, the third RHS (integral) term captures the present-value effect of the temporal interdependence between the focal and adjacent stands. It describes whether complementarity or substitutability of the stands becomes stronger, weaker, or remains unchanged when the rotation age of the focal stand changes.

The response of the focal rotation age, T^H, to a change in the adjacent stand, τ, can be summarized by the following proposition:

Proposition 3.5   The focal rotation age shortens, remains constant, or lengthens in response to a longer rotation age for the adjacent stand when temporal dependence is decreasing, constant, or increasing, respectively; that is, when

$$T_\tau^H \begin{Bmatrix} < \\ = \\ > \end{Bmatrix} 0 \quad \text{as } F_{\tau T} \begin{Bmatrix} < \\ = \\ > \end{Bmatrix} 0.$$

The full proof of proposition 3.5 is given in Koskela and Ollikainen (2001b). This shows that the response of the focal stand rotation age will depend only on how temporal dependence between the stands is affected by a change in the focal rotation age. Therefore, it is critical to know the sign of the term $F_{\tau T}$ alone, not the notion of spatial dependence per se.

In the following three cases we apply the notion of temporal interdependence developed in definition 3.3 to provide a further interpretation of proposition 3.5. Before doing this, we should point out that under spatial independence, both $F_\tau = 0$ and $F_{\tau T} = 0$, so that the rotation age of the adjacent stand has no effect on that of the focal stand. A more interesting case can be obtained when $F_\tau \neq 0$.

Case 1 (temporal independence): $F_{\tau T} = 0$: Temporal independence means that the spatial complementarity or substitutability relationship is merely due to site-specific characteristics and does not depend on rotation ages. Thus a change in the rotation age of the adjacent stand affects neither the marginal valuation nor the opportunity costs of delaying harvest for the focal stand. Thus the landowner of the focal stand has no reason to change the stand's rotation age.

Case 2 (increasing temporal dependence): $F_{\tau T} > 0$: Increasing temporal dependence implies that spatial complementarity becomes stronger, or that spatial substitutability becomes weaker, as the rotation age of the adjacent stand changes. For complements, a rise in the rotation age of the adjacent stand increases both the marginal amenity valuation at harvest time and the opportunity cost effect of the future amenity valuation, but the former effect is stronger. For substitutes, a longer rotation age for the adjacent stand decreases both the marginal valuation and the opportunity cost of amenity services, and the latter effect is stronger. Therefore, in both cases the landowner lengthens the focal stand's rotation age.

Case 3 (decreasing temporal dependence): $F_{\tau T} < 0$. If temporal dependence decreases, then spatial complementarity becomes weaker or spatial substitutability becomes stronger as the rotation age of the adjacent stand increases. For weakening complements, a higher rotation age for this stand increases both the marginal valuation and the opportunity costs of delaying harvest, but the latter effect is stronger. For increasing substitutability, a higher rotation age for the adjacent stand decreases both the marginal valuation of amenities and the opportunity costs, the former effect being stronger. In both cases the landowner shortens the focal stand's rotation age.

In practice, the problem here is more complicated once we extend our thinking to an endogenous adjacent stand, where landowners may interact with each other and possibly behave strategically. This possibility has been analyzed in Amacher et al. (2004).

3.6 Modifications

There are many modifications of the Hartman model outside of stand interdependence, such as preservation of old-growth forests, biodiversity conservation, water quality issues, and wildlife management.[5] We return to some of these topics in later chapters. In this section we extend the land allocation and life-cycle con-

---

5. For example, Conrad and Ludwig (1994) examine the optimal stock of old-growth forest. Strange et al. (1999) focus on the joint production of timber, forage, and water protection. Bowes et al. (1984) analyze forest management for increased timber and water yields using a numerical simulation. Steinkamp and Betters (1991) and Swallow et al. (1990) examine joint production of timber and forage, the latter focusing on the nonconvexities that joint production may cause. Strange et al. (1999) include an additional nontimber benefit and examine joint production of timber and forage and water protection.

sumption models of chapter 2 to include amenities, and we show how carbon policies can be studied with this framework.

3.6.1 Competing Land Uses

The land allocation model of chapter 2 can be adapted to accommodate amenity benefits in the spirit of Hartman. As long as the Hartman rotation age remains finite, regular forest management prevails and land will be allocated between forestry and agriculture just as we saw earlier. However, if the amenity valuation function is such that some parcels have an infinite rotation age, then these parcels will always be devoted entirely to amenity production and not forest harvesting. Thus devoting some parcels to pure amenity production provides an additional land-use form.

As with chapter 2, assume that a single landowner holds a plot of land of varying quality, divided into a continuum of parcels for which quality is uniform within a parcel but varies across parcels. The indirect net present value profit function for agricultural production remains the same, $\pi^*(\,p_a, w, r, q)$ (recall that p_a is the price of agricultural product and w is the input price). Let forest growth depend on land quality in the same way as before. Amenity benefits may behave in many ways; they may be tied to land quality or may be independent of it. In the former case, high amenity services may be associated with better or less desirable land qualities (e.g., wetlands may belong to low qualities in terms of forest growth for some species, yet they sustain high biological diversity). Fast-growing forests are usually found on higher-quality land, and these can also provide considerable amenity services in terms of recreation. To keep our discussion simple, we assume here that high amenity benefits are associated with better land qualities, so that we express amenity valuation as $F(T, q)$ with $F_q > 0$. The net present value of forest revenue and amenity benefits for any land quality q is given by

$$W = (1 - e^{-rT})^{-1}\left[\,p_f f(T, q)e^{-rT} + \int_0^T F(s, q)e^{-rs}\,ds - c\right]. \tag{3.25}$$

Finally, assume that the landowner values amenity benefits from old stands, and let this valuation be strong enough that for some better-quality parcels, devoting land to amenity production forever is more profitable than commercial plantation-based forestry. The indirect net present revenue function per parcel associated with (3.25) has the following properties: $W^* = W^*(\,p_f, c, r, q)$.

As before, the total amount of the land is given by

$$G = \int_0^1 g(q)\,dq, \tag{3.26}$$

where $g(q)$ is the density function for G; i.e., $g(q) = G'(q)$. The landowner allocates area G to agriculture, commercial forestry, and amenity production. We retain our previous assumption that agriculture has higher profits on higher-quality lands and forestry has higher profits on lower-quality lands. However, we add an assumption that on some of the parcels of highest-quality land, amenity production is the most profitable land use. Denote the share of land devoted to amenity production by h_{ap}, agriculture by h_a, and commercial forestry by $h_f = 1 - h_{ap} - h_a$. The landowner now solves the following problem:

$$\max_{h_f, h_{ap}, h_a} PV = \int_0^1 \{\hat{W}^*(q)h_{ap}(q) + \pi^*(q)h_a(q) + W^*(q)[1 - h_{ap}(q) - h_a(q)]\}g(q)\,dq.$$

$$(3.27)$$

The objective function is the sum of the rents from all land uses indexed by quality, summed over all land quality levels. The necessary first-order conditions for land allocation are

$$\frac{\partial PV}{\partial h_a} = \pi^*(q) - W^*(q) = 0,$$

$$(3.28a)$$

$$\frac{\partial PV}{\partial h_{ap}} = W^*(q) - \hat{W}^*(q) = 0.$$

$$(3.28b)$$

These conditions define upper and lower critical qualities, q_i^c, $i = 1, 2$, which define the margins of land allocation between amenity production and agriculture, and between agriculture and commercial forestry; $q_1^c : \hat{W}^*(q) - \pi^*(q) = 0$ and $q_2^c : \pi^*(q) - W^*(q) = 0$. Using these margins, the choices of land area are given by

$$h_f = \int_0^{q_2^c} g(q)\,dq \quad \text{and} \quad h_a = \int_{q_2^c}^{q_1^c} g(q)\,dq \quad \text{and} \quad h_{ap} = \int_{q_2^c}^1 g(q)\,dq.$$

$$(3.29)$$

The comparative statics of land allocation could now be solved in a similar way as before by differentiating equations (3.28a) and (3.28b) with respect to exogenous variables, accounting for the margins given by (3.29).

3.6.2 Life-Cycle Models

Recall from chapter 2 our simple life-cycle model where the landowner's problem was to simultaneously choose consumption and rotation age. Suppose in addition to what was assumed in chapter 2 that this landowner now has a separable utility function that depends on both consumption and amenities. Assume that this utility function is concave and that definition 3.1 holds for amenities. Once again, the landowner finances his consumption by using an exogenous income stream and harvest revenue:

$$\underset{(c,T)}{\max} \ U = \int_0^\infty u[c(t)]e^{-\rho t} \, dt + E, \ \text{s.t.} \tag{3.30a}$$

$$\int_0^\infty c(t)e^{-rt} \, dt = \int_0^\infty m(t)e^{-rt} \, dt + V. \tag{3.30b}$$

In this model, the amenity benefit, E, is part of the utility function, while the Faustmann part, V, is part of the intertemporal budget constraint because it represents timber revenues. The Lagrangian function for this problem is

$$L = \int_0^\infty u[c(t)]e^{-t} \, dt + E + \lambda \left[\int_0^\infty m(t)e^{-rt} \, dt + V - \int_0^\infty c(t)e^{-rt} \, dt \right],$$

where λ is the Lagrange multiplier, which has the following first-order conditions for optimal consumption and harvesting:

$$L_c = u'[c(t)]e^{-\rho t} - \lambda e^{-rt} = 0 \tag{3.31a}$$

$$L_T = E_T + \lambda V_T = 0 \Leftrightarrow F(T) - rE + \lambda[pf'(T) - rpf(T) - rV] = 0 \tag{3.31b}$$

$$L_\lambda = \int_0^\infty m(t)e^{-rt} \, dt + V - \int_0^\infty c(t)e^{-rt} \, dt = 0. \tag{3.31c}$$

Condition (3.31a) characterizes the optimal consumption path at the margin; it is similar to what we found in chapter 2; i.e., the path is constant, increasing, or decreasing over time, depending on the relationship between the rate of time preference and real interest rate. The difference here compared with the model in chapter 2 can be seen in the choice of rotation age in equation (3.31b). If λ were equal to one, then we would obtain the Hartman solution. However, from (3.31a) we have $\lambda = u'[c(t)]e^{(r-\rho)t}$, and thus (3.31b) becomes

$$F(T) - rE + u'[c(t)]e^{(r-\rho)t}[pf'(T) - rpf(T) - rV] = 0. \tag{3.32}$$

Clearly, the rotation age now depends, not only on amenity benefits, but also on the marginal utility of consumption. This means that consumption choices and forest management decisions depend on each other, and so Fisherian separation no longer holds.

To be more precise, the Hartman model relies on the assumption of quasi-linearity; that is, the model is linear in harvest revenue but nonlinear in amenity benefits. Linearity in income means that the landowner's marginal utility of income is equal to one, so that preferences over consumption do not matter. A natural consequence of this is that income effects are zero. Conversely, in equation (3.32) the concave utility function links forest management to consumption choices via decreasing marginal utility. Now all variables affect forest management. We would expect, for instance, that an increase in exogenous income will affect the optimal rotation age. It can be shown that this income effect on rotation age is positive (or negative) if the landowner values amenities from old (or young) sands. Not surprisingly, the effects of stumpage price and the interest rate on rotation age are generally ambiguous. Regeneration cost has an ambiguous effect on rotation age when the landowner values amenities from old stands, but it increases rotation age if young stands are valued. Finally, a higher rate of time preference increases (decreases) the rotation age if old (young) stands are valued (Tahvonen and Salo 1999).

3.6.3   Forests and Carbon Sinks

Climate policy and the role of forests in reducing global warming has been a major research agenda since the 1990s. In terms of forest economics theory, researchers have addressed many questions, including whether forest plantations provide an efficient means of sequestrating carbon in trees; the management of old-growth forests, a particularly large existing carbon sink; and what role second- and third-growth stands should have in climate policy debates. Furthermore, the role of emerging carbon markets on stand and forest management remains an active research area.

By interpreting the amenity valuation function as the ability of forest stocks to sequester carbon, these studies focus on the optimal rotation age of a stand when timber and carbon reductions are jointly produced. A series of articles by Englin and Callaway (1993), Gutrich and Howarth (2007), Plantinga and Birdsey (1994), Hoen and Solberg (1994), and van Kooten et al. (1995) focused on these issues using the Hartman framework (for a broader review and assessment of these issues, see Sedjo et al. 1995, 1997). Plantinga and Birdsey (1994) note that carbon in a stand of trees is stored in four categories: trees, soil, litter, and understory vegetation. In the Hartman framework, these site characteristics are most easily combined into a single carbon index. In addition, it is important to specify how much carbon will be released by harvesting and whether forests should be placed into reserves and not harvested (Parks and Hardie 1995).

We consider a model in this section that is based on van Kooten et al. (1995). Carbon benefits are usually assumed to be a function of the change in biomass and the amount of carbon per cubic meter of biomass. As trees grow, they sequester carbon, but once the capacity of trees to store carbon is reached, no further benefits accrue. Naturally, one cannot define carbon benefits without making an assumption about the social valuation of these benefits. All the papers mentioned here postulate the existence of an implicit carbon price to describe the social valuation of carbon benefits.[6]

Following van Kooten et al. (1995), but using our previous notation, let $f(t)$ denote the growth function of trees at time t and let α denote the tons of carbon sequestered per cubic meter of timber biomass. The carbon sequestered is then given by $\alpha f'(t)$, i.e., by the growth rate of timber. The present value of carbon uptake benefits $F(T)$ over a rotation period T can then be written as

$$F(T) = q\alpha \int_0^T f'(s)e^{-rs}\,ds, \tag{3.33}$$

where q denotes the constant implicit social value of carbon that is removed from the atmosphere, r is the discount rate, and all other variables are as previously defined in the chapter. Equation (3.33) is an amenity valuation function written in terms of carbon sequestered. Integrating it by parts yields

$$F(T) = q\alpha \left[f(T)e^{-rT} + r \int_0^T f(s)e^{-rs}\,ds \right]. \tag{3.34}$$

One now has to define the carbon release that is due to harvesting. Let β denote the fraction of carbon that remains in long-term storage after harvesting, such as in structures and landfills. The share $(1 - \beta)$ denotes the amount of carbon released into the environment by harvesting. This can be expressed as $(1 - \beta)\alpha f(T)$ at the rotation age, T.

We first analyze the case of second- and third-rotation stands; that is, we assume the landowner starts with bare land and some regeneration (which is assumed to be costless). The social net present value of timber harvested is then given by the difference between the harvest revenue and the discounted external cost of the carbon that is released. Over one rotation this is

$$\hat{V} = pf(T)e^{-rT} - q\alpha(1 - \beta)f(T)e^{-rT}, \tag{3.35}$$

---

6. The existence of a price is typically motivated by either the nonmarket values that society places on stored carbon, or the assumption of an explicit carbon market where storage capacity is traded at some price. Current emissions trading programs in the United States (Regional Greenhouse Gas Initiative) and in the European Union (Emissions Trading Scheme) provide emission permit prices as market-based estimates for the marginal costs of carbon reductions and thereby provide an estimate for the benefits obtained by sequestering carbon in forests.

where the first RHS term denotes the present value of harvest revenue for one rotation period and the second term defines the present value of external costs (social costs) that are due to carbon release. The present value of harvest revenue and carbon sequestration benefits, $\hat{W}$, over all future rotations is then given by the sum of (3.34) and (3.35), so that we can write

$$\hat{W} = \left\{ [p - q\alpha(1-\beta)]f(T)e^{-rT} + q\alpha \left[f(T)e^{-rT} + r \int_0^T f(s)e^{-rs}\, ds \right] \right\} (1 - e^{-rT})^{-1}.$$

(3.36)

It makes sense here that both carbon uptake and carbon release are valued according to the same social price (the marginal benefit from sequestering a unit of carbon equals the marginal cost of releasing one unit). Differentiating (3.36) with respect to rotation age yields

$$\hat{W}_T = 0 \Leftrightarrow (p + q\alpha\beta)f'(T) + q\alpha f(T) = r\hat{W}.$$

(3.37)

The LHS of equation (3.37) is the marginal benefit of delaying harvesting. It consists of the value of harvested timber as a sum of stumpage price plus the value of the share of sequestered carbon not released by harvesting. The RHS is the opportunity cost of delaying harvesting expressed as interest lost by not harvesting plus the interest rate multiplied by rent on land plus sequestration. According to equation (3.37), the optimal rotation age is to be chosen so that the marginal benefits of delaying harvest equal the opportunity costs of delaying harvest.

The question of an old-growth forest can be approached by postulating an initial stand following the ideas of Strang (1983). Assume further that the age of this initial stand is greater than the optimal rotation age defined by (3.37). Van Kooten et al. (1995) demonstrate that in this case the condition for harvesting an "over-aged" stand is

$$pf(T) - q(1-\beta)\alpha f(T) + \hat{W} > q\alpha \int_0^\infty f'(s)\, ds.$$

(3.38)

According to equation (3.38), the marginal benefits of cutting the overmature stand (LHS) must exceed the marginal costs (RHS). The marginal benefits are given as the sum of the stumpage value of the trees, net of the costs of carbon release, plus the discounted value of all future harvests. The marginal cost consists of forgone carbon benefits. Weitzman (2003) in his book provides a similar condition but assumes that the future consequences of global warming are uncertain.

In their empirical application, van Kooten et al. (1995) find that for a specific estimate of the social valuation of carbon sequestration, rotation age becomes longer than the Faustmann rotation age, and under some values it becomes optimal not to harvest stands at all.

3.7 Summary

The model first proposed by Hartman was the earliest attempt to include amenity values of forest stocks in landowner decision making. It turns out that the importance of amenity benefits in the rotation age decision is more complicated than Hartman's original analysis and depends on whether young or old stands are valued by the landowner, and on the conditions under which preservation of forests is optimal. In this model, with its quasi-linear separability of timber income and amenity valuation, we find that interest rate and regeneration cost parameters affect the optimal rotation age in a way similar to that of the Faustmann model, but the effect of price can be to either lower or raise rotation age, depending on the properties of the amenity valuation function. These properties also make the effects of these parameters on supply much more complicated than in the Faustmann case.

Although these findings were all established by the mid-1980s, the model could still be the basis of new and interesting developments. There is an obvious need to describe amenity services in a more detailed way, and especially to account more adequately for interactions between the stands in amenity production. The obvious future importance of policies targeting biodiversity conservation and climate (carbon) will mean a continuing need to describe the processes of the forest ecosystem on both unit and landscape scales. From a social value point of view, the life-cycle approach reveals that landowner preferences may considerably affect the choice of rotation age and harvesting behavior. Despite a large body of empirical work on harvesting behavior, we still do not have a great understanding of how various amenities drive this behavior in these situations. We imagine that the Hartman model and its variants will feature importantly in continuing investigations of this type for years to come.

4   Two-Period Life-Cycle Models

In the 1980s, Swedish economists Karl-Gustav Löfgren and Per-Olov Johansson proposed a new model for analyzing forest economics.[1] They considered the problem of a risk-averse landowner who maximizes the present value of harvest revenue over two periods in the presence of uncertainty for the future price of the timber. The landowner begins with an initial endowment of forest and chooses how much to harvest in the current and future periods. Their model turned out to combine two approaches: a classical fishery model in that it made reference to the biomass of forest stands, and a traditional two-period (life-cycle) model, which has been widely applied elsewhere in economics. The first complete presentation of the forestry two-period life-cycle harvesting model was by Koskela (1989a,b), who relaxed Johansson and Löfgren's assumption that period one consumption and period two consumption are perfect substitutes.

These models greatly simplified the analysis of uncertainty, capital market imperfections, and amenities that invalidate the Fisherian separability. They also provided a theoretical framework for understanding how landowner preferences affect production decisions and timber supply (Montgomery and Adams 1995 provide a short review of these models). The two-period model was developed to cover joint production of timber and amenities (Max and Lehman 1988, Ovaskainen 1992), and during the 1990s it was extended to problems in which generations of landowners overlap through time. Perhaps most important, the life-cycle model has also been quite useful in motivating econometric studies of forest landowner behavior, providing a means to derive testable hypotheses concerning preferences of nonindustrial private forest (NIPF) landowners. Binkley (1981) was one of the first to point out that predictions of the basic Faustmann model consistently failed in empirical tests of harvesting behavior for nonindustrial private forest owners, among other things because it does not allow owner-specific variables (preferences) to affect harvesting. For some of the more

---

1. See Johansson and Löfgren (1985), chapter 12, "The Properties of Timber Supply Function in a Risky World."

important studies in this area, see Conway et al. 2003, Pattanayak et al. 2002, Brännlund et al. 1985, Newman 1987, Hultranz and Aronsson 1989, Kuuluvainen and Salo 1991, Hetemäki and Kuuluvainen 1992, Toppinen 1998, and Kuuluvainen et al. 1996).

In this chapter we explore at length the analytical structure of the two-period framework. We first assume that a representative landowner values only harvest revenue and then we allow the landowner to value amenity services from forest stocks. The basic model is then extended to problems of overlapping generations to determine how timber bequests between generations and forest management decisions affect long-run forest stocks. Throughout the chapter we are careful to compare comparative statics and other qualitative results of the two-period model with those of Faustmann- and Hartman-based rotation models. This comparison has often been confused within the forest economics literature. We will use these results to identify the strengths and weaknesses of two-period models relative to rotation models.

In section 4.1 we analyze the problem of a landowner who maximizes utility by choosing consumption and harvesting, but without amenity services. We call this the two-period timber production model. We introduce amenities in section 4.2, calling this the two-period amenity model. In section 4.3 we show how the two-period model gives rise to a new way of thinking about long-run forest stocks and bequests made across generations in models of overlapping generations, which proves to be a relatively new and useful framework for long-run policy analysis. Section 4.4 presents some modifications.

4.1   Two-Period Timber Production Model

4.1.1   Harvesting Possibilities

We start by describing a landowner's harvesting possibilities defined by his initial endowment of forest stock and its biological growth properties. Denote the initial endowment of forest stock by Q. The landowner can harvest x_1 at the beginning of the first period. The remaining timber stock, $k_1 = Q - x_1$, grows for the rest of the first period. At the beginning of the second period the forest stock consists of k_1 plus growth in the stock; i.e., $k_2 = k_1 + g(k_1) = Q - x_1 + g(Q - x_1)$. The function g denotes forest growth and is conventionally assumed to be concave; i.e., $g' > 0$ but $g'' < 0$. A typical representative of concave forest growth is the logistic function $g(k_1) = bk_1[1 - (k_1/K)]$, where $b > 0$ and K is biological carrying capacity. There are two values of k_1 where the growth rate is zero and one value of k_1 that gives maximum growth. More formally, there exists $k_1 = 0$ and $k_1 = K$, at which $g(0) = g(K) = 0$, so that there is a unique value k_1^* for which $g'(k_1^*) = 0$. These assumptions correspond to the plausible notion that at low levels of forest stock there is rapid growth, but eventually growth slows as the stock decays (see e.g., Clark 1990, p. 268).

A graph of the growth as a function of the resource stock is given in figure 4.1 of box 4.1. Figure 4.1 is well known and can be found in almost any textbook on natural resources. The timber stock at k_1^* is the somewhat famous, but economically inefficient, point of the maximum sustained yield we discussed at length in chapter 2.

Denote second-period harvesting by x_2 and assume for now that the landowner harvests his entire remaining forest stock in the second period. This is always the case when there is no salvage value for the stock in the second period.[2] Thus we can express harvesting possibilities in terms of second-period stock, or equivalently, in terms of harvesting flows and the initial endowment in equation (4.1):

$$x_2 = k_2 = (Q - x_1) + g(Q - x_1). \tag{4.1}$$

Equation (4.1) highlights the biological tradeoff between current and future harvesting. Totally differentiating it with respect to x_1 and x_2 yields $dx_2/dx_1 = -(1 + g') < 0$. This derivative indicates that any unit of timber harvested "today" reduces future harvesting by that unit plus forgone growth. The frontier of harvesting possibilities for an arbitrary initial forest stock is shown in figure 4.3. The vertical axis indicates future harvesting and the horizontal axis indicates current harvesting.

The frontier measures the tradeoff between current and future harvesting for a given forest endowment with a slope $-(1 + g')$. If a landowner harvests his entire stock during the first period, then $x_1 = Q$ and nothing is left for the second period. Not harvesting at all in the first period yields the entire forest stock plus its growth for future harvesting in the next period; i.e., $x_2 = Q + g(Q)$.

We have not explicitly assumed that forest regeneration is present in defining the frontier of harvesting possibilities. The implicit assumption is that the stock regenerates naturally. Montgomery and Adams (1995) suggest that the two-period life-cycle harvesting model reflects uneven-aged harvesting, with continuous or selective harvesting. However, if we make the assumption that second-period harvesting exhausts the entire stock, then the model merely explains short-term timber supply behavior for a given landowner. The justification for the focus on short-term timber supply rests on the fact that short-term fluctuations in market parameters are often more important to the functioning of timber markets than long-term ones (see e.g., Hyde 1980, pp. 79–80).

---

2. An important salvage value exists if the landowner values amenity services or if he gives timber bequests to his heirs. This landowner will never harvest the entire forest stock during the second period. We will see how the model works with endogenous second-period harvesting in the next section and we take up the issue of bequests more formally later in this chapter.

Box 4.1
Short-Term Harvest Volumes in a Two-Period Model

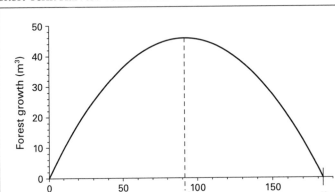

Figure 4.1
Forest growth as a function of stock.

The two-period model is seldom applied numerically, mainly because it is more useful in econometric work. However, since the two-period model relies on a biomass-based growth function, it is easy to develop a parametric model. We can formulate a logistic growth function, g, indicating growth as the function of the forest stock, Q. This type of function is well known in natural resource economics. In this box we examine the general properties of the following logistic forest growth function: $g(Q) = b[1 - (Q/K)]Q$.

In this function parameter b defines the natural regeneration rate of the forest and K is the carrying capacity. The carrying capacity defines the maximum stock that the forest ecosystem can sustain forever on any given unit of land. For forest stock, we can plausibly set $b = 1$ and we choose, for example, $K = 190$ m^3 to reflect conservative boreal growth conditions. Figure 4.1 presents the resulting (parabolic) growth as a function of the forest stock. The growth peaks at a volume of 95 m^3, yielding a maximum sustained growth of 47.5 m^3 per period. These points of MSY stock and growth have been the subject of many debates, as we have seen earlier in this book.

Referring back to the stand growth functions presented in chapters 2 and 3, we can link the growth function in this box to the stand growth function in box 2.1 quite easily. To do this, note that the logistic growth function with the stock as its argument is simply a differential equation of the following form: $\dot{Q}(t) = g(Q)$. As is well known, this can be solved for time to give forest growth at each point in time. For logistic growth we have $\dot{Q}(t) = dQ/dt = bQ(K - Q)/K$. By arranging we obtain $1/Q + 1/(Q - x) = b dt$. Next we can integrate the LHS of this equation as follows:

$$\int_{Q_0}^{Q(t)} 1/Q + 1/(Q - x)\, dQ = b dt,$$

Box 4.1
(continued)

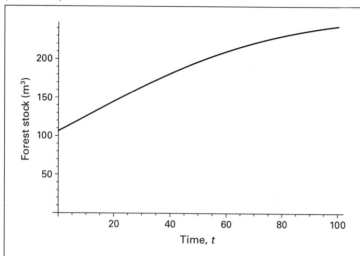

Figure 4.2
Logistic growth.

where Q_0 refers to initial stock. Noting that the integrand is the product of a logarithmic function, which yields by integration $\log[Q/(K-Q)] - \log[Q_0/(K-Q_0)] = bt$. Taking the antilogarithm and denoting $c = Q_0/(K-Q_0)$, we obtain $Q/(K-Q) = ce^{bt}$, or simply $Q(t) = K(1 - ce^{-bt})^{-1}$. For arbitrary parameter values $b = 0.03$ and $K = 260$, we obtain the growth function over time shown in figure 4.2. This function starts around 100 m^3 and increases up to the level of carrying capacity within the next 100 years.

4.1.2   Landowner Preferences and Consumption

Assume that the representative forest owner has a preference ordering over present and future consumption, denoted by c_1 and c_2. This preference ordering is conveniently represented by a utility function that is additively separable across periods and concave in each argument (consumption):

$$U = u(c_1) + \beta u(c_2), \tag{4.2}$$

where $\beta = (1 + \rho)^{-1}$ describes the landowner's rate of time preference. Denote the timber price in periods one and two by p_i, $i = 1, 2$. The primary source of the landowner's income is revenue from harvesting in each period. (One can easily assume the presence of other exogenous nonharvest income, but we omit it here

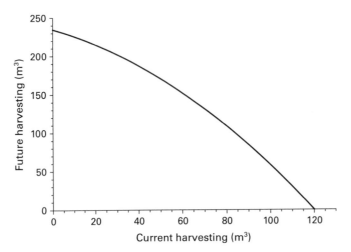

Figure 4.3
Frontier of harvesting possibilities as a function of the initial stock.

because it does not affect the basic properties of timber supply.) During the first period, the landowner allocates revenue from harvesting between current consumption (c_1) and saving (s), so that current consumption is expressed as

$$c_1 = p_1 x_1 - s, \tag{4.3}$$

where s is positive if the landowner is a net saver and negative if he is a net borrower.

In the second period, the landowner receives harvest revenue and interest income from any savings (if the landowner borrows, then he must pay back the loan and any interest that has accrued). The real interest rate, r, is constant and determined in a perfect capital market. Thus future consumption is defined as the sum of future revenue from harvesting plus investment income:

$$c_2 = p_2 x_2 + Rs, \tag{4.4}$$

where $R = (1 + r)$. Solving for s using (4.3) and substituting this into (4.4) produces the landowner's intertemporal budget constraint:

$$c_2 = p_2 x_2 + R(p_1 x_1 - c_1). \tag{4.5}$$

The way equations (4.1), (4.3), and (4.4.) are defined implies that once the landowner chooses current consumption and harvesting, forest growth then uniquely defines future consumption and harvesting.

4.1.3   Short-Term Harvesting Behavior

The landowner's economic problem is now to choose current consumption and harvesting so that utility is maximized; i.e., $\max_{c_1, x_1} U = u(c_1) + \beta u(c_2)$; s.t. (4.5).

The first-order conditions for current consumption and harvesting at an interior solution are, respectively,

$$U_{c_1} = u'(c_1) - (1+r)\beta u'(c_2) = 0, \tag{4.6a}$$

and

$$U_{x_1} = \beta u'(c_2)[(1+r)p_1 - p_2(1+g')]. \tag{4.6b}$$

The interpretation of these conditions is straightforward. Equation (4.6a) defines consumption over both periods under perfect capital markets. It implies that the marginal rate of substitution for consumption equals the ratio of market interest rates to the rate of time preference:

$$\frac{u'(c_1)}{u'(c_2)} = \frac{(1+r)}{(1+\rho)}. \tag{4.7}$$

Equation (4.7) is a "Euler equation" characterizing the intertemporal choice of consumption. Here the marginal rate of substitution between period 1 and period 2 consumption equals the ratio of the real interest rate to the rate of time preference.

Equation (4.6b) characterizes the landowner's harvesting decision. Since marginal utility $\beta u'(c_2)$ is positive by assumption, the term in brackets must be zero at the interior solution. This implies that the landowner's preferences for consumption are separable from the harvesting decision, in that harvesting does not depend on utility (see Hirshleifer 1970, pp. 53–66). Thus the optimal condition for current harvesting can be expressed from (4.6b) as

$$Rp_1 - (1+g')p_2 = 0. \tag{4.8}$$

Equation (4.8) provides the following harvesting rule:

Proposition 4.1   Harvesting rule in two-period timber production model   The landowner chooses current harvesting by equating the marginal revenue from the last unit of forest stock to its opportunity cost defined by the revenue that this unit would provide in the second period if the stock were left to grow.

Figure 4.4 is a graph of this result. It shows the marginal revenue of harvesting curve (MR), which is given by p_1, and the opportunity cost of harvesting curve

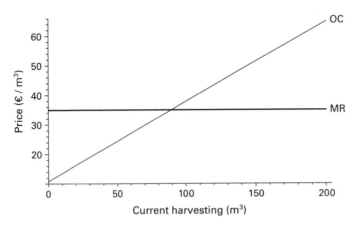

Figure 4.4
Harvesting decision of the landowner in the two-period model.

(OC), given by $R^{-1}(1+g')p_2$. Marginal revenue and opportunity costs are expressed as a function of x_1.

The marginal revenue curve is constant, but the opportunity cost curve is rising because the growth function is assumed increasing and concave. An interior solution is obtained at the point where the MR and OC curves cross. This intersection point defines optimal current harvesting, x_1^*, and given the initial stock, it also determines future harvesting, x_2^*.

If prices are constant over the two periods, such as in a steady state, we would write $p_1 = p_2 = p$, and timber prices would no longer be present in the first-order conditions (4.6b) and (4.8). Thus in the steady state, prices no longer affect optimal harvesting in current and future periods. Instead, harvesting is undertaken so that the following condition holds (see 4.6b): $g' = r$. This has something in common with rotation age models. Recall from chapter 2 (see equation 2.6) that the single-rotation von Thunen-Jevons optimal rotation period was defined by the condition $f'(T)/f(T) = r$. Rotation and two-period models can therefore be explicitly linked because $f'(T)/f(T) = g'$. We can conclude that two-period harvesting models yield outcomes qualitatively equivalent to those of a single rotation-period model.

We can also compare the steady-state harvesting rule with the harvesting rule obtained using the Faustmann formula when regeneration costs are zero. Recall equation (2.4.a) describing the Faustmann rotation age solution. Setting the replanting cost c equal to zero, we can re-express equation (2.4a) in chapter 2 as $f'(T)/f(T) = r/(1 - e^{-rT})$. Recall that the denominator of the RHS measures the discount factor for an infinite sequence of identical rotations. It is interesting that timber price has cancelled out of this equation and therefore no longer affects the optimal rotation period. We can therefore conclude that the main qualitative difference between solving the rotation age using a Faustmann formula and solving for the harvesting rule a two-period harvesting model rests on the absence of a requirement for ongoing (sustainable) production in the two-period model.

The optimal amount of current and future harvesting can also be used to define the representative landowner's timber supply function. Because preferences are separable from harvesting, we can study the properties of timber supply using only the comparative statics results obtained from the optimal harvesting rule. Comparative statics are derived by totally differentiating (4.8) to obtain $R\,dp_1 - (1+g')\,dp_2 + p_1\,dr - g''p_2\,dQ + g''p_2\,dx_1 = 0$. We denote current (or future) timber supply by x^{1S} (x^{2S}). Expressing the derivative of this using subscripts, we then arrive at the following supply effects from the differential:

$$x_{p1}^{1S} = -\frac{R}{g''p_2} > 0; \quad x_{p2}^{1S} = \frac{(1+g')}{g''p_2} < 0; \quad x_r^{1S} = -\frac{p_1}{g''p_2} > 0; \quad x_Q^{1S} = 1 > 0. \quad (4.9)$$

According to (4.9), a higher first-period timber price p_1 increases the marginal return on harvesting in the first period, so that it becomes optimal to increase current harvesting. A higher p_2 in turn increases the opportunity cost of current harvesting, thus reducing x_1. A higher interest rate increases the marginal return on harvesting and therefore increases current harvesting. Finally, a higher initial timber stock Q increases current harvesting by the entire amount of its increase. The reason for this is the absence of any constraints on harvesting capacity, which allows the landowner to immediately adjust his stock to the optimum level.[3]

Given the comparative statics of current harvesting in (4.9), the tradeoff between current and future harvesting can be obtained by differentiating (4.1), $dx_2/dx_1 = -(1+g') < 0$. Hence the effects of exogenous parameters on future harvesting can be determined directly as

$$x_{p1}^{2S} = -(1+g')x_{p_1}^s < 0; \quad x_{p2}^{2S} = -(1+g')x_{p_2}^s > 0;$$

$$x_r^{2S} = -(1+g')x_r^s < 0; \quad x_Q^{2S} = 0. \quad (4.10)$$

The exogenous parameters affect future harvesting in an opposite direction relative to that of current harvesting because the entire forest stock is harvested in the second period. The only exception to this rule is for the parameter Q, which has no effect on future harvesting.

---

3. This idea corresponds to a bang-bang solution in dynamic biomass (fishery) models in the absence of constraints in iharvesting capacity, demonstrating that the two-period model can capture this crucial feature of a more general dynamic model.

We summarize the general properties of short-term and future timber supply in the next proposition.

Proposition 4.2   For a landowner, under certainty and perfect capital markets:

(a) Current timber supply depends positively (negatively) on current (future) timber price, and positively on the real interest rate and initial forest endowment.

(b) Future timber supply depends negatively (positively) on current (future) timber price, negatively on the real interest rate, and is not affected by an increase in the initial timber endowment.

Proposition 4.2 reveals that the effects of exogenous variables on timber supply in the basic version of the two-period model indeed are qualitatively close to those found in the Faustmann model studied in chapter 2. The way in which short-term and long-term timber supplies depend on the real interest rate is similar across models. Timber price is constant in the basic Faustmann model. In the two-period model, a higher current timber price affects timber supply in a manner similar to that of the short-run effect of a higher timber price in the Faustmann model. The qualitative similarity between the Faustmann and two-period models occurs simply because landowner preferences are separable from production decisions in the two-period model. As we will see, once we relax the assumptions concerning certainty in later chapters, differences between the two models will emerge. Box 4.2 presents a numerical analysis of short-term harvesting behavior in the two-period model of this section.

4.1.4   Forest Taxation and Timber Supply

We return to the question of how forest taxes affect harvesting decisions and timber supply. Consider the following three alternative tax forms: a yield tax, τ; a unit tax, u; and a site productivity tax, a. As in previous chapters, the yield tax is a proportional tax levied on harvest revenues and the unit tax is proportional to harvest volume. Using these taxes, periodic after-tax prices can be expressed as $\hat{p}_i = p_i(1 - \tau) - u_i$. The site productivity tax a is a lump-sum tax based on the natural productivity of the forest site. We will consider cases where forest taxes are constant or periodic.

The presence of harvesting taxes changes the landowner's consumption flow equations as follows: $c_1 = \hat{p}_1 x - a_1 - s$ and $c_2 = \hat{p}_2 z - a_2 + Rs$. Maximizing the landowner's utility in (4.2) using these new flow equations gives the following condition at an interior solution:

$$R\hat{p}_1 - (1 + g')\hat{p}_2 = 0. \tag{4.11}$$

Box 4.2
Short-Term Harvesting Behavior

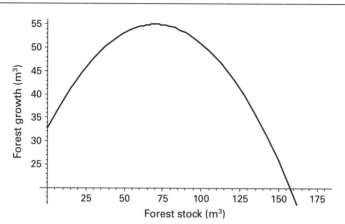

Figure 4.5
Forest growth as a function of its harvest.

In this box we analyze a landowner's short-term harvesting behavior in a parametric
two-period model. Before doing this, we need to have values for all economic parameters (current and future timber price, initial endowment of forest stock, and the real
interest rate), and we must also decide on the length of period two in calendar years.

We must first rewrite the logistic growth function so that it includes harvesting in
the current period, x_1. Recall from the text that the growth function in the two-period
model was generally expressed as $g(Q - x_1)$. The logistic counterpart to this is the
following function:

logistic growth under harvesting: $g(\bar{Q} - x_1) = \left[1 - \dfrac{(\bar{Q} - x_1)}{K}\right](\bar{Q} - x_1).$

The bar above Q denotes a given initial volume of timber out of which an amount of
x_1 is harvested during the first period. Using $K = 220$ and $\bar{Q} = 180$, figure 4.5 shows
how increasing x_1 affects growth.

As the figure reveals, the initial stock $\bar{Q} = 180$ produces a growth of 32.73 m³. Low
levels of harvesting increase the growth of the stock as it gradually shifts toward the
MSY point, which is obtained by $x_1 = 70$ m³. When current harvesting goes beyond
the MSY point, the stock becomes smaller. The figure demonstrates effectively how
current harvesting changes future harvesting possibilities.

To use this growth function to determine current and future harvesting levels, we
assume in the baseline case that the current stumpage price is €35, the future price
is €40, and the real interest rate is 3%. To apply this, we assume that the length of
period two is 10 years; that is, the second harvesting takes place in calendar time 10
years after the first harvesting. When the landowner values his harvest for revenue
only, the parametric objective function, V, becomes

Box 4.2
(continued)

Table 4.1
Harvesting and Growth in Two Period Timber Production Model (m^3)

	Current harvesting	Future harvesting	Forest stock, k_1	Forest growth	Total harvested volume
Baseline	89.4	143.9	90.6	53.3	233.3
Sensitivity Analysis					
Current price €45	126.3	94.3	53.7	40.6	220.6
Future price €50	63.5	171.3	116.5	54.8	234.8
Interest rate 4%	102.5	127.7	77.5	45.1	230.2

$$V = 35x_1 + (1 + 0.03)^{-10} * 40\left[180 - x_1 + \frac{180 - x_1}{220}(180 - x_1)\right].$$

The baseline results and comparative statics for this objective function are given in table 4.1. In addition to periodic harvesting, we report forest growth between periods one and two and forest stock in the first period. The latter facilitates comparison with the amenity production model in box 4.3 (recall that future stock is always equal to zero in the timber production model because it is always optimal to harvest 100% of the remaining stock at the beginning of the second period). The reported volumes demonstrate that the amounts of harvested volume may change considerably between periods one and two owing to changes in exogenous variables. Increases in current and future timber prices effectively shift harvesting between the periods, illustrating the short-term usefulness of this model. Interestingly and in accordance with figure 4.5, the total harvested volume varies according to how current harvesting and the resulting growth are chosen.

Totally differentiating (4.11) and assuming that taxes are periodic (nonconstant) yields $0 = g''\hat{p}_2\, dx_1 - Rp_1\, d\tau_1 - R\, du_1 + p_2(1 + g')\, d\tau_2 + (1 + g')\, du_2$. Using the same notation as earlier, we have the following effects:

$$x^{1S}_{a_1} = 0; \quad x^{1S}_{a_2} = 0, \tag{4.12a}$$

$$x^{1S}_{\tau_1} = \frac{Rp_1}{g''\hat{p}_2} < 0; \quad x^{1S}_{\tau_2} = -\frac{(1 + g')p_2}{g''\hat{p}_2} > 0, \tag{4.12b}$$

and

$$x^{1S}_{u_1} = (p_1)^{-1}x^s_{\tau_1} < 0; \quad x^{1S}_{u_2} = (p_2)^{-1}x^s_{\tau_2} > 0. \tag{4.12c}$$

According to (4.12a), a site productivity tax has no effect on harvesting because the tax is independent of any actions of the landowner (except the decision to grow trees). We will see in chapter 9 that once uncertainty about future parameters is introduced, the site productivity tax ceases to be neutral. Periodic harvest taxes distort the relative profitability of timber production over the periods, so that higher current (future) yield and unit taxes decrease the marginal return (opportunity cost) of current harvesting. This increases future (current) harvesting.

If the tax rates are constant ($\tau_i = \tau$, $a_i = a$, $u_i = u$) but timber prices are different over time, then both the site productivity tax and yield tax are neutral, but the unit tax continues to be distortionary, i.e.,

$$x_\tau^{1S} = x_a^{2sS} = 0, \tag{4.13a}$$

and

$$x_u^{1S} = \frac{(r - g')}{\hat{p}_2 g''} \geq (<) \, 0 \quad \text{as } (r - g') \leq (>) \, 0. \tag{4.13b}$$

Why does the yield tax have no effect on harvesting while the unit tax turns out to be distortionary? The answer lies in the way that the taxes are introduced into the model. The yield tax is a proportional-value tax, changing both the marginal return and the opportunity cost of harvesting by equal amounts. Therefore the yield tax does not discriminate between periods. The unit tax is a quantity tax, so that it changes the marginal revenue and opportunity cost curves differently, depending on the magnitudes of harvested volume. Only in the case where current and future timber prices are equal, implying $r = g'$, is the unit tax neutral. We collect our findings for this discussion into the following proposition:

Proposition 4.3   Under certainty and perfect capital markets when timber prices vary over time, a constant site productivity tax and yield tax are neutral, while a unit tax is distortionary in that it decreases or increases timber supply depending on the relative size of growth and interest rates.

The results of proposition 4.3 differ from the Faustmann equivalent in chapter 2. A constant yield tax is neutral in the two-period model, but in the Faustmann model it decreases short-term timber supply and increases long-term supply. The reason for the difference is the presence of regeneration costs in the Faustmann model. Recall that in the absence of regeneration costs, prices do not affect the optimal rotation period in the Faustmann model—and in this case, the yield tax is a neutral tax because it affects only net price. Incidentally, the yield tax is always distortionary when harvesting costs are tax-deductible, as they are in many countries. For further discussion on the differences between two-period and

Faustmann models, see Amacher (1997) and Ollikainen (1996a). If timber prices are constant over time, then all taxes mentioned in proposition 4.3 are neutral.

4.2   Two-Period Amenity Production Model

We next introduce amenity benefits in the two-period life-cycle model by assuming that the landowner values both harvest revenue and amenities. This model was originally formulated in Ovaskainen (1992), who provided a refined version of the model by Max and Lehman (1988). The two-period amenity model is in many ways equivalent to the Hartman rotation model. However, in the two-period version, the landowner decides how much to harvest now and in the future to generate both income and amenities from forest stock. In order to enjoy amenity services, the landowner will generally not harvest the entire stock at the beginning of the second period. Consequently, both current and future harvesting become endogenous decisions.

We start this section by clarifying how amenities enter into our biological description of the harvesting technology, and then we redefine landowner preferences to include amenity valuation. We continue with the assumption that the landowner harvests forest stock at the beginning of both periods. The harvesting decision in the absence and presence of taxes is then analyzed. The reader is encouraged to check the references for more thorough analyses.

4.2.1   Joint Production of Timber and Amenity Services

As in chapter 3, we continue to assume that amenity services are closely tied to the size of the standing forest stock. Therefore we can use the periodic stock as a proxy for amenity services, much in the way that Hartman used time as a proxy for amenity services during a rotation. The landowner's harvesting and amenity production possibilities can be described using the following two equations of motion:

$$k_1 = Q - x_1, \tag{4.14a}$$

and

$$k_2 = (Q - x_1) + g(Q - x_1) - x_2. \tag{4.14b}$$

By choosing x_1, the landowner defines the first-period stock, and this stock provides current amenity services during the first period. By choosing the size of future harvesting x_2, the landowner determines the size of the future forest stock k_2 and the resulting flow of amenity services in the future period. The biological tradeoff in terms of the effect of current and future harvesting on amenities is now given by the expression $(dk_2 + dx_2)/dx_1 = -(1 + g') < 0$. An increase in current harvesting means that the sum of future harvesting plus future amenity services provided by the future forest stock will decrease by an amount that depends on the growth function of the forest.

4.2.2 Landowner Preferences

Following the same logic as in chapter 3, assume that the landowner values both harvest revenue and the amenities provided by the forest stock. Also assume that the landowner has preferences through a modification of (4.2). Denoting the valuation of current and future flow of amenity services by $v(k_1)$ and $\beta v(k_2)$, with $v'(k_i) > 0$ and $v''(k_i) < 0$, $i = 1, 2$, the landowner's preferences over consumption and amenities can be expressed in a general form as

$$U = u(c_1) + \beta u(c_2) + v(k_1) + \beta v(k_2). \tag{4.15}$$

4.2.3 Harvesting and Amenity Production

The landowner maximizes utility in (4.15) by choosing current and future harvesting subject to (4.14a) and (4.14b). The first-order conditions at the interior solution are

$$U_{c_1} = u'(c_1) - \beta R u'(c_2) = 0 \tag{4.16a}$$

$$U_{x_1} = p_1 \beta R u'(c_2) - v'(k_1) - (1 + g')\beta v'(k_2) = 0 \tag{4.16b}$$

$$U_{x_2} = p_2 \beta u'(c_2) - \beta v'(k_2) = 0. \tag{4.16c}$$

From (4.16c) we have that $U_{x_2} = 0 \Leftrightarrow p_2 u'(c_2) = v'(k_2)$. Plugging this in (4.16b) and rearranging yields

$$p_1 - (1 + g')p_2 R^{-1} = \frac{v'(k_1)}{\beta u'(c_2)}. \tag{4.17}$$

From (4.17), the harvesting decision is no longer separable from landowner preferences because marginal utility terms from consumption and amenity benefits are present on the right-hand side of the harvesting rule. On the left-hand side, we have a measure of the tradeoff between current and future harvest revenue, which is positive because the right-hand term is the (positive) marginal rate of substitution between consumption and amenity services. Because the forest growth function is concave, the landowner harvests less than in the case where amenity services are not valued. Finally, if prices are constant, $p_1 = p_2 = p$, then equation (4.17) would be written as $p(r - g') = Rv'(k_1)/\beta u'(c_2)$. Prices would still affect harvesting, and now we find that an interior solution requires $(r - g') > 0$, reflecting the fact that the timber stock is higher, owing to amenity valuation; thus the forest growth rate is less than the interest rate at the optimal harvesting solution. We now summarize this new harvesting rule for the case of amenity valuation.

Proposition 4.4 In the two-period amenity model, the landowner chooses current harvesting so that the difference between the marginal return and the opportunity cost of current harvesting is equal to the marginal rate of substitution between consumption and amenity services.

This shows that landowner preferences now affect timber supplies. One of the implications of proposition 4.4 is that the entire stock is never exhausted because the landowner derives utility from amenities during the second period. In effect, amenities represent the salvage value of the forest stock at the end of the future period.

For the rest of this section we assume that the landowner's utility function is quasi-linear so that $u'(c_1) = u'(c_2) = 1$. This assumption ensures that the landowner's objective function resembles the Hartman model's objective function. Using this assumption, the two-period objective function equivalent to the Hartman model is

$$U = p_1 x_1 + R^{-1} p_2 x_2 + v(k_1) + \beta v(k_2). \tag{4.15'}$$

For utility function (4.15'), the harvesting rule reduces to $p_1 - (1 + g') p_2 R^{-1} = v'(k_1)$.

The optimal levels of x_1 and x_2 define both the landowner's current and future supply of timber as well as the periodic stocks of standing timber. Provided that the second-order conditions hold, we can solve for the comparative statics of x_1 and x_2 using the two-equation system defined by the first-order conditions (4.16a) and (4.16b). It is easy to check that the second-order conditions hold (the reader should verify this). Using the implicit function theorem, applying Cramer's rule, and expressing the periodic supplies in the amenity model by x^{1A} and x^{2A}, we obtain the following effects:

$$x^{1A}_{p_1} = -\Delta^{-1}\{\beta v''(k_2)\} > 0; \quad x^{2A}_{p_1} = -(1 + g')x^{1A}_{p_1} < 0, \tag{4.18a}$$

$$x^{1A}_{p_2} = R^{-1}x^{2A}_{p_1} < 0; \quad x^{2A}_{p_2} = -(1 + g')x^{1A}_{p_1} - R^{-1}N > 0, \tag{4.18b}$$

$$x^{1A}_r = R^{-1}p_2(1 + g')x_{p_1} > 0; \quad x^{2A}_r = -(1 + g')x^{1A}_r + N < 0, \tag{4.18c}$$

and

$$x_Q^{1A} = 1; \quad x_Q^{2A} = 0, \tag{4.18d}$$

where $\Delta > 0$ is the determinant of the Hessian matrix of the second-order derivatives in the equation system given by (4.16a) and (4.16b), and $N = \Delta^{-1}\{v''(k_1) + g''\beta v'(k_2)\} < 0$.

From (4.18a), a higher current timber price increases current supply. A higher future timber price in (4.18b) decreases current harvesting because it increases the opportunity cost of harvesting. A higher future timber price increases future harvesting through two reinforcing effects: First, reduced current harvesting increases future harvest possibilities and second, amenity production becomes relatively less profitable compared with timber production. From equation (4.18c), a higher interest rate increases current harvesting but decreases future harvesting; the same two reinforcing effects apply here, but in the opposite direction. Finally, an increase in the initial endowment of forests increases current harvesting by the entire amount of the increase but it has no effect on future harvesting (see 4.18d). These are summarized in proposition 4.5.

Proposition 4.5   When the landowner values both harvest revenue and amenity services from forests:

(a) Current timber supply depends positively (negatively) on current (future) timber price, and positively on the real interest rate and initial forest endowment.

(b) Future timber supply decreases (increases) with current (future) timber price, decreases when interest rates increase, and remains unchanged when initial timber endowment increases.

A comparison of propositions 4.2 and 4.5 reveals that the effects of exogenous parameters on current and future supply are qualitatively (but not quantitatively) the same in the presence and in the absence of amenity values.

Naturally, we are also interested in how amenity production (i.e., standing timber stock) is affected by market parameters (see Amacher and Brazee 1997a and Koskela and Ollikainen 1989). Recall that equations (4.14a) and (4.14b) govern periodic forest stocks as a function of harvesting. Differentiating them with respect to exogenous parameters and utilizing the comparative static effects (4.18a)–(4.18d), we can establish that

$$k_1 = k_1(\underset{-}{p_1}, \underset{+}{p_2}, \underset{-}{r}, \underset{0}{Q}) \quad \text{and} \quad k_2 = k_2(\underset{0}{p_1}, \underset{-}{p_2}, \underset{+}{r}, \underset{0}{Q}). \tag{4.19}$$

We have now proved the following:

Proposition 4.6   When the landowner values both harvest revenues and amenity services, the standing forest stock is independent of initial forest endowment

when there are no restrictions on harvesting capacity. Moreover, timber prices
and interest rates change the profitability of timber relative to amenity production, as well as the temporal tradeoff between them, as follows:

(a) A higher current timber price reduces the current forest stock but leaves the
future forest stock unchanged.

(b) A higher future timber price increases the current stock and reduces the future forest stock.

(c) A higher interest rate reduces the current forest stock but increases the future
forest stock.

Proposition 4.6 makes it clear that the general state of amenity production by a
forest landowner is always a function of timber prices and the real interest rate.
Box 4.3 explores some of these results and offers a comparison of two-period
models for amenity and timber production using a parametric example.

4.2.4 Forest Taxation

A government might be concerned with how tax instruments affect the relative
profitability of harvesting and amenity production, particularly if it is interested
in modifying the behavior of forest landowners by using tax instruments. Consider again the three alternative tax forms used in this section: the yield tax, τ;
unit tax, u; and the site productivity tax, a. Assume that these tax rates are constant over periods one and two. Even though they are constant, they still affect
the relative profitability of timber and amenities.

The flow equations for current and future consumption are $c_1 = \hat{p}_1 x_1 - a - s$
and $c_2 = \hat{p}_2 x_2 - a + Rs$. Maximizing the landowner's utility in the presence of
these taxes yields the following first-order conditions at the interior solution:

$$U_{x_1} = \hat{p}_1 - v'(k_1) - (1 + g')\beta v'(k_2) = 0, \tag{4.20a}$$

$$U_{x_2} = \hat{p}_2 R^{-1} - \beta v'(k_2) = 0. \tag{4.20b}$$

These conditions can be used to form a harvesting rule in the presence of taxes:

$$\hat{p}_1 - (1 + g')\hat{p}_2 R^{-1} = v'(k_1). \tag{4.21}$$

Again, the site productivity tax does not appear in the harvesting rule, indicating that it is neutral. The yield and unit taxes clearly decrease the profitability of
timber production in favor of amenity services while keeping the intertemporal
tradeoff between first- and second-period harvesting unchanged.

Deriving the comparative statics for these taxes yields the following effects on
current timber supply:

Box 4.3

Two-Period Amenity Model: Short-Term Harvest Volumes

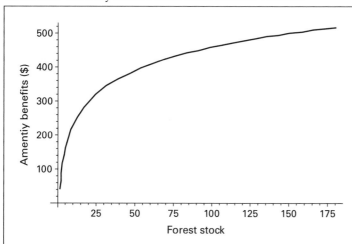

Figure 4.6

Concave amenity benefit function.

The two-period model in this chapter requires an additively separable amenity valuation function that describes a landowner's preferences for standing forest stock. In this box we develop a parametric specification for such an amenity valuation function and combine it with the timber production returns of box 4.2 to ultimately solve for harvested volumes and standing stocks in periods one and two. The amenity benefit functions have periodic stocks, k_i, $i = 1, 2$ as their arguments. We adopt here a logarithmic amenity benefit function for periods one and two from the literature (Kuuluvainen and Uusivuori 2003):

Amenity benefit function: $v(k_i) = b \log[k_i]$, $i = 1, 2$.

We set parameter $b = 100$ in the logarithmic amenity function. Recalling the empirical specification of the logistic growth function in box 4.2, we express the arguments of this logarithmic function as follows:

$$k_1 = \bar{Q} - x_1$$

$$k_2 = 180 - x_1 + \frac{\bar{Q} - x}{K}(180 - x_1) - x_2,$$

where $K = 220$ denotes the carrying capacity and $\bar{Q} = 180$ is the initial stock.

Figure 4.6 shows a graph of the amenity benefits derived as a function of the size of the forest stock k_i. We use the same baseline parameter values for the timber production as in box 4.2 (current and future stumpage prices equal €35 and €40; the interest rate is 3%; and the length of period two is 10 years). When the landowner values harvest revenue and amenity benefits, the parametric objective function, W, is given by

$$W = 35x_1 + (1 + 0.03)^{-10} * 40x_2 + 100 * \log[k_1] + (1 + 0.03)^{-10} * 100 * \log[k_2].$$

The baseline of the model and comparative statics are reported in table 4.2.

Box 4.3
(continued)

Table 4.2
Harvesting and Forest Stock in the Two Period Amenity Model (m^3)

	Current harvesting	Future harvesting	Current stock, k_1	Future stock, k_2
Baseline	85.0	146.0	95.0	2.5
Sensitivity Analysis				
Current price €45	120.1	100.9	59.9	2.5
Future price €50	60.1	171.6	119.0	2.0
Interest rate 4%	97.5	131.3	82.5	2.8

Table 4.3
Comparison of Timber and Two Period Amenity Production Models (m^3)

	Current harvesting $x_1^S - x_1^A$	Future harvesting $x_2^S - x_2^A$	Current stock $k_1^S - k_1^A$
Baseline	4.4	-2.1	-4.4
Sensitivity Analysis			
Current price €45	6.2	-6.6	-6.2
Future price €50	3.4	-0.3	-2.5
Interest rate 4%	5.0	-3.6	-5.0

The reported volumes in table 4.2 demonstrate that the amount of harvesting may change considerably between periods one and two. An interesting feature is that the future forest stock remains positive, as the theory shows, but it is very small. This results from the specification of our chosen amenity benefit functions and the fact that the model is a short-run model with the time horizon ending in period two.

To compare the amenity production model with the timber production model, table 4.3 reports differences in current and future harvesting and current forest stocks. We denote by superscripts A and S variables in the amenity and timber production models, respectively.

The differences between harvesting behavior in this table match the results shown in the theory. The most dominant feature is that the presence of amenity benefits reduces current harvesting in a way that increases the volume of the current forest stock, so that the landowner captures higher amenity benefits. Quantitatively, the differences are quite small owing to our assumption of a conservative amenity benefit function. An interesting detail here is that the comparative statics results differ most with regard to the interest rate. While higher prices increase the relative profitability of harvesting over amenity valuation in both models, a higher interest rate affects future harvest revenue and amenity benefits alike. Thus it does not distort the relative weight of amenity and timber production.

$$x_a^{1A} = 0; \quad x_\tau^{1A} = \Delta^{-1}\{v'(k_1)\beta v''(k_2)\} < 0;$$

$$x_u^{1A} = \Delta^{-1}\{(r - g')R^{-1}\beta v''(k_2)\} < 0 \tag{4.22}$$

Thus, while the site productivity tax has no impacts on current supply, higher yield and unit taxes unambiguously decrease current harvesting. Recall, however, that for a unit tax, $r > g'$ holds for constant prices or for the case where current and future prices are "close enough"; if this does not hold, we have $x_u^{1A} > 0$. For future harvesting, we have $x_a^{2A} = 0$, but $x_\tau^{2A} \gtreqless 0$, and $x_u^{2A} \gtreqless 0$. This ambiguity results from two countervailing factors (see the equations in note 4). The landowner takes into account reduced current harvesting and increased future harvesting possibilities through the term $-(1 + g')x_\tau^{1A}$. This effect tends to increase future harvesting. However, the tax also makes amenity production more profitable relative to timber production in the second period, tending to decrease future harvesting.

The effects of taxes on the forest stocks turn out to be similar in both periods, so we have:[4]

$$k_i = k_i(\underset{0}{a}, \underset{+}{\tau}, \underset{+}{u},), \quad i = 1, 2. \tag{4.23}$$

Both types of harvest taxes tend to increase periodic forest stocks and therefore they could be used to increase the amenity benefits produced from forests by distorting landowner behavior. These results give rise to the following proposition:

Proposition 4.7 If the landowner values harvest revenue and amenity services, then higher yield and unit taxes always increase periodic forest stocks, whereas the site productivity tax has no effect on periodic forest stocks.

The harvest taxes are clearly powerful instruments in the hands of a social planner wishing to modify production of amenity services. However, the site productivity tax remains a neutral instrument, indicating it is an excellent means of raising tax revenue if the government must raise revenue but has no other reason to intervene in forest management. In chapter 5 we study these issues in detail.

4.3 Overlapping Generations Models

We now turn to an important extension of two-period life-cycle models. Overlapping generations models allow the study of how forest stocks are linked over time

---

4. The reader can verify this by differentiating equations (4.14a) and (4.14b) with respect to the tax parameters, using the equations in (4.22) for the effects on timber supply. To solve for effects on the future stock, notice that the exact tax effects on future supply are $x_\tau^{2A} = -(1 + g')x_\tau^{1A} + p_2 M \gtreqless 0$ for yield tax and $x_u^{2A} = -(1 + g')x_u^{1A} + M \gtreqless 0$ for a unit tax, where $M = \Delta^{-1}\{p_2 R^{-1}[v''[k_1 + (1 + \rho)^{-1}v'(k_1)g'']\} < 0$.

through forest landowners whose lives overlap. Two interesting questions emerge from this analysis: First, how do landowners make bequest choices regarding their forests and second, how are these bequests affected by policy instruments? Ultimately, landowner bequests determine the size of the long-run forest stock in the economy, which in and of itself is of interest to policy makers concerned about sustaining forest cover over time.

4.3.1 General Features

As for many other resources in an economy, generations of forest resource owners overlap through time. Forest stocks can be transferred between these generations either by forestland sales or bequests from the old generation to the young. Timber bequests are most important for private nonindustrial landowners, the largest landowner class in the world. The Faustmann model and its variants studied in chapter 2 are based on a series of infinite rotations that do not make explicit the ways that timber capital can be transferred across generations.[5] In fact, the implicit assumption of perfect land markets effectively makes timber transfers redundant. Nonetheless, bequests are important because they in part determine the size of the future forest stock and amenities present in the economy.

It is well known that government policies can modify landowner bequests. Heavy inheritance taxes on timber bequests are likely to make monetary bequests relatively more profitable, and this may encourage landowners to harvest forest stocks when they inherit them. This timber capital is often liquidated by the next generation when parents die and leave their forests to their children, who then face the burden of relatively high taxes. As a consequence of these actions, the amount of forest stock and amenities may decline over time in an economy. In the United States, fragmentation of forest land has been caused in part by current-generation landowners selling inherited parcels of timber to avoid paying inheritance taxes.

A convenient framework for studying these issues is the overlapping generations model (OLG), which has a recent history in forest economics, with the earliest work due to Löfgren (1991) and Hultkrantz (1992), and more recently Ollikainen (1998), Amacher et al. (1999, 2002, 2003), and Koskela et al. (2002).[6] The basic structure of an OLG model is relatively simple. Agents in any generation live for two periods; they are typically labeled as "young" in the first period

---

5. A notable exception is in Tahvonen (1998), where the landowner decides upon consumption or saving, rotation age of the stands within his lifetime, and the amount and form of bequest to leave to the next generation.
6. The economics literature on these models dates back to Allais in 1947 (see Malinvaud 1987) and Samuelson (1958) and is widely used in general economics (see Azariadis 1993 for a comprehensive review of OLG models through the early 1990s and de la Croix and Michelle 2002 for a recent exposition).

and "old" in their second period of life. At any period of time there are two agents alive; when an agent of generation t is young, an agent of generation $t - 1$ is old. Agents of two generations therefore overlap at any point in time. The two generations are linked through assumptions concerning utility, budget constraints, and bequests.

Building an OLG model requires modifying the timing of landowner actions in order to close the model properly. While we previously assumed that the decisions are made at the beginning of each period, now we assume they are made at the end of each period. Among other things, this implies that each generation makes a harvesting decision only once in its lifetime.

4.3.2   Bequests

The OLG model requires some other precise assumptions about altruism and bequests. Bequests can be unintentional (accidental), as in the case of unexpected death, or intentional. The latter case, which we follow here, dates back to Barro (1974). In his model and the many that followed, altruism takes place within the family, i.e., parents care for their children. For simplicity, all generations are assumed to have identical preferences and perfect foresight, and parents are also assumed to enjoy the well-being of their children by recognizing their own effect on the maximum utility attainable by their children. Given that their children in turn derive utility from their children and so on, the solution of the model leads to outcomes that are equivalent to the solution of an infinite time-horizon problem. These models have one strong condition, which is that bequests can only be positive or zero. Thus the possibility for negative bequests, or gifts from children to parents, does not exist. The case where parents leave bequests and children give gifts to their parents (monetary or otherwise) is called "two-sided altruism" (see Kimball 1987, Abel 1987, and Bergstrom 1999), which will not concern us here.

Consider a one-sided altruistic OLG economy in which the population grows exogenously over time and agents live for two periods and maximize the following intertemporally additive concave utility function:

$$U^t = u(c_{1t}) + \beta(c_{2t}) + v(k_{1t}) + \beta v(k_{2t}) + \phi V_{t+1}. \tag{4.24}$$

Here c_{it} and k_{it} denote the periodic consumption and forest stocks for a landowner born at time t in period $i = 1$ and $i = 2$ of their life. The same assumptions concerning the properties of utility and amenity benefit functions earlier continue to hold. A new term in the utility function is V_{t+1}; this denotes the indirect utility of the representative landowner in the next generation. This term introduces altruism into the model. The parameter ϕ is known as the intergenerational discount rate, which measures the weight that the current-generation landowner attaches to the welfare of the future-generation landowner. This is a measure of the degree of altruism. The welfare of the next generation is partly determined by the decisions of the current-generation landowner, but it is also a function of initial conditions determined by bequests made to them from the previous generation. The explicit specification of V_{t+1} comes from maximizing equation (4.24) defined for the generation born at $t+1$, i.e, U^{t+1}, so that V_{t+1} is the maximized utility of the generation born at $t+1$.

Bequests modify the previous equations for consumption and harvesting presented earlier in this chapter. Denote the volume of timber bequests by b and the value of money bequests by m. Timber and money bequests received by the landowner in the current (t) generation are b_t and m_t, respectively. Likewise, timber and money bequests given by the current generation to the next generation are denoted m_{t+1} and b_{t+1}, respectively. Both forms of bequests are subject to respective tax rates τ_b and τ_m. Finally, let the number of heirs of the next generation be defined through a population growth rate equal to n. Hence $(1+n)$ is the number of heirs for each generation. All heirs are assumed to receive a bequest from the previous generation. The growth rate n is assumed to be exogenous (for more details, see Carmichael 1983 and Blanchard and Fischer 1990, chapter 3).

The periodic consumption of the representative landowner in each generation is now defined by the following equations:

$$c_{1t} = w_t - T + p_t^* x_{1t} + (1 - \tau_m)m_t - \tau_b p_t^* b_t - s_t \tag{4.25a}$$

$$c_{2t} = p_{t+1}^* x_{2t} + (1 + r_{t+1})s_t - (1 + n)m_{t+1}. \tag{4.25b}$$

In equation (4.25a) w_t is the wage and all harvest prices are net of taxes; that is, $p_t^* = (1 - \tau_x)p_t$ and $p_{t+1}^* = (1 - \tau_x)p_{t+1}$ where τ_x is a tax on harvest revenue. The variable s_t denotes saving, and $R = (1 + r_{t+1})$ is the (exogenous) real interest rate. T is a lump-sum tax, such as an income tax paid when the landowner is working in the first period of life. Equations (4.25a) and (4.25b) are familiar from our previous discussions, except that now they include inherited and given money and timber bequests.

All harvesting decisions continue to be made at the end of each period. Hence during the first period of the current generation t, the size of the forest stock has been determined by the harvesting decision of the previous generation, $t-1$ at the end of the previous period. At the end of the first period, the landowner in the current generation t inherits a forest stock b_t and harvests a share x_{1t}. This in turn determines the increase in the second period's forest stock through forest growth, denoted by the function g. At the end of the second period the landowner harvests an amount x_{2t} and bequeaths to the next generation an amount of the forest stock equal to $(1+n)b_{t+1}$. In accordance with this timing of decisions, the equations for periodic forest stock and second-period harvesting are given by

$$k_{1t} = (b_{t-1} - x_{1t-1}) + g(b_{t-1} - x_{1t-1}) = k_{2t-1} \tag{4.26a}$$

$$k_{2t} = (b_t - x_{1t}) + g(b_t - x_{1t}) = k_{1t+1} \tag{4.26b}$$

$$x_{2t} = (b_t - x_{1t}) + g(b_t - x_{1t}) - b_{t+1}(1+n). \tag{4.26c}$$

The economic problem of the landowner in generation t is now to maximize (4.24) subject to (4.26a)–(4.26c) by choosing saving, harvesting, and money and timber bequests. Substituting these constraints into the utility function and differentiating, we obtain the following first-order conditions:

$$U_{s_t}^t = u'(c_{1t}) + R\beta u'(c_{2t}) = 0 \tag{4.27a}$$

$$U_{x_{1t}}^t = u'(c_{1t})p_t^* - \beta u'(c_{2t})p_t^*(1+g') = 0 \tag{4.27b}$$

$$U_{m_{t+1}}^t = -(1+n)\beta u'(c_{2t}) + \phi(\partial V_{t+1}/\partial m_{t+1}) \leq 0 \Leftrightarrow m_{t+1} \geq 0 \tag{4.27c}$$

$$U_{b_{t+1}}^t = -\beta u'(c_{2t})p_{t+1}^*(1+n) + \phi(\partial V_{t+1}/\partial b_{t+1}) \leq 0 \Leftrightarrow b_{t+1} \geq 0. \tag{4.27d}$$

In (4.27c)–(4.27d), the last terms depend on the next generation's indirect utility function. Using the envelope theorem (see appendix A.4), we have $\partial V_{t+1}/\partial m_{t+1} = u'(c_{1t+1})(1 - \tau_m)$ and $\partial V_{t+1}/\partial b_{t+1} = u'(c_{1t+1})\tau_b p_{t+1}^* + (1+g')\beta[u'(c_{2t+1})p_{t+2}^* + v'(k_{2t+1})]$. Equation (4.27a) defines a Euler equation for consumption. Using it in (4.27b) we obtain the following harvesting rule under altruistic preferences: $Rp_t^* - (1+g')p_{t+1}^* = (1+g')[\beta v'(k_{2t}) + \phi v'(k_{1t+1})]/\beta u'(c_{2t})$. Recalling that $k_{2t} = k_{1t+1}$ by definition, we can reduce the harvesting rule to

$$Rp_t^* - (1+g')p_{t+1}^* = (1+g')\frac{(\beta + \phi)v'(k_{2t})}{\beta u'(c_{2t})}. \tag{4.28}$$

According to (4.28), our previous harvesting rule derived here in the two-period model is slightly different in the OLG model. Altruism now implies that the utility of the next generation becomes important to the harvesting decisions of the current generation. The current-generation landowner chooses harvesting so that the difference between the marginal return to harvesting and its opportunity cost equals the marginal rate of substitution between future amenities of the current and next generation and future consumption of the current generation.

Here altruism implies that a greater weight is given to amenities. Hence, while $r - g' > 0$ both in the presence and absence of altruism, this difference is greater in the altruistic case. Moreover, another new feature of the harvesting rule is that the growth rate shows up in the RHS of (4.28), serving to reinforce the above effect. Note, finally, that while the harvest tax affects the harvesting rule, bequest taxes do not.

The first-order conditions here implicitly define optimal levels of timber and money bequests. Clearly, a tax on money bequests is likely to decrease the bequest at the margin. According to (4.27d), timber bequests occur to the point where the marginal cost of leaving a bequest to the next generation, expressed as lost consumption for the current generation, is equal to the marginal benefits of the bequest. Marginal benefits include amenities to the landowner and future consumption and amenities accruing to future generations who inherit the timber bequest. Again, a tax on timber bequests decreases the amount of forest stock transferred to heirs at the margin.

The OLG literature has traditionally devoted much attention to whether or not money bequest motives are operative in a long-run steady state. Weil (1987) first studied this by comparing steady states in the presence and absence of bequests. To follow his approach, we assume a steady-state equilibrium exists in our model here (for more details, see Amacher et al. 2002). Using a "hat" notation to define the steady state, a steady state exists when for each generation, $\hat{c}_{1t} = \hat{c}_{2t}$, and also $\hat{c}_{2t} = \hat{c}_{1t+1}$. Before proceeding, we also need the following definition (Blanchard and Fisher 1990):

Definition 4.1 Efficiency in an OLG economy with population growth If the real interest rate equals the population growth rate, then the economy resides at the golden rule, which corresponds to the socially optimal level of the long-run capital stock. If the real interest rate is lower (higher) than the population growth rate, then the economy is dynamically inefficient (efficient).

The higher interest rate case implies that the long-run steady-state level of forest capital falls below the golden rule level, whereas a lower interest rate implies overinvestment of capital in the long run. Since consumption is tied to capital, the golden rule is the highest sustainable consumption path for the economy and thus is of interest.

The conditions for operative money and timber bequests can now be formally developed from the first-order conditions given here using the following procedure: First, we re-express the flow of consumption equations in both the absence of bequests and in the steady state. This is solved to allow us to define an implicit real interest rate $(1 + \bar{r}) = u'(\hat{c}_{1t})/\beta u'(\hat{c}_{2t})$, or equivalently $(1 + \bar{r})\beta = 1$. We then assume that no taxes are levied. From the first-order conditions for money and timber bequests, (4.27c) and (4.27d), we arrive at

$$m_{t+1} > 0 \Leftrightarrow \phi > \frac{(1+n)}{(1+\bar{r})} \tag{4.29a}$$

$$b_{t+1} > 0 \Leftrightarrow \phi > \frac{(1+n)}{(1+\bar{r})(1+g')\left[1 + \dfrac{v'(\hat{k}_2)}{pu'(\hat{c}_2)}\right]}. \tag{4.29b}$$

According to (4.29a), the propensity of landowners to give money bequests depends on the real interest rate and population growth. Referring to definition 4.1, we conclude that money bequests cannot be operative (i.e., positive) in a dynamically inefficient economy since $n > \bar{r}$ because α is less than one. According to (4.29b), the propensity to leave a timber bequest also depends on the forest growth rate and the next generation's marginal rate of substitution between non-timber and consumption (harvesting) benefits, $v'(\hat{k}_2)/u'(\hat{c}_2)$. The greater this marginal rate of substitution, the more the current landowner is willing is to forgo his consumption by leaving a timber bequest. It is interesting that timber bequests can also be operative in dynamically inefficient economies because the denominator of (4.29b) would be greater than in the case of money bequests. This is a different result than that found in the economics literature on bequests, which, focusing only on money, finds that such bequests are operative only in dynamically efficient economies.

A comparison of these two conditions reveals that conditions for operative timber bequests are more easily fulfilled than those for operative money bequests. Hence timber bequests dominate money bequests when amenity benefits from forests are important. This means that if there is no penalty (i.e., taxation) for bequests, then there is less incentive for a private landowner to convert timber to money (through harvesting) prior to leaving a bequest. If the landowner does not value amenities, then he would still prefer leaving timber bequests because from (4.29a) and (4.29b) we have $(1+n)/(1+\bar{r}) > (1+n)/(1+\bar{r})(1+g')$. This establishes proposition 4.8.

Proposition 4.8   When the rate of time preference is equal to the interest rate, timber bequests always dominate money bequests in the absence of policies regardless of whether amenities are valued. Finally, while money bequests cannot be operative in dynamically inefficient economies, timber bequests can be.

The fact that timber bequests can be positive in a broader range of situations than money bequests is reassuring for long-run forestry and clearly results because forests provide amenity services that go beyond harvesting values; these translate into additional long-run benefits of giving timber bequests to the next generation.

The case where taxes are present is quite interesting. Now the conditions for positive money and timber bequests become

$$m_{t+1} > 0 \Leftrightarrow \phi > \frac{(1+n)}{(1+\bar{r})(1-\tau_m)} \tag{4.30a}$$

$$b_{t+1} > 0 \Leftrightarrow \phi > \frac{(1+n)(1-\tau_x)}{(1+\bar{r})(1+g')\left[(1-\tau_x) + \dfrac{v'(\hat{k}_2)}{u'(\hat{c}_2)p} - \tau_b\right]}. \tag{4.30b}$$

A tax on money bequests decreases a landowner's propensity to give them, as we would expect. While the condition on money bequests is affected only by a money bequest tax, the condition for timber bequests depends on the timber bequest tax as well as the harvest tax; both of these taxes tend to decrease the propensity to leave timber bequests. Hence equations (4.30a) and (4.30b) demonstrate that the threshold level of altruism certainly is dependent on the tax structure of the economy. Whether or not timber bequests continue to dominate money bequests now depends on the relative size of the bequest and harvest taxes. From the equations we can see that a sufficient but by no means necessary condition for this to hold is that

$$(1+g')\left[(1-\tau_x) + \frac{v'(\hat{k}_2)}{u'(c_2)p} - \tau_b\right] \geq (1 - \tau_m).$$

Clearly, this condition need not hold when all landowner choices are taxed. Still, it is evident that a subsidy on timber bequests (or a reduction in the bequest tax) strongly increases the dominance of timber bequests. Hence, we have proposition 4.9.

Proposition 4.9 When the rate of time preference is equal to the interest rate, a timber bequest subsidy reinforces the condition for operative bequests in the presence of harvest and money bequest taxes. If timber bequests are also taxed, then it is unclear whether timber bequests are reinforced relative to money bequests.

Proposition 4.9 lends support to claims that timber bequests are sensitive to government policies such as inheritance taxes. In chapter 5 we will return to the OLG model and ask how a tax structure can be designed to balance bequests and amenities in an economy.

In closing, the reader should note that it is straightforward to extend the OLG model here to include management effort. From chapter 2, this choice captures

various timber stand improvements, such as fertilization or more intensive planting efforts. In the OLG economy, it can be shown that management effort affects
the optimal timber capital present in the steady state and also the bequest decision
of the landowner. In such a model, we can again show that harvest and inheritance taxes potentially reduce the long-run forest capital stock relative to the stock
that would exist in the absence of these taxes.

4.4 Modifications

Two-period models have been extended to study the case where there is uncertainty in future economic parameters, such as prices, or where there are no longer
perfect markets concerning capital for borrowing by forest landowners attempt-
ing to finance consumption. We now comment on several extensions and then
present one concerning the incidence of forest taxation.

4.4.1 Some Extensions of Two-Period Models

As in the Faustmann model, management intensity can also be introduced into
the forest growth function as either regeneration or silvicultural effort (see e.g.,
Ovaskainen 1992 or Amacher and Brazee 1997). Timber market equilibria in both
perfectly and imperfectly competitive markets have also been studied (see e.g.,
Johansson and Löfgren 1985, and Koskela and Ollikainen 1998a,b). These models
have allowed an analysis of the incidence of forest taxation, as found in Koskela
and Ollikainen (1998a). Others have used this framework to explore socially optimal design of forest taxation when externalities or fiscal federalism are present
(see e.g., Amacher and Brazee 1997a, Amacher 2002).

Finally, the two-period model has been extended to an analysis of landowner
behavior under various forms of uncertainty. Although we will not discuss it
here, a modification of two-period models that deserves mention concerns cases
where landowners are credit constrained. Johansson and Löfgren (1985) were the
first to propose a theory for this problem in which landowners harvest timber to
finance purchases when borrowing is restricted. Their idea was couched as one of
buying a Volvo, and since then the "Volvo theorem" was born in forest economics. Fina et al. (2001) and Brazee (2003) provide a relatively recent analysis
and discussion of the history behind this idea in addition to deriving some generalizations about it.

Capital market imperfections have also been introduced in many forms. For instance, saving and borrowing rates have been assumed to differ, or the interest
rate was considered to be a nonlinear function of the size of a loan a landowner
wished to borrow in financing consumption (Ollikainen 1996b). The easiest form
of credit rationing to model is a binding loan ceiling in which the landowner

would like to borrow more than financial institutions are willing to lend him at a constant interest rate (Koskela 1989b). Kuuluvainen (1990) used a two-period model to consider credit rationing in the form of a shadow price using the same starting point.

Finally, Ovaskainen (1992) examined short- and long-term timber supply in a three-period model. In Ovaskainen's model, the landowner has two decision variables—harvesting and silvicultural effort—and the third period is modeled as the long run. If we omit effort from his model, we can arrive at exactly the same harvesting rule we obtained earlier for short-term harvesting. Thus the short-term timber supply function is robust in terms of increasing the number of periods under certainty and perfect capital markets. Furthermore, assuming prices are constant over time, the long-run timber supply solution from this model results in a long-run optimal forest stock that is always lower than the maximum sustained yield stock (see figure 4.1). The rest of this chapter is devoted to one modification that embeds the two-period landowner decision model within a market equilibrium to examine the incidence of forest taxation.

4.4.2   Incidence of Forest Taxation

The incidence of forest taxes has interested forest economists for years. The first modern studies in this area were done by Gamponia and Mendelsohn (1987) and Aronsson (1990). Their main objective of evaluating tax incidence was to determine how taxes imposed on forest landowners are reflected in market prices. Ultimately this tells us where the burden of taxation falls in the market.

Most of the work in forest economics concerning tax incidence has been derived using two-period models. Drawing on Koskela and Ollikainen (1998a) we present a simple derivation of market equilibrium and tax incidence. Suppose firms in the forest industry produce final products (pulp, paper) using a timber input over two periods, now and the future. Let current and future timber inputs be denoted by x_1 and x_2, respectively. In order to simplify the analysis, assume that the production functions are identical for both periods and quadratic in terms of the timber input: $Q_1 = [a - (1/2)bx_1]x_1$ $Q_2 = [a - (1/2)bx_2]x_2$. We also make a small open-economy assumption that firms are price takers and that the price of the final product is exogenous. Normalizing this price to equal one, the decision problem of the firms is to choose x and z to maximize the present value of expected profits:

$$\max_{\{x,z\}} \pi = [a - (1/2)bx_1]x_1 - p_1x_1 + R^{-1}\{[a - (1/2)bx_2]x_2 - p_2x_2\}. \tag{4.31}$$

This solution yields the current and future demand for timber as functions of the parameters of the production function and timber prices. Applying previously adopted notation for timber supply, we can express timber demand as follows:

$$x^{1d} = \alpha - \beta p_1; \quad x^{2d} = \alpha - \beta p_2, \tag{4.32}$$

where $\alpha = a/b$ and $\beta = 1/b$. Two notable features of these demand functions are that they depend negatively on timber prices and are characterized by temporal separability. This results from a lack of interdependence here between current and future production functions.

Adopting a logistic specification for the forest growth function, we can define growth as

$$(Q - x_1)\left[f - \frac{k}{2}(Q - x_1)\right],$$

where f is a linear component, k is a concavity factor, and Q is the original stock of timber at the beginning of the first period. Future harvesting is then defined by

$$x_2 = (Q - x_1) + (Q - x_1)\left[f - \frac{k}{2}(Q - x_1)\right]. \tag{4.33}$$

Using the same symbols as before, we define the period consumption flow equations for a representative landowner as $c_1 = \hat{p}_1 x_1 - a - s$ and $c_2 = \hat{p}_2 x_2 - a + Rs$. In what follows we allow the yield tax (and unit tax) rate to differ across periods.

The landowner again generates timber supply through maximizing utility by choosing current consumption c_1 and current harvesting x_1. It suffices here to solve only for current harvesting. Under certainty and perfect credit markets, separability holds, so that x^s is solely determined by the harvesting condition $(1 + r)\hat{p}_1 - \hat{p}_2(1 + g') = 0$. Using the derivative of equation (4.33), this condition becomes $R\hat{p}_1 - \hat{p}_2[1 + f - k(Q - x_1)] = 0$, and we can write the following explicit current and future timber supply functions as

$$x^{1S} = Q + \frac{R\hat{p}_1 - \hat{p}_2(1 + f)}{k\hat{p}_2}; \quad x^{2S} = Q(1 + f) - x^s(1 + f). \tag{4.34}$$

According to (4.34), short-term timber supply depends positively on the initial forest stock, the real interest rate, and current timber price, and negatively on future timber price. Our explicit solution for x also indicates that growth factors ("fertility" f and concavity k) negatively affect current harvesting. As for forest taxes, we know that the site productivity tax is neutral with respect to timber supply. For the effects of current unit and yield taxes in the case when timber prices are not constant over time, we have

$$x^{1S}_{\mu_1} = -\frac{R}{k\hat{p}_2} < 0 \quad \text{and} \quad x^{1S}_{\tau_1} = p_1 x^{1S}_{\mu_1} < 0. \tag{4.35}$$

Applying the implicit function theorem, we can now analyze qualitatively how equilibrium timber prices change when taxes change. For an increase in the current yield tax, it must be true that $x^{1d} - x^{1S} = 0$ and $x^{2d} - Q(1+f) + x^{1S}(1+f) = 0$ at the market equilibrium. Differentiating this system with respect to taxes and current and future timber prices, and solving the differential for price effects of the current yield tax gives

$$\frac{dp_1}{d\tau_1} = \Delta^{-1}\{x_{\tau_1}^{1S} x_{p_2}^{2d}\} > 0; \quad \frac{dp_2}{d\tau_1} = -\Delta^{-1}\{(1+f)x_{\tau_1}^{1S} x_{p_2}^{2d}\} < 0, \tag{4.36}$$

where Δ is the Jacobian matrix (see appendix A for a discussion of Jacobian matrices).

Equation (4.36) shows that a current yield tax increases the current equilibrium market timber price but decreases the future equilibrium price. This makes sense because a higher current yield tax makes the landowner shift some harvesting to the second period. Since current timber supply decreases, future supply increases. Because the demand for timber remains unaffected by the changes in the yield tax, shifts of the supply function uniquely determine the new timber price subject to the elasticity of the demand function. We leave it to the reader to show that an increase in the future yield tax causes an opposite effect on equilibrium timber prices. In a similar way, we can analyze the effects of the unit tax on current and future equilibrium timber prices.

Market equilibrium is defined by an equality of demand and supply in both periods. Equilibrium timber prices will be of second-order magnitude, so that we will have two pairs of solutions in general. These can be solved numerically using a program such as Mathematica. In the simple case where the concavity term of the growth function is negligible, and taxes are absent, timber prices equal the following expressions:

$$p_1 = \frac{(1+f)^2 + 2kQ(1+f) + R - \beta^{-1} + 3\alpha kD}{(1+f)2k\beta}$$

$$p_2 = \frac{-(1+f)^2\beta + R - k\alpha + D\beta^{-1}}{2k\beta},$$

where

$$D = \pm\sqrt{-4kR(1+f)Q - (2+f)\alpha + [1 + f(2+f) + R - k\alpha]^2\beta^2}.$$

The general effects of the yield tax on timber market equilibrium and its incidence for suppliers and demanders depend on the price elasticity of demand and supply functions. In the case of a horizontal timber demand function and upward-sloping timber supply function, the entire increase in the yield tax is passed on to the supply side, i.e., to landowners (this is the case where the comparative statics of timber supply is valid for the entire timber market). For a vertical demand function and an upward-sloping supply function, the entire tax increase is passed onto the (inelastic) demand side (to consumers). In the typical case of downward-sloping demand and upward-sloping supply functions, both parties pay a share of the tax according to their price elasticities.[7]

Application of the yield tax also causes a deadweight loss from reduced use of timber by the forest industry. The deadweight loss describes a distortion that is due to taxation, and it implies that the welfare decrease that higher taxation causes is larger than the increase in government revenue from the higher taxation. Thus unless the goal of the government is actually to create distortions (there may be some reasons to do this, as we will see in chapter 5), a neutral forest tax is desirable because it minimizes the deadweight loss of taxation. It also therefore minimizes the total burden of the tax imposed on landowners and forest firms. We can approximate the deadweight loss as the area of a triangle abc formed by a price increase and quantity decrease as $abc \approx -(1/2)\Delta x_1 \tau_1$, where $\Delta x_1 = x_1^0 - x_1^1$. If timber supply is infinitely elastic, then by using the elasticity equation for timber demand,

$$\varepsilon = -\frac{\Delta x_1/x_1}{\Delta p_1/p_1} = -\frac{\Delta x_1}{x_1}\frac{p_1}{\tau_1},$$

where $abc \approx (1/2)(x_1^0/p_1^0)\varepsilon\tau_1^2$. The deadweight loss here depends positively on the elasticity and the tax rate. The relationship between the deadweight loss and the tax rate is also now convex; i.e., a higher tax rate will increase deadweight loss proportionately more.

4.5 Summary

The two-period life-cycle model approaches harvesting from a different perspective than the rotation framework. The aim of two-period modeling is to develop a theory of short-run timber supply. Unlike in the rotation model, forest production technology in the two-period model is described in the form of the standing forest stock or biomass. The landowner may value only harvest revenue, or both harvest revenue and amenity benefits. Under certainty and perfect markets, the properties of short-run timber supply are easy to solve and interpret. Higher current and future timber prices and interest rates will increase and decrease timber supply,

---

7. The elasticity of timber demand and supply is affected by many market features and new developments (see, for example, forest fragmentation discussed in Decoster 2002).

respectively, from the representative landowner. Amenity valuation serves to decrease current harvesting and implies that some of the stock is left standing in the second period. These timber supply functions can be used as a component of an analysis of more general timber market equilibrium and policy, and it is straightforward to study how uncertainty concerning economic parameters affects supply. The strengths of this model will become clearer in later chapters when we take up uncertainty and forest resources.

The two-period model also provides the basic building block in analyzing forest economics for overlapping generations. Such a framework allows one to examine how forest stocks are transferred from one generation to another. An especially interesting issue here involves the motives of landowners in leaving timber bequests, something the basic rotation framework cannot inform. The OLG model also allows a study of how policy instruments affect the long-run forest stock in an economy. These questions will have an increasingly important role as forestry is further integrated into more general economic models.

In chapter 8, we will provide a formal discussion of the relationship between two-period models and dynamic uneven-aged management problems. This qualifies the points originally made by Montgomery and Adams (1995). In the two-period interpretation of uneven-aged modeling, the first-period harvest choice moves the forest to a steady state, while future harvesting (in the second and subsequent periods) is then defined by the resulting sustained steady-state forest growth.

II   Policy Problems

5 Design of Forest Policy Instruments

The taxation of forest and land has captured the interest of foresters and economists for more than a century, encouraging Martin Faustmann to write his celebrated contribution in 1849. Faustmann set out in that famous article to define the principles by which the value of forest land could be accurately determined, so that forest taxes could be levied more efficiently and fairly in order to raise funds for government programs.

To design any forest tax system, a policy maker must answer two basic questions: what types of tax instruments are "best," and what is the appropriate level for each instrument? Obviously, the answers to these questions depend on the objective function chosen by the policy maker and any constraints on the problem. The policy maker must also be able to anticipate the reaction of landowners to choices of instrument.

The most common way of structuring forest policy problems in economics is to consider the benchmark of a benevolent social planner. This planner is assumed to choose instruments to maximize a social welfare function for the forest sector given by the sum of consumer and producer surplus. For any forest sector, social welfare consists of the net economic surplus of forest landowners, which may include amenities enjoyed privately by the forest landowner plus amenities derived from private forests that are consumed as public goods by nonlandowner members of society. The optimal design of taxation will differ, depending on whether these public goods are important enough for the social planner to promote them through the design of instruments. It will also depend on the nature of any binding revenue constraints the government faces and revenue targets that the government may set for the forest sector.

There are three possible definitions of policy design problems. A "first-best" taxation problem is solved when an instrument set, or single instrument, is chosen to eliminate distortions in landowners' forest management decisions that cause distortions in the level of public goods consumed by nonlandowners. A "second-best" taxation problem arises when the government faces binding constraints on taxation choices; for instance, when a government has no reason to intervene in the forest sector other than to collect tax revenue using distortionary taxes. Often truly neutral taxes are not possible under current legislation in many countries, and thus taxation problems are often second best in nature. Finally, we should note that there are sometimes combinations of first- and second-best problems. For example, a government might seek to raise funds using tax instruments but do so in a way that provides an efficient level of public goods from forests. We will cover all three types of problems here, but the reader interested in more general treatments is referred to the public economics texts by Tresch (2002), Myles (1995), and Salanie (2003).

When characterizing first- and second-best problems, some terminology is used from public finance. If a tax does not affect landowner decisions, such as harvesting, then we say that it is a neutral instrument. Taxes that do affect landowner behavior are called either "nondistortionary" or "distortionary" instruments. A nondistortionary tax is one that does not change relative prices faced by the landowner but does potentially change landowner income. Thus, in welfare economics language, these taxes induce an income effect but not a substitution effect in the landowner's decision making. Distortionary taxes affect the behavior of forest landowners by changing both relative prices and income. Thus, these taxes have both income and substitution effects on landowner decisions.

As we will see, a neutral tax is always optimal when the social planner does not wish to distort the decisions of private landowners but does need to meet certain revenue collection needs. This reflects a well-established result, called a "Ramsey rule" (Ramsey 1927), which states that a government should seek to raise tax revenue by causing as little distortion as possible. Conversely, a distortionary, or sometimes called "corrective" or "Pigouvian" tax, dates back to Pigou (1932). This type of tax is optimal if private landowner decisions must be adjusted because of the presence of externalities. We will return to these definitions later.

Another convention in modeling optimal forest policy is to assume that the social planner acts as a "Stackelberg" leader when choosing instruments. Landowners and consumers take these instruments as given and behave as followers. The government therefore chooses forest tax rates by anticipating the reactions of consumers and landowners. Necessarily, the social planner must be able to credibly commit to an optimal tax design for a long term once one is chosen. While commitment has been the common assumption in the 50 plus years of research on optimal taxation theory, it is clear that commitment may not hold in a dynamic context (see Ljungqvist and Sargent 2004, chapter 22, for a detailed discussion). Potential lack of commitment remains an open question in forest economics policy problems.

The history of optimal taxation in the public economics literature is long, but its use in forestry problems is recent. Gamponia and Mendelsohn (1987), Englin and Klan (1990), and Koskela and Ollikainen (2003) examine these problems in a rotations framework. Gamponia and Mendelsohn (1987) focused on the excess burden of yield and timber taxes in a Faustmann model, while Englin and Klan (1990) studied tax policy in the absence of a binding tax revenue requirement. In their models, amenities arise when landowners forgo harvesting forest stocks, but landowners are assumed to value only harvesting revenues; this makes their problems a pure first-best one. Koskela and Ollikainen (2003) studied optimal forest taxation in both Faustmann and Hartman models when the government faces a revenue collection constraint and amenities are produced by forests but not necessarily valued by forest landowners. This is similar to a combination of the first and the second-best problems.

Amacher (1999), Amacher and Brazee (1997a), and Koskela and Ollikainen (1997a, b) study optimal taxation in life-cycle models. These authors also generally assume that the policy maker faces revenue constraints and that there is a need to correct certain externalities arising from private landowner behavior that diverges from socially optimal behavior. Finally, first-best taxation problems have been considered in models with what is called an "Austrian" sector, where forest capital appreciates, and an ordinary sector. Here the issue has been how to design instruments that can achieve both intrasectoral and intersectoral efficiency, with the forest sector existing in equilibrium with other sectors. Some examples include Kovenock and Rothschild (1983), Kovenock (1986), Koskela and Ollikainen (1997a), and Uusivuori (2000).

This chapter is organized as follows: In section 5.1 we examine the first-best design of forest taxation within a Faustmann framework. In section 5.2 we allow for the presence of amenity benefits accruing to society from not harvesting forest stocks, making use of the Hartman framework and specifying a level of public goods produced in the forest sector. Section 5.3 contains the same analysis as sections 5.1 and 5.2, but here we make use of the two-period models discussed in chapter 4. In section 5.4 we return to policy design in overlapping-generations models, and in the last section we discuss several modifications.

5.1 Optimal Taxation—Faustmann Interpretations

We start by examining a first-best forest tax problem using the Faustmann framework. We assume the existence of a social welfare function that describes social preferences for the use of forests, and then we compare socially optimal and privately optimal rotation age decisions in order to define the optimal tax structure.

5.1.1   First-Best Taxation—Absence of Government Revenue Constraint

Assume that both a private landowner and society value only the net present value of harvest revenue from ongoing forest rotations. As in chapter 2, assume the following objective function for the landowner in the absence of any taxes:

$$V = \frac{pf(T) - ce^{rT}}{e^{rT} - 1},$$
(5.1)

where p is the timber price, $f(T)$ is the forest growth technology, c is harvesting and replanting costs, and r is the real interest rate. The notation of this Faustmann problem is slightly different than that in chapter 2. The objective function (5.1) is derived from (2.1) simply by multiplying both numerator and denominator by e^{rT}. This restates the Faustmann form in future-value terms and is more convenient for deriving taxation results. If (5.1) is the social value function, it would imply that society only valued harvest revenues and the social welfare function, denoted SW^F, is equivalent to the landowner's objective function, $SW^F = V$. Maximizing this function yields a rotation age that is both privately optimal (i.e., optimal for a private landowner) and socially optimal (i.e., optimal for society) (see Johansson and Löfgren 1985 and Hellstien 1988). Suppose that the government wants to choose a tax system that will maximize this social welfare function. Because the privately and socially optimal rotation ages coincide, if there were no collection target for tax revenues, then it would simply be optimal to set all forest tax rates equal to zero, since harvesting returns decrease with taxes.

5.1.2   First-Best Taxation—Presence of Neutral Tax and Revenue Constraint

Beyond the trivial problem just discussed, suppose that the social planner faces a binding revenue constraint but has a neutral forest tax available. Now the interaction between the government and the representative landowner is described as a two-stage Stackelberg game. In the first stage, the government, acting as a leader, chooses its forest taxation policy and credibly commits to using it. In the second stage, private landowners choose their harvesting levels conditional on the announced tax policy. The response of the landowner to the government's choice of policies is given by the comparative static results already derived in chapter 2. The resulting policy game is solved using backward induction.

The nature of the Stackelberg game now requires us to define the social welfare function in terms of the indirect net revenue function of the landowner, V^*. This is obtained by substituting the landowner's optimal rotation age choice, $T = T[\tau, u, \theta, \alpha]$, solved in the presence of taxes, into his objective function in (5.1) to obtain

$$SW^F = V^*[\tau, u, \theta, \alpha, a(i)],$$
(5.2)

where $a(i)$ is the site productivity tax; τ and u are yield and unit taxes, respectively; α is a property tax levied on stumpage value, called a timber tax; and θ is the site value tax.

This V^* function represents the best in present-value terms that the landowner can do making his decision in the face of taxes. This is the maximum welfare received by the landowner for a given tax structure.

The social planner's objective is to choose tax rates that meet its revenue constraint while decreasing the welfare of landowners in (5.2) as little as possible. The optimal choices of the landowner represent reaction functions to the government's choice of tax instruments. To introduce a binding exogenous tax revenue target, $\bar{G}$, assume that short-run government debt or surplus is not regarded as an important factor in tax revenues. Suppose also that the government responds only to the discounted sum of tax revenue collected from the forest sector, perhaps because it assigns a revenue target to it. Revenue for all of the taxes defined before in chapter 2 (using the same notation) is given by

$$G = \frac{(p\tau + u)f(T) + \alpha e^{rT} \int_0^T pf(s)e^{-rs}\,ds}{e^{rT} - 1} + \frac{a(i)}{r} + \theta V, \qquad (5.3)$$

with $V = [pf(T) - ce^{rT}](e^{rT} - 1)^{-1}$.

In spite of expressing social welfare as a function of all in (5.2) taxes, it is rarely assumed that the policy maker can use all possible taxes at the same time. Thus in analyzing the problem we have to indicate what taxes are assumed to be operative. Recall from chapters 2 and 3 that in the Faustmann model, the site productivity tax and site value tax are neutral, but in the Hartman model only the site productivity tax is neutral. Therefore we will use the site productivity tax as our benchmark neutral tax for both Faustmann and Hartman models. We will then ask whether it is socially optimal to introduce a distortionary tax at the margin. We obviously also need to make the assumption that tax revenues are positively related to tax rates and negatively related to any tax exemptions.[1]

The government's problem is now to maximize (5.2) subject to G in (5.3) equaling some revenue target $\bar{G}$. The Lagrangian for this problem is

$$\Omega = V^* - \lambda(\bar{G} - G). \qquad (5.4)$$

To examine the optimal site productivity tax, we differentiate (5.4) with respect to $a(i)$. However, because V^* is a maximized value of the landowner's objective

---

1. The dependence of the tax revenue on the tax rate is usually described by the Dupuit-Laffer curve, according to which there is a positive (or negative) relationship between tax revenue collections and tax rate (tax exemption). Fullerton (1982) provides a survey of the empirical literature regarding the relationship between tax rates and tax revenues, while Malcomson (1986) provides a self-contained theoretical analysis of the Laffer curve.

function, the envelope theorem must be used to obtain this first-order condition, which implies that the derivative of V^* with respect to the choice variable of the landowner, T, is equal to zero (see the appendix at the end of this book for a review of this theorem). Furthermore, recalling that the site productivity tax is neutral, there will be no landowner response to it through the budget constraint since $T_a^F = 0$. The first-order condition for the tax is therefore

$$\Omega_{a(i)} = -\frac{1}{r} + \lambda \frac{1}{r} = 0. \tag{5.5}$$

According to (5.5), it is optimal to set the site productivity tax $a(i) = a^*$ so as to equate the marginal welfare loss of landowners that is due to the tax (first term) with the marginal cost of public funds (defined as the shadow price of the revenue constraint, λ). This shadow price measures the marginal decrease in the value of social welfare for an incremental increase in government revenue that follows from a tax increase (see Sandmo 1975 for a detailed explanation). In (5.5), λ is equal to one at the optimum, meaning that shifting one dollar from the landowner to the government costs nothing. Thus, the neutral site productivity tax provides an ideal tax instrument to collect revenue because it does not distort the landowner's privately optimal behavior. Another interpretation is that the deadweight loss for this tax is zero.

Given that the social planner sets the site productivity tax at its optimal level, since this collects the required tax revenue, it is evident that no other taxes are needed. Thus, we have proposition 5.1.

Proposition 5.1 Optimal forest taxation in the presence of neutral tax If society values only harvest revenue from forests and wishes to collect a given revenue requirement from the forest sector, then it should employ only a neutral tax and set all distortionary taxes equal to zero.

Proof Consider first the yield tax τ. Differentiating the Lagrangian Ω with respect to τ and making use of the envelope theorem gives

$$\Omega_{\tau|a=a^*} = \tau \left\{ \frac{(e^{rT} - 1)pf'(T) - re^{rT}pf(T)}{(e^{rT} - 1)^2} \right\} T_\tau^F = 0 \quad \text{with}$$

$$\frac{(e^{rT} - 1)pf'(T) - re^{rT}pf(T)}{(e^{rT} - 1)^2} < 0 \quad \text{and} \quad T_\tau^F > 0.$$

Thus this expression can be zero only if the yield tax is zero. The respective first-order condition for the timber tax is

$$\Omega_{\alpha|a=a^*} = \frac{\alpha}{(e^{rT} - 1)} \left\{ \frac{1}{(e^{rT} - 1)} [pf'(T) - rU] \right\} T_\alpha^F = 0,$$

where U was previously defined in (2.22b) of chapter 2, with $[pf(T) - rU] > 0$ $T_\alpha^F < 0$, and $[pf(T) - rU] > 0$. Hence the timber tax rate is also optimally equal to zero. Q.E.D.

The message of proposition 5.1 is clear. In the absence of externalities, a neutral tax is optimal and follows the Ramsey rule by minimizing deadweight losses, as pointed out by Gamponia and Mendelsohn (1987).

5.1.3   Second-Best Taxation—Distortionary Tax and Revenue Constraint

The site productivity tax may not be feasible, and indeed such taxes are rare in practice, given the complexity inherent in defining site classes for each landowner. It is therefore logical to next ask what happens if the government does not have a site productivity tax or for that matter any neutral tax available. The question then becomes which tax instruments meet the government's revenue target with minimal decreases in landowner welfare.

Differentiating the Lagrangian Ω in (5.4) with respect to the unit and timber taxes and assuming there are no other taxes, and again using the envelope theorem, we obtain the following first-order conditions by slight rearranging:

$$\Omega_u = (\lambda - 1) \frac{f(T)}{(e^{rT} - 1)} + \frac{\lambda}{(e^{rT} - 1)^2} \{uA + \alpha B\} T_u^F = 0 \tag{5.6}$$

$$\Omega_\alpha = (\lambda - 1)U + \frac{\lambda}{(e^{rT} - 1)^2} \{uA + \alpha B\} T_\alpha^F = 0, \tag{5.7}$$

where $A = (e^{rT} - 1)f'(T) - rf(T)e^{rT} < 0$ using the first-order condition (2.3) given in chapter 2, $B = (e^{rT} - 1)[pf(T) - rU] > 0$, $T_u^F > 0$ and $T_\alpha^F < 0$. From (5.6) and (5.7) one can see immediately that these conditions are satisfied with $\lambda = 1$ and $uA + \alpha B = 0$. The latter implies that the privately optimal rotation age of the stand is determined by the following first-order condition:

$$V_T = 0 \Leftrightarrow p[(e^{rT} - 1)f'(T) - re^{rT}f(T)] + cre^{rT} = 0.$$

This is exactly the same condition for harvesting that holds without forest taxes. Solving for an optimal combination of unit and timber taxes that makes $uA + \alpha B = 0$, and using the definitions of A and B yields the optimal ratio of unit and timber taxes as

$$\frac{\hat{u}^*}{\hat{\alpha}^*} = -\frac{(e^{rT} - 1)[pf(t) - rU]}{(e^{rT} - 1)f'(T) - rf(T)e^{rT}} > 0. \tag{5.8}$$

The hat notation is used to indicate that these rates are solved under a binding tax revenue requirement. This tax mix is nondistortionary with respect to the rotation age decision of the private landowner. This result holds for any combination of the yield tax and timber tax as long as the above condition is satisfied. We therefore have proposition 5.2.

Proposition 5.2 Optimal forest taxation in the absence of a neutral tax If a neutral tax is not available, and if society values only harvest revenue and seeks to collect a given revenue target, then a combination of a unit tax and timber tax (or, equivalently, a combination of a yield tax and timber tax) is socially optimal and can be used to collect tax revenue in a nondistortionary way.

The economic intuition behind this result lies in the use of two corrective taxes that affect rotation age in opposing ways. Thus an appropriate combination can be constructed to ensure the instrument mix is nondistortionary. This result is derived analytically by Koskela and Ollikainen (2003a), and Gamponia and Mendelsohn (1987) derive a similar result using numerical simulations. Whether the government uses a unit or yield tax does not matter as long as the timber market is competitive. While proposition 5.2 provides an interesting possibility of collecting tax revenue without distortions, its practical relevance is questionable. Typically, governments do not levy multiple taxes on natural resource use, and a plan such as this may not be obvious to forest landowners, who would most likely oppose any plan of using two taxes. Still, from a theoretical standpoint, proposition 5.2 is perfectly in line with the Ramsey rule of taxation.

5.2 Optimal Taxation—Hartman Interpretations

We now consider a case where potential externalities arise either because private landowners do not value public goods, or they do, but not as much as society does.[2] As in the previous section, we proceed by first defining the social welfare maximization problem, and then we analyze the optimal design of taxation in both the presence and the absence of neutral taxation.

5.2.1 First-Best Taxation—Absence of Government Revenue Constraint
When amenity benefits from forests are important, there are two possible ways of defining the social welfare function: amenity services might be valued by both landowners and nonlandowners, or, alternatively, the landowner may value only harvest income while nonlandowners value only amenities from forests. An easy but still informative way to frame the difference in the first case is to consider a stylistic forest economy where the number of (nonlandowner) citizens is n and there is one (representative) landowner. Thus the number of nonlandowners valuing amenity benefits is $(n-1)$. Furthermore, it is easiest to assume that all non-

---

2. Tresch (2002) provides an excellent discussion of externalities in public economics problems.

landowners have identical preferences over amenity benefits. The landowner may or may not value amenity services. However, if he does, then assume that the landowner's valuation coincides with the values of nonlandowners. Thus it is only the number of the nonlandowners that leads to the key difference in private and social valuation of amenities.[3]

In terms of chapter 3 notation, equations (5.9a) and (5.9b) represent two potential social welfare functions:

$$SW^H = V + E + (n - 1)E, \tag{5.9a}$$

$$SW^F = V + (n - 1)E. \tag{5.9b}$$

Under both target functions, the preferences of the society and the landowner differ, so we immediately expect that the privately and socially optimal rotation ages cannot coincide. We are therefore now studying a first-best policy choice situation.

The case in (5.9b) was first used by Englin and Klan (1990) under the implicit assumption that $n = 1$ and later by Koskela and Ollikainen (2003a) for $n > 1$. These authors first solved for the socially optimal and privately optimal rotation ages in the presence of taxes. Equalizing the corresponding first-order conditions, they then solved for tax rates so that the first-order conditions of society and the private landowner coincided. We apply this approach to define first-best instruments under the social welfare function (5.9a). It turns out that the first-best instruments under welfare function (5.9b) are identical, so that we only need to work with equation (5.9a). Choosing rotation age T to maximize the social welfare function (5.2) gives the socially optimal rotation age

$$SW_T = pf'(T) - rpf(T) - rV + F(T) - rE + (n - 1)[F(T) - rE] = 0. \tag{5.10}$$

Recall from chapter 3 that the first-order conditions of the private rotation age in the presence of distortionary taxes are defined by equation (3.10) for harvest taxes, equation (3.16) for the site value tax, and equation (3.18) for the timber tax. Setting the privately optimal first-order condition in the presence of harvest taxes equal to the socially optimal first-order condition, and solving for the first-best optimal forest tax rates where these conditions are equal gives the following solutions:

$$\tau^{**} = -\frac{(n - 1)[F(T) - rE]}{pf'(T) - rpf(T)(1 - e^{-rT})^{-1}} \leq (>) 0 \quad \text{as } F'(T) \geq (<) 0 \tag{5.11a}$$

$$u^{**} = p\tau^{**} \leq (>) 0 \quad \text{as } F'(T) \geq (<) 0, \tag{5.11b}$$

---

3. Alternatively, we could have postulated a general public amenity valuation function that differs from the private landowner's valuation of amenities, but the analysis would produce the same qualitative results, albeit with more complicated mathematics.

where the two asterisks refer to optimal tax rates under the assumed Hartman model.

In (5.11a) the sign of both the denominator and numerator depend on the valuation of amenities by society. The nature of this valuation determines the conclusions we reach about the optimal yield tax. First, if amenities do not depend on management decisions, that is, $F'(T) = 0$, and $F(T) - rE = 0$, then the optimal yield tax equals zero. Second, if amenities increase with rotation age, $F'(T) > 0$, then using the first-order condition (5.10) we have $F(T) - rE > 0$ but $pf'(T) - rpf(T)(1 - e^{-rT})^{-1} < 0$ and the optimal yield tax is positive (this induces the landowner to increase the private rotation age; see chapter 3). Finally, if we have the case where $F(T) - rE < 0$ and the sign of $pf'(T) - rpf(T)(1 - e^{-rT})^{-1}$ is ambiguous, then the optimal yield tax can either be positive, in which case it is a net tax, or negative, in which case it is a net subsidy paid to landowners. A similar analysis applies to the unit tax in (5.11b). Hence when amenity services from young stands are valued, the information requirement for use of the yield tax becomes high.

Equations (5.11a)–(5.11b) define what we call first-best Pigouvian harvest taxes. The point of using these instruments is to induce the private landowner to choose the socially optimal rotation age. Another way of saying this is that the tax rates in (5.11a) and (5.11b) internalize the externality implicit in different private and social rotation ages by forcing the landowner to pay a higher marginal cost for deviating from the socially optimal rotation age.

Consider next the optimal first-best timber and site value taxes. Setting the privately optimal first-order conditions in the presence of taxes (i.e., equations 3.14 and 3.18) equal to the socially optimal condition, and solving for the optimal tax rates gives

$$\alpha^{**} = -\frac{(n-1)[F(T) - rE]}{[pf(T) - rU]} \leq (>)\, 0 \quad \text{as } F'(T) \geq (<)\, 0 \tag{5.11c}$$

$$\theta^{**} = -\frac{(n-1)[F(T) - rE]}{pf'(T) - rpf(T) - rV} = \frac{(n-1)}{n} > 0. \tag{5.11d}$$

The denominator is always positive, so that the solution to this tax rate depends on the nature of the amenities. When amenities from older stands are valued, it is socially optimal to use a timber subsidy, but when younger stands are preferred to older stands with regard to amenities, a positive timber tax becomes optimal. The optimal size of the timber subsidy or tax depends on the ratio of the size of the externality generated by harvesting, measured by forgone marginal amenity benefits, and the effect of the timber subsidy or tax on timber production.

Equation (5.11d) defines the optimal site value tax. The simplification of this formula reflects the observation that owing to the landowner's first-order conditions, we have $E_T^* = -V_T^*$, where the asterisk refers to the value of the partial derivative at the optimal rotation age. The LHS of this equality is in the numerator and the RHS is in the denominator of equation (5.11d). Thus the first-best site value tax is a classical Pigouvian tax. The precise choice of the tax rate depends on the value of amenities lost by harvesting; these in turn depend on the share of citizens who value these amenities in the economy. Further insight can be obtained by looking more closely at (5.11c). The sign of the denominator depends on the marginal valuation of the amenities. If amenities from old (or young) stands are preferred, then this term is negative (or positive). The numerator under these valuations is positive (or negative), respectively. When $F'(T) > 0$, the site value tax will have a positive effect on the private rotation age, but when $F'(T) < 0$, the site value tax will have a negative effect. In the former case, the privately optimal rotation age is too short relative to the social optimum, and it is too long in the latter case. In both cases society should use the site value tax to lengthen and shorten the private rotation age, respectively. The findings here can be summarized in proposition 5.3.

Proposition 5.3   When amenity benefits are public goods, the first-best tax policy depends on the nature of the amenity benefits. Optimal corrective forest taxes should be designed as follows:

(a) Under increasing marginal valuation of amenity benefits over time, the socially optimal yield tax is positive, but under decreasing amenity benefits either a yield-based subsidy or tax is optimal, depending on the exact relative size of net marginal amenity benefits and regeneration costs.

(b) Under decreasing amenity valuation over time, a positive timber tax is optimal, but under increasing marginal valuation over time a positive timber subsidy is optimal.

(c) While the optimal site value tax depends on the size of the externality, it does not depend on the marginal valuation of amenity benefits.

5.2.2   First-Best Taxation—Presence of Neutral Tax and Revenue Constraint
As we did earlier, we can re-express the social welfare function (5.9a) in terms of the indirect net present value function of the private landowner, augmented by the direct utility of citizens in (5.12):

$$SW^H = V^*[\tau, u, \theta, \alpha] + E^*[\tau, u, \theta, \alpha] + (n-1)E. \tag{5.12}$$

The economic problem of the social planner is again to choose forest taxes so that the social welfare function is maximized subject to the tax revenue requirement (5.3). We start by choosing the optimal site productivity tax. The Lagrangian is

$$\Omega^H = V^* + E^* + (n-1)E - \lambda(\bar{G} - G), \tag{5.13}$$

where the superscript H refers to the Hartman case and λ is the marginal cost of public funds. It is trivial to show that the optimal site productivity tax, $a(i)$, is neutral, yielding $\lambda = 1$ given that there is no rotation age effect, $T_a^H = 0$. Hence even in the presence of amenities, the site productivity tax is an ideal instrument for collecting tax revenues without causing any distortions to landowner behavior. As for the other taxes, referring to proposition 5.1 and its proof, we can immediately give the following result.

Proposition 5.4 Optimal forest taxation in the presence of a neutral tax If society values both harvest revenue and forest amenities, needs to meet a given tax revenue target, and amenities are pure public goods, then the socially optimal taxation mix includes a site productivity tax and the following:

(a) no other taxes when amenity valuation is site specific

(b) a corrective first-best site value tax if amenities depend on stand age, with the tax rate reflecting the size of the externality; a first-best harvest (yield or unit) tax when marginal valuation of amenities is increasing with the age of the stand; a timber tax if amenity valuation decreases with the age of the stand; or a timber subsidy if amenity valuation increases with the age of the stand.

Proof See Koskela and Ollikainen (2003a).

Because the site productivity tax takes care of collecting tax revenues, the government should apply site value and harvest taxes in a way to correct the externalities that emerge when amenity valuation is either increasing or decreasing with the stand's age. The properties of the first-best instruments were discussed in proposition 5.3.

Box 5.1 illustrates the results of some of the propositions presented here. It presents the socially optimal tax solutions under first- and second-best assumptions for both Faustmann and Hartman models (discussed later).

5.2.3 Optimal Forest Taxation—Absence of Neutral Tax

Neutral taxes (such as the site productivity tax) are generally not available for use in the forest sector. We therefore must consider an optimal combination of other taxes. Recall in the Hartman model of chapter 3 that Hartman and Faustmann rotation ages were equal when amenities were site-specific $[F'(T) = 0]$. Thus, for our case, the private and social rotation ages are equal. For this case, a neutral tax is the socially optimal instrument. In the other cases, we can show that the socially optimal tax mix consists of a corrective tax scheme that internalizes the externality caused by private harvesting. Such a scheme is provided by a mix of unit and timber taxes. Whether each tax is positive or negative depends on whether amenities

Box 5.1
Socially Optimal Forest Tax Rates

Table 5.1
Socially Optimal Corrective Pigouvian Taxes

Tax instrument	Optimal tax rates	Tax revenue (€)
Yield tax rate	0.207	246.95
Unit tax payment	€16.7/m^3	246.95
Timber tax (subsidy) rate	−0.00137	−84.19

Policy makers tax forest landowners either to raise money or to affect landowners' incentives to undertake certain management decisions. The presence of externalities calls for corrective first-best Pigouvian taxes, while tax revenue collection requires only neutral taxes. If a distortionary tax must be used to raise revenue in a second-best problem, then it should distort landowner decisions and welfare as little as possible.

Consider an economy in which society values amenity benefits, but landowners maximize only harvest revenue. Boxes 2.2 and 3.2 indicated rotation ages of 58.1 and 59.3 years for the Faustmann and Hartman models, respectively. Thus the optimal tax is one that lengthens the private rotation age by 1.2 years. Obviously, we need a distortionary tax, and we know from the theory that such taxes are the yield tax, unit tax, or timber tax (other taxes were neutral in the Faustmann model). The second-best tax rates based on the information presented in boxes 2.2 and 3.2 are reported in table 5.1. Also given in the table are the impacts on government tax revenues from the optimal application of each tax instrument.

Table 5.1 confirms proposition 5.1. That is, it is either optimal to use a negative timber tax (i.e., a subsidy), or positive yield or unit taxes. Yield and unit taxes generate the same tax revenue, while a timber subsidy requires outlays from the government to the landowner. However, these outlays are more than three times smaller than the tax revenue collections paid by private landowners to the government.

Now suppose that the government must collect a given sum of tax revenue, $\bar{G}$, from the forest sector. Suppose this revenue requirement is a net present value of €265 per hectare, which exceeds the tax revenue reported in table 5.1 and ensures that the tax rates become distorted beyond the first-best solution. For the timber subsidy, we assume that the maximum sum that the government can allocate to subsidy payments is €45, which is less than that paid in table 5.1. Suppose finally that society solves a Hartman model based on the problem in box 3.2 for determining rotation ages, but private landowners can follow either the Faustmann (box 2.2) or the Hartman model. If landowners base their decisions on the Faustmann model, then we have both an externality that needs correction and a need to raise tax revenue. If they base their decisions on the Hartman model, then tax instruments are used only to raise revenues. Finally, suppose that society does not have a neutral tax available. The second-best taxation solutions are reported in table 5.2.

Box 5.1
(continued)

Table 5.2
Socially Optimal Taxes under Binding Tax Revenue Requirement

Tax instrument	Tax rate	Rotation age (years)
Faustmann model		
Yield tax rate	0.222	59.40
Unit tax payment	€7.784/m^3	59.340
Timber subsidy rate	−0.000059	−58.81
Hartman model		
Yield tax rate	0.225	60.99
Unit tax payment	€7.860/m^3	60.99
Timber subsidy rate	−0.000058	−60.01

The second-best yield tax is slightly higher in table 5.2 than those found in table 5.1, while the timber subsidy is lower. The unit tax in table 5.2 is now much lower than in table 5.1. Second-best tax rates in the Hartman case are higher than those solved in the Faustmann model, while the subsidy rate is lower.

are decreasing with rotation age $[F'(T) < 0]$, or increasing with rotation age $[F'(T) > 0]$. We leave this proof to the reader, but it follows the same lines as the one that led to proposition 5.2. The main results from this exercise are summarized in the following proposition:

Proposition 5.5 If society values both harvest revenue and forest amenities, needs to meet a given revenue target, and amenities are pure public goods, then in the absence of any neutral taxes, the government should use the following:

(a) a neutral combination of a unit tax and timber tax (or combination of a yield tax and timber tax) for site-specific amenities, and

(b) a corrective combination for the case of increasing marginal amenity valuation.

The exact nature of the corrective combination of taxes for the case of decreasing marginal amenity valuation cannot be defined a priori.

Proposition 5.5 is the Hartman counterpart to proposition 5.2. The main difference is that now the tax combination is used for the dual purpose of raising revenue and correcting externalities. This leads to a more difficult applications of taxes, especially for the case of decreasing amenity benefits.

5.3   Optimal Taxation—Life-Cycle Interpretations

Optimal forest tax policy issues can also be studied in the two-period model of chapter 4. An advantage of these models is that they allow amenities to be defined as a function of the forest stock.

5.3.1   Timber Production

Assume that both the private landowner and society value only net harvest revenue from forestry and maximize the following utility function familiar from chapter 4:

$$U = u(c_1) + \beta u(c_2), \tag{5.14}$$

where now U refers to utility in the two-period model, c_1 and c_2 denote consumption in periods one and two, and $\beta = 1/(1 + \rho)$ where ρ is the rate of time preference. The government again must choose a design for forest taxation that maximizes social welfare. Recalling our earlier analysis, it is evident that private and socially optimal harvesting are identical here, and so taxes should be zero if the government faces no revenue constraint. However, assume there is a binding revenue constraint $\bar{G}$. Define the social welfare function using the indirect utility function of the landowner, $U^* = U^*(a, \tau, u)$, which can be obtained by substituting the current and future harvesting choices as functions of taxes, $x_1 = x_1(a, \tau, u)$ and $x_2 = x_2(a, \tau, u)$, into equation (5.14). The economic problem of the social planner is then to maximize (5.15) by choosing forest taxes subject to the present-value revenue requirement (5.16), but now we allow for timber prices that are not constant; i.e.,

$$\max SW = U^*(a, \tau, u) \tag{5.15}$$

$$\text{s.t. } G = (1 + R^{-1})a + (\tau p_1 + u)x_1 + R^{-1}(\tau p_2 + u)x_2, \tag{5.16}$$

where $R^{-1} = (1 + r)^{-1}$ and r is the discount rate. The Lagrangian for this problem is $\Omega = U^*(\cdot) - \lambda(\bar{G} - G)$.

The optimal choice for the neutral site productivity tax using the envelope theorem applied to U^* yields

$$SW_a = U_a^* + \lambda(1 + R^{-1}) = 0, \tag{5.17}$$

where $U_a^* = -\beta u'(c_2)(1 + R^{-1}) < 0$. Hence we find that the optimal level of the site productivity tax requires that $\lambda = \beta u'(c_2)$; i.e., the marginal cost of public funds should equal the marginal disutility the landowner faces from having to pay the tax. Note the equivalence to optimal policies in the Faustmann model. If the landowner maximizes only the net present value of harvest revenue, then we would have $(1 + \rho)^{-1} u'(c_2) = 1$ at an interior solution. In this case, the optimal tax structure requires $\lambda = 1$ just as in the Faustmann case.

Proposition 5.6 If society values only harvest revenue and has a binding tax revenue requirement, then the optimal design of forest taxation consists of only a neutral site productivity or yield tax.

In this model the government could use either the site productivity tax a or the constant yield tax, τ. The neutrality of the site productivity tax is obvious. Also, a constant yield tax affects both the marginal return and opportunity costs of harvesting equally and thus it ends up being neutral in the two-period timber production model (see chapter 4). We leave it to the reader to prove this proposition. The proof entails first deriving the optimal yield tax rate or site productivity tax and then showing that given the optimal site productivity or yield tax rate, the distortionary unit tax rate must equal zero to satisfy its first-order condition.

The intuition of proposition 5.6 is the same with the Faustmann model. Since public and private preferences are identical, the government has no reason to introduce distortionary taxes and change landowner behavior on the margin. In the absence of externalities, the only function of tax instruments is to collect revenue, which is accomplished most efficiently using neutral taxes following the Ramsey rule.

5.3.2 Joint Production of Timber and Amenity Services

The case where amenity services from forests are valued by the representative landowner and by nonlandowners is more interesting. Assume as before that the amenity valuation function is the same for the representative landowner and nonlandowners. Then we can express landowner preferences and public preferences for amenities, respectively, as

$$U = p_1 x + R^{-1} p_2 z + v(k_1) + \beta v(k_2) \tag{5.18a}$$

$$V^N = (n - 1)[v(k_1) + \beta v(k_2)], \tag{5.18b}$$

where n continues to denote the number of citizens and $(n - 1)$ is the number of nonlandowners. When citizens have full access to amenity services from private forests and there are no congestion externalities, the social welfare function is written

$$SW = U + (n - 1)[v(k_1) + \beta v(k_2)]. \tag{5.19}$$

First-best corrective instruments can be solved by the same approach we used here in the Hartman model. The socially optimal choice of current and future harvesting, x_1 and x_2, is characterized by the following first-order conditions:

$$SW_{x_1} = p_1 - v'(k_1) - \beta v'(k_2)(1 + g') - (n - 1)[v'(k_1) + \beta v'(k_2)(1 + g')] = 0 \quad (5.20a)$$

$$SW_{x_2} = R^{-1}p_2 - \beta v'(k_2) - (n - 1)\beta v'(k_2) = 0. \quad (5.20b)$$

Using $R^{-1}p_2 = \beta v'(k_2) + (n - 1)\beta v'(k_2) = 0$ from the future harvesting condition in (5.23a) yields the following socially optimal harvesting rule:

$$p_1 - R^{-1}p_2(1 + g') - nv'(k_1) = 0. \quad (5.21)$$

Thus, socially optimal harvesting accounts for the amenity valuation of all consumers. From (5.21) we have that $p_1 - R^{-1}p_2(1 + g') > 0$. Moreover, condition (5.21) is larger in absolute value than the respective privately optimal condition (see equation 4.15′ in chapter 4). Given concavity of the growth function, the socially optimal harvesting level is lower than the level a private landowner will choose.

The first-best taxes needed to correct private harvesting can be determined by setting the privately optimal harvesting rule in the presence of taxes (equations 4.20a and 4.20b in chapter 4) equal to the socially optimal harvesting rule in (5.21). Doing this and rearranging the resulting condition, we obtain the following first-best tax rates:

$$\tau^* = \frac{n - 1}{n}; \quad u^* = \frac{n - 1}{n}\left[\frac{p_1 - R^{-1}p_2(1 + g')}{R^{-1}(r - g')}\right]. \quad (5.22)$$

The optimal corrective-yield tax rate is directly related to the size of the externality (i.e., to the difference in private and social harvesting choices). The same holds true for the unit tax in a steady state where timber prices are constant over the two periods. When prices differ over periods, the optimal unit tax includes a correction term to reflect the fact that as a tax on quantity, it asymmetrically affects the marginal return and opportunity cost of harvesting.

Suppose next that the government is not free to choose tax instruments and in addition faces a binding tax revenue requirement. As we discussed before, this fits the classic second-best problem of policy design. We again express social welfare in terms of the landowner's indirect utility function:

$$SW = U^*(a, \tau, u) + (n - 1)[v(k_1) + \beta v(k_2)]. \quad (5.23)$$

The government chooses forest taxes to maximize (5.23) subject to (5.16). We already know from the previous analysis what the government should do. If the government has a nondistortionary tax available, then it should use it and set the available corrective Pigouvian instruments at their first-best levels.

Proposition 5.7   Under a binding tax revenue requirement, the socially optimal design of forest taxation consists of a neutral site productivity tax and a corrective harvest, yield, or unit tax set at the first-best optimal level.

Clearly, from this proposition, if the government cannot apply a neutral tax, the Ramsey rule dictates again that the choice of taxes should minimize distortions in private landowner decisions.

So far in this chapter we have seen that the optimal design of forest policy instruments depends greatly on whether standing forests provide public goods in the economy and whether the government faces a binding tax revenue requirement. Moreover, we know that any optimal instrument design depends on the nature of available taxes, that is, on whether they are distortionary or nondistortionary instruments. Our analysis demonstrates that the logic of the optimal tax analysis and the results remain the same no matter what type of model is used to frame the problems. Depending on the taxes available to the government, forest tax designs can consist of Pigouvian corrective taxes, revenue-raising Ramsey-type taxes, or combinations of these. As we will see later in this book, once uncertainty is introduced into landowner decision making, tax instruments will take on additional roles.

5.4   Optimal Taxation—Overlapping Generations Interpretations

In chapter 4 we saw that OLG models can provide a fruitful framework for studying the timber bequest behavior of private landowners and examining determinants of long-run forest stocks in an economy. This framework also affords study of a new type of instrument, the timber bequest tax, which is commonly applied in developed countries. The optimal timber bequest tax is closely associated with two new features of forest taxation made visible by the OLG framework. First, an externality arises that is caused by bequest-induced changes in public goods; now there is a possibility for intergenerational externalities. That is, the current-generation landowner may not leave the socially optimal volume of forest stocks to the next generation. Hence we must analyze how these bequests modify optimal policy choices. The second new aspect concerns the market-driven level of the long-run forest capital stock. As discussed in chapter 4, the long-run capital stock may or may not coincide with a golden rule level; that is, it may end up as dynamically efficient or inefficient in the long run. This implies that harvest and timber bequest taxes can be used to internalize externalities and shift the economy to the golden rule level (Amacher et al. 2005).

To make the analysis as simple as possible, we assume that only timber bequests are operative. The government is assumed to have available a lump-sum tax, timber bequest (inheritance) tax, and a harvest tax, the rates of which are constant over time. The government again acts as a Stackelberg leader by choosing tax rates to maximize social welfare subject to an exogenously given tax revenue requirement and the reaction function of landowners. Note that we solved for landowners' responses to harvest and timber bequest taxes in section 4.3.2 of chapter 4. Furthermore, we will focus on the steady state in this model. This implies that we employ constancy for consumption and other time-based variables in the same way we defined the steady state in section 4.3.2. The government's choice of policy instruments in this steady state will be examined by first assuming that amenity services from forest stocks are private goods, so that the social welfare function consists only of the representative landowner's indirect utility function. We then consider the case where amenity services are public goods.

5.4.1   Amenities as Private Goods

The government's exogenous revenue target, denoted by G, must be met in the steady state and represents a constraint on the choice of policies. This steady-state revenue target can be derived by adding tax revenue collections over the two overlapping generations that are alive at any point in time:

$$G = T + \tau_b pb + \tau_x p \frac{(2+n)}{(1+n)} x, \tag{5.24}$$

where x and b denote steady-state levels of harvesting and timber bequests. This equation is written in per capita form. The first term is the lump-sum tax. The second term is the bequest tax paid by the old generation, while the young generation harvests and receives revenues (measured by the third term on the RHS).

The government chooses policies to maximize a welfare function given by the indirect utility of overlapping generations, $V^*(T, \tau_x, \tau_b, \ldots)$, subject to its revenue requirement (and the optimal behavior of the representative landowner in each generation,[4]

$$\max_{\{T, \tau_x, \tau_b\}} V^*(T, \tau_x, \tau_b, \ldots), \quad \text{s.t. (5.24).} \tag{5.25}$$

The Lagrangian for this problem is

$$\Omega = V^* - \lambda \left[G - \tau_b pb - T - \tau_x \frac{(2+n)}{(1+n)} px \right],$$

---

4. See Ihori (1996) for a discussion of the basic OLG optimal taxation problem.

where λ is the shadow price of the government's budget constraint. The first-order condition for the optimal lump-sum tax, T is

$$\Omega_T = V_T^* + \lambda\left[1 + \tau_b p b_T + \tau_x \frac{(2+n)}{(1+n)} p x_T\right] = 0, \tag{5.26a}$$

where $V_T^* = -[u'(c_{1t}) + \alpha u'(c_{2t+1})] < 0$. Equation (5.26a) implies that the government should set the lump-sum tax rate so that the negative marginal welfare effect of the tax ($V_T^* < 0$) equals the marginal cost of collecting revenue (second RHS bracketed term). Obviously, the revenue collection effect depends on population growth rates and timber prices, as well as the response of timber bequests and harvesting to the lump-sum tax (b_T and x_T). The first-order conditions for the bequest and harvest taxes are

$$\Omega_{\tau_b} = V_{\tau_b}^* + \lambda[pb + \tau_b p b_{\tau_b} + \tau_x p x_{\tau_b}] = 0, \tag{5.26b}$$

$$\Omega_{\tau_x} = V_{\tau_x}^* + \lambda[\tau_b p b_{\tau_x} + px + \tau_x p x_{\tau_x}] = 0, \tag{5.26c}$$

where $V_{\tau_b}^* = pbV_T^* < 0$ and $V_{\tau_x}^* = (1 + R^{-1})pxV_T^* < 0$. These conditions have the same interpretation as we had for the optimal lump-sum tax, T^*. Using (5.26a) for the optimal lump-sum tax, we can re-express equations (5.26b) and (5.26c) as follows:

$$\tau_{x|T=T^*}^* = \frac{b_{\tau_b}^c \frac{(2+n)}{(1+n)} \varepsilon}{(x_{\tau_x}^c b_{\tau_b}^c - x_{\tau_b}^c b_{\tau_x}^c)}; \quad \tau_{b|T=T^*}^* = -\frac{x_{\tau_b}^c \varepsilon}{(x_{\tau_x}^c b_{\tau_b}^c - x_{\tau_b}^c b_{\tau_x}^c)}, \tag{5.27}$$

where $\varepsilon = -[1/(1+n) - 1/(1+r)]x$. The fact that the denominators of these solutions are positive can be proven using the Slutsky equations for this problem (see Amacher et al. 2005).

Developing an economic interpretation of the optimal policies requires us to consider the golden rule, dynamically inefficient, and dynamically efficient steady states discussed in chapter 4.

Golden Rule (r = n) Now we have $\varepsilon = 0$ and the economy resides at the golden rule level of capital and consumption. Therefore τ_b^* and τ_x^* are equal to zero in (5.27). The optimal tax policy design now requires the government to use only nondistortionary lump-sum taxes to raise revenues. This is a result equivalent to those we found in the two-period timber production and Faustmann models of chapters 4 and 2.

Dynamic Inefficiency (r < n) Now $\varepsilon > 0$, and the optimal policy mix requires non-zero bequest and harvest taxes. Because $x_{\tau_b}^c < 0$ and $b_{\tau_b}^c < 0$, the numerators of

(5.27a) and (5.27b) are always positive. Thus the government should issue a subsidy for forest harvesting, $\tau_x < 0$, and a tax on forest bequests, $\tau_b > 0$. The economy here is inefficient owing to overinvestment of forest capital because of low interest rates. In forestry terms, the volumes of bequeathed forest stocks are too large, so the steady-state forest stock is too high. To correct this inefficiency, the government should choose a policy mix that leads to a reduction of forest capital in the steady state. Revenue shortfalls from the harvest subsidy can be satisfied by taxing timber bequests. Moreover, as shown earlier, this combination makes the operative timber bequest less likely.

Dynamic Efficiency (r > n) Now we have $\varepsilon < 0$ and there is underinvestment in forest capital and too few bequests in the long run, which reverses the signs of the optimal tax–subsidy mix. By taxing harvesting and subsidizing timber bequests, the government can move the economy closer to the golden rule. Furthermore, we know from chapter 4 that this instrument mix also reinforces the condition for operative timber bequests. Because the steady-state forest stock is too low, harvest taxes and bequest subsidies encourage intergenerational transfers and long-run investments in forests. We collect these findings in proposition 5.8.

Proposition 5.8   If the economy exists at a golden rule level of forest capital in the long run, then it is optimal to use a lump-sum tax to collect revenue without introducing any distortionary instruments. If the economy is dynamically inefficient, then the optimal policy mix also includes a subsidy on harvesting and a tax on timber bequests. However, for a dynamically efficient economy, a harvest subsidy and bequest tax are optimal in order to push the capital stock to the golden rule level.

Proposition 5.8 represents a departure from the results derived in OLG models that do not explicitly include forests. These other models have shown that capital taxes are optimally equal to zero in the long run, and that these results are relatively robust. Here, although the harvest tax can be interpreted as a capital tax, in some cases we find that it should be used in the steady state to achieve the desired level of long-run bequests and forest stocks. For optimal forest policy design, it is critical to know where the economy resides relative to the golden rule. This suggests a new information requirement in the design of forest policies.

5.4.2   Amenities as Public Goods

Consider now the case where amenity services from timber stocks represent public goods in the economy. The government should respond to these values when making policy choices by accounting for how conservation of the timber stock, through bequests, produces amenity benefits to all citizens in both current and future generations. Thus society has two targets for choosing policy design now: the long-run forest stock and the level of amenity services. Since society values public goods from forests, the policy problem in (5.25) is modified accordingly.

$$\max_{\{T, \tau_b, \tau_x\}} \ V^*(T, \tau_b, \tau_b) + (n+2)(q-1)\{v(k_{1t}) + \beta v(k_{2t}) + \phi[v(k_{1t+1}) + \beta v(k_{2t+1})]\},$$

(5.28)

where $(n+2)(q-1)$ represents the number of nonlandowners in the old and young generations. As before, we assume that the tax choices in (5.28) are restricted by (5.24). Following the procedure given here, we first solve for the optimal lump-sum tax T^a, and then, assuming it has been set at the optimal level, we solve for the optimal harvest and timber bequest tax (see appendix 5.1 for details):

$$\tau^a_{x|T=T^a} = \tau^*_{x|T=T^*} + \frac{(q-1)(1+g')[(\phi+\beta)v'(k_{2t}) + \phi\beta v'(k_{2t+1})]}{\lambda p}$$

(5.29a)

$$\tau^a_{b|T=T^a} = \tau^*_{b|T=T^*} - \frac{(n+2)(q-1)[(\phi+\beta)v'(k_{2t}) + \phi\beta v'(k_{2t+1})]}{\lambda p},$$

(5.29b)

where the superscript a refers to the public goods case and the superscript asterisk refers to the optimal policy defined in equation (5.27). Note that the government's revenue multiplier describing the marginal cost of public funds, λ, remains positive.

Equations (5.29a) and (5.29b) imply that the optimal tax instrument mix is composed of a policy solving the previous government's problem (first term on each RHS), plus a new additive externality term (second term on each RHS), which represents an adjustment that reflects public goods lost through harvesting of the steady-state forest stock. The presence of this new term ensures that the efficient level of public goods (timber stock) is achieved in the long run. Again we consider the golden rule, dynamically inefficient, and dynamically efficient steady states separately.

Golden Rule Using $\varepsilon = 0$ implies $\tau^*_{b|T=T^*} = 0$ and $\tau^*_{x|T=T^*} = 0$. The optimal tax rates in (5.29) are not zero, however, because of the presence of amenities. Now we find that a tax–subsidy mix is optimal, $\tau^a_b < 0$ and $\tau^a_x > 0$. In this case the government corrects the externality through an adjustment to harvesting and bequests while raising revenues using the lump-sum tax. The adjustment in private decisions should be made using a subsidy on timber bequests and a tax on harvesting. This result is another case where (timber) capital taxes are optimally positive in the long run, even though the capital stock equates with its golden rule level. In contrast to the standard result in the economics literature, our new result occurs because overaccumulation of forest capital relative to the golden rule might actually be efficient if public goods are produced from the stock.

Dynamic Inefficiency Recall that in the absence of public goods we had $\tau^*_{x|T=T^*} < 0$ and $\tau^*_{b|T=T^*} > 0$ because the aim was to decrease the forest stock, which is too high from a social point of view. Introducing public goods aspects of the forest stock tends to increase the size of the optimal stock in the long run, leading to opposing effects on this policy mix and making the design of instruments a priori ambiguous.

Dynamic Efficiency In the absence of public goods, we saw earlier that $\tau^*_{x|T=T^*} > 0$ and $\tau^*_{b|T=T^*} < 0$, and the forest stock was lower than socially optimal. Now, public goods being present tends to strengthen this policy choice; that is, it increases the size of the optimal subsidy and tax rates, $\tau^a_b < 0$ and $\tau^a_x > 0$, because this further increases the long-run forest stock and corrects inefficient equilibrium underinvestment of capital. Hence the capital restoration and amenity aspects of the policy instruments reinforce each other.

Proposition 5.9 If the economy resides at the golden rule forest capital stock, then in addition to a lump-sum tax, both a harvesting tax and timber bequest subsidy are optimal. In a dynamically inefficient economy, the policy mix is ambiguous in terms of using a harvesting and bequest tax, but in a dynamically efficient economy, a subsidy on timber bequests and a tax on harvesting are optimal.

Collectively, propositions 5.8 and 5.9 add an interesting wrinkle to the debate about forest tax structure and the ability of a government to influence long-run forest stocks and public goods in an economy. We saw in chapter 4 that timber bequests can be operative in a range of situations, even in dynamically inefficient steady states. For most governments facing the need to raise revenue yet desiring to provide public goods from forests in the long run, clearly policies targeting harvesting should not be used without also considering policies targeting timber bequests. The usual case in North America and Europe is to tax both timber harvesting and bequests. However, there is no case here where we find such a policy mix to be the efficient approach for raising revenues and ensuring the right levels of public and private goods produced from forests in the long run, regardless of the nature of the steady state. It is critical to follow the optimal policy mix when one considers that it is often claimed that sustainable use of a forest leads to the right level of long-run forest stocks. The economy will not achieve these first- and second-best stocks if the best policy mix is not applied simultaneously to timber bequests and to harvesting.

5.5   Modifications

Modifications of optimal policy design in forest sector models have followed a number of directions. First, the harvesting and preservation of public forests have been added as an additional policy instrument (Amacher 1999, Koskela and Ollikainen 1999a). Amacher (2002) also examines policy design under fiscal federalism arising from many layers of governments. Second, the effects of forest taxes on the forest sector have been analyzed when there are preexisting distortions from other policies in the economy. This has often been called the "Austrian sector problem," and originated with work by Fairchild (1909, 1935). Bentick (1980) and Chisholm (1975) refocused his analysis to deal specifically with forest issues. Kovenock and Rothschild (1983) and Kovenock (1986) provided a comprehensive analysis within a rotation framework, while Uusivuori (2000) used an N-period framework in which agents in the Austrian sector can have a bequest motive. Koskela and Ollikainen (1997a) introduced future price uncertainty in the Austrian sector. Central questions in this collection of research are how a capital income tax affects the decision making of forest landowners among these other distortions and how to design a tax system so that the right amount of forest capital exists in the forest sector.

In this section we concentrate on topics studied most recently by Aronsson (1993), Mendelsohn (1993), Koskela and Ollikainen (2003b), and Amacher and Brazee (1997b). In practice, most tax systems are either progressive or regressive. This means that the average tax rate is either increasing or decreasing in the tax base, respectively (i.e., in income) (see e.g., Musgrave and Thin 1948 and Lambert 2002, chapters 7–8). Furthermore, forest tax exemptions are common in many countries, and often their presence makes the forest tax system de facto progressive (see e.g., Uusivuori 2000). Here we briefly examine the extent to which a progressive forest tax is socially optimal. We will find, again, that the answer depends on the presence of amenities.

5.5.1   Progressive Taxation

Assume that the forest tax rate is constant and that there is a tax exemption, i.e., a threshold below which there is no taxation and above which a constant tax rate is charged. The threshold depends on a target, such as income, that defines where in the tax base the exemption occurs. With such a system, the average tax rate increases with the income of the taxpayers so that taxation is linearly progressive. This fits many countries, where landowners who harvest less and capture less than a certain amount of income are not taxed. Suppose we consider an exemption m that is present with a timber tax α and yield tax τ. Using previous notation, the objective function of the landowner with these instruments, defined for each one separately, is given by

$$\hat{W}(\alpha, m) = V - \alpha[U - X] + E, \tag{5.30a}$$

and

$$\hat{W}(\tau, m) = V - \tau[Y - X] + E, \tag{5.30b}$$

where $U = (1 - e^{-rT})^{-1} \int_0^T pf(s)e^{-rs}\, ds$, $X = m(1 - e^{-rT})^{-1}$ and $Y = pf(T)e^{-rT} \times (1 - e^{-rT})^{-1}$. The second RHS terms in brackets describe the taxation part. The first-order conditions for the privately optimal rotation age under each tax are

$$\hat{W}_T(\alpha, m) = pf'(T) - rpf(T) - rV - \alpha[pf(T) - r(U - X)] + F(T) - rE = 0 \tag{5.31a}$$

$$\hat{W}_T(\tau, m) = p(1 - \tau)[f'(T) - rf(T)] - r[V - \tau(Y - X)] + F(T) - rE = 0. \tag{5.31b}$$

Forest taxes have the same qualitative effects here as those found in chapter 3. Totally differentiating the first-order conditions, we can solve for the comparative statics of the exemption level on rotation age as follows:

$$T_m^H(\alpha) = -\frac{\hat{W}_{Tm}(\alpha)}{\hat{W}_{TT}(\alpha)} \begin{Bmatrix} < \\ > \end{Bmatrix} 0, \quad \text{as } \alpha \begin{Bmatrix} > \\ < \end{Bmatrix} 0 \tag{5.32a}$$

$$T_m^H(\tau) = -\frac{\hat{W}_{Tm}(\tau)}{\hat{W}_{TT}(\tau)} \begin{Bmatrix} < \\ > \end{Bmatrix} 0, \quad \text{as } \tau \begin{Bmatrix} > \\ < \end{Bmatrix} 0, \tag{5.32b}$$

where $\hat{W}_{Tm}(\alpha) = -\alpha r(1 - e^{-rT})^{-1}$ and $\hat{W}_{Tm}(\tau) = -\tau r(1 - e^{-rT})^{-1}$. A higher tax exemption with the timber or yield tax (subsidy) will shorten (lengthen) the rotation age.

We will need to use Slutsky equations because they define how rotation age depends on taxes in the presence of an exemption. The Slutsky equations are used to decompose the total effects of taxes on the rotation age decision into substitution and income effects (see the mathematical review at the end of this book for a description of these equations). We can treat the tax exemption as a lump-sum reduction in income (a pure income effect), given how it enters income of the landowner (see 5.30a and 5.30b). We then look for the combination of a tax rate and tax exemption that will keep the utility of the representative landowner constant. The Slutsky decompositions are

$$T_\alpha^{Hc} = \underbrace{T_\alpha^H}_{-} + \underbrace{\frac{(1 - e^{-rT})(U - X)}{\alpha} T_m^H}_{-} \tag{5.33a}$$

$$T_\tau^{Hc} = \underbrace{T_\tau^H}_{?} + \underbrace{\frac{(1 - e^{-rT})(Y - X)}{\tau} T_m^H}_{-}.$$ (5.33b)

The first part is the substitution effect while the second part is the compensated income effect. Drawing on total effects, the substitution effects are

$$T_\alpha^{H^C} = \frac{pf(T)}{\hat{W}_{TT}} < 0$$ (5.34a)

$$T_\tau^{H^C} = \frac{p[f'(T) - rf(T)]}{\hat{W}_{TT}} \begin{Bmatrix} > \\ = \\ < \end{Bmatrix} 0, \quad \text{as } r[V - \tau(Y - A)] \begin{Bmatrix} < \\ = \\ > \end{Bmatrix} F(T) - rE.$$ (5.34b)

Not surprisingly, the substitution effect of the timber tax on harvesting is unambiguously negative while the substitution effect of a yield tax is more complicated. For an amenity valuation with $F'(T) > 0$, the substitution effect of the yield tax on rotation age is ambiguous, but for $F'(T) < 0$ it is negative.

Consider now a reform of the forest tax system, asking how a tax revenue-neutral change in progression of the timber tax and yield tax will affect the privately optimal rotation age. In the presence of the timber tax, tax revenue is given by $G = \alpha(U - X)$. Differentiating this function with respect to α and m and setting the derivative equal to zero gives us a condition that ensures that revenue is constant as tax parameters are adjusted, i.e., $dG = 0 = G_\alpha \, d\alpha + G_m \, dm$. Solving this equation for the differential describing the change in tax progression, we have

$$dm_{|dG=0} = -\frac{G_\alpha}{G_m} d\alpha.$$ (5.35)

Assuming that there is a positive (or negative) relationship between the tax rate (tax exemption) and tax revenue, we can write

$$G_\alpha = \alpha(U_T - X_T)T_\alpha^H + U - X > 0,$$ (5.36a)

and

$$G_m = \beta(U_T - X_T)T_m^H - \frac{\alpha e^{rT}}{1 - e^{-rT}} < 0,$$ (5.36b)

where the terms are derived from the response of rotation age to changes in the tax rate and the exemption level. An increase in the timber tax and tax exemption means that the average tax rate increases in the tax base. A compensation rule between the tax rate and the tax exemption, defined by equation (5.35), represents a pure change in the progressivity of the tax system in an ex post sense.

The total rotation age effect can be expressed as $dT^H = T_\alpha^H \, d\alpha + T_m^H \, dm$. Substituting the RHS of (5.35) for dm in the above equation and rearranging yields

$$\left. \frac{dT^H}{d\alpha} \right|_{dG=0} = G_m^{-1} [T_\alpha^H G_m - T_m^H G_\alpha].$$

Finally, applying the Slutsky decomposition for the timber tax rate, we obtain after some manipulation

$$\left. \frac{dT^H}{d\alpha} \right|_{dG=0} = -G_m^{-1} \frac{\alpha}{1 - e^{-rT}} T_\alpha^{HC} < 0. \tag{5.37}$$

Hence a rise in timber tax progression that holds tax revenues constant will shorten the rotation age, owing to the negative substitution effect of the timber tax.

To consider the progressivity of the yield tax, note that government tax revenue is given by $G = \tau(Y - X)$. Differentiating this with respect to τ and m and again setting the derivative equal to zero gives $dG = G_\tau \, d\tau + G_m \, dm = 0$. Assuming there is a positive (negative) relationship between the tax rate (tax exemption) and tax revenue, we have

$$dm_{|dG=0} = -\frac{G_\tau}{G_m} d\tau, \tag{5.38}$$

where $G_\tau = \tau(Y_T - X_T)T_\tau^H + Y - X > 0$ and $G_x = \tau(Y_T - X_T)T_m^H - \tau/(1 - e^{-rT}) < 0$.

Changing the tax rate and exemption affects the optimal rotation age in the following manner: $dT^H = T_\tau^H \, d\tau + T_m^H \, dm$. Substituting for dm from (5.38) in this equation and rearranging yields

$$\left. \frac{dT^H}{d\tau} \right|_{dG=0} = G_m^{-1} [T_\tau^H G_m - T_m^H G_\tau].$$

Applying the Slutsky decomposition gives

$$\left. \frac{dT^H}{d\tau} \right|_{dG=0} = -G_m^{-1} \frac{\tau}{1 - e^{-rT}} T_\tau^{HC} = ? \tag{5.39}$$

An increase in tax revenue-neutral progression of the yield tax has an ambiguous effect on rotation age owing to the indeterminacy of the substitution effect.

A decreasing marginal amenity valuation $F'(T) < 0$ is a sufficient, but not a necessary, condition for the effect on rotation age to be negative.

5.5.2 Optimality of Progressive Forest Taxation

The next obvious question is whether a progressive forest tax can be socially optimal. As before, we have to examine the choice of tax rates and the exemption to maximize social welfare, $SW^H = V^*[a(i), \alpha, m] + E^* + (n-1)E$. This assumes that the private landowner does not value amenities but society does. We will assume that the government has a neutral tax available, such as a site productivity instrument $a = a(i)$ used earlier.

The tax revenue requirement for the timber tax is $\bar{G} \leq G = (a/r) + \alpha(U - X)$, where U and X are defined in the previous section. Choosing a to maximize the Lagrangian $\Omega^H = V^* + E^* + (n-1)E - \lambda(\bar{G} - G)$ when this tax is applied optimally ($\lambda = 1$), and using this in the first-order conditions for the timber tax α and the tax exemption m gives

$$\Omega^H_{\alpha|a=a^*} = T^H_\alpha[(n-1)E_T + \alpha(U_T - X_T)] = 0 \tag{5.40a}$$

$$\Omega^H_{m|a=a^*} = T^H_m[(n-1)E_T + \alpha(U_T - X_T)] = 0. \tag{5.40b}$$

From previous discussion, both $T^H_\alpha \neq 0$ and $T^H_m \neq 0$. Therefore they imply that $\Omega^H_{\alpha|a=a^*} = 0 \Leftrightarrow [(n-1)E_T + \alpha(U_T - X_T)] = 0$. Utilizing the expressions for E_T, U_T, and X_T and rearranging, we obtain the following expression for the optimal timber tax rate:

$$\alpha^*_{|a=a^*} = -\frac{(n-1)[F(T) - rE]}{pf(T) - r(U - X)}. \tag{5.41}$$

As before, the policy maker should use a timber tax (or subsidy) when the marginal valuation of amenities decreases (or increases) with the age of the stand, i.e., when $F'(T) > (<) 0$. The denominator of equation (5.41) reveals the relationship between the timber tax rate α and the tax exemption m (present in the X term). If the marginal valuation of amenities increases with the age of the stand, then the policy maker should use a net subsidy (a negative value for α is optimal). This subsidy should be inversely related to the size of the tax exemption. However, if the marginal valuation of amenities decreases with rotation age, then the policy maker should use a net timber tax (a positive value for α); again, this is inversely related to the size of the tax exemption. Comparing equations (5.40a) and (5.40b) gives the following result:

Proposition 5.10 When the government has two objectives, collection of tax revenues and correction of any externalities arising from a divergence in privately and socially optimal rotation ages, and it can use both a site productivity tax and a proportional timber tax, then progression in the tax system is not socially optimal.

The results show that progression is not optimal if the government already has two instruments at its disposal (equity issues aside). In this case, a third type of instrument (progression) becomes redundant. The same analysis would hold for the yield tax.

Proposition 5.10 is based on the assumption that the government has a neutral tax available. We end by asking whether these results would change if the government did not have such a tax available. The question is certainly more relevant in practice, where site productivity taxes and any other truly neutral taxes for that matter, are hard to find. In this case, following the earlier procedures it can be shown that a positive exemption is socially optimal, along with a timber or yield tax rate that is set so that the marginal cost of public funds is equal to one; i.e., the required tax revenue is raised in a nondistortionary fashion. This allows targeting of a corrective tax to remove any externality associated with differences between private and social rotation ages when amenities are present (for a proof, see Koskela and Ollikainen 2003a).

5.6   Summary

This chapter demonstrates the ways in which a government can design the best tax systems for the forest sector under a variety of objectives. We examined how the optimal structure of forest taxation depends on the goals of the social planner and the wedge between private and social decisions in both the short and the long run. Possible goals of the planner are to raise tax revenues to finance public spending or to promote amenity benefits by influencing the harvesting decisions of private landowners through changes in the tax structure. The principles of Pigouvian taxation (correcting of externalities) apply in the latter case, while Ramsey principles of taxation (neutrality through minimum distortions) apply in the former case.

The results in this chapter provide a starting point for empirical comparisons concerning what economic theory suggests is the best tax structure, and taxes that are actually present in forest sectors. The best taxation systems are simply not applied in practice, meaning there are social costs present that would be removed through tax reform. Forest taxes are a surprisingly powerful tool for affecting the harvesting behavior of landowners and the public goods provided by forests, yet tax instruments are rarely used in the ways we have shown are suggested by economic theory. The results in this chapter set the stage for much further empirical work evaluating the inefficiency of actual policy instruments targeting forests.

The methods outlined in this chapter also have implications that go far beyond the problems we have studied. We return to many of the ideas here in the policy-oriented chapters, 6–9. As these chapters will show, the optimal policy framework is useful for framing the design of tropical forest concessions, maintaining biodiversity in commercial forests, and producing climate benefits through forest management. We will also see that once uncertainty is introduced into landowner decision making, tax instruments often take on additional roles besides correcting externalities and raising revenues for the government.

Appendix 5.1   Derivation of Tax Formulas in Section 5.3

A.   Optimal Taxes When Amenity Services Are Private Goods

The first-order conditions for the optimal choice of taxes are

$$\Omega_T = V_T^* + \lambda \left[1 + \tau_b p b_T + \tau_x \frac{(2+n)}{(1+n)} p x_T \right] = 0 \tag{A5.1a}$$

$$\Omega_{\tau_b} = V_{\tau_b}^* + \lambda \left[pb + \tau_b p b_T + \tau_x \frac{(2+n)}{(1+n)} p x_{\tau_b} \right] = 0 \tag{A5.1b}$$

$$\Omega_{\tau_x} = V_{\tau_x}^* + \lambda \left[\tau_b p b_{\tau_x} + \frac{(2+n)}{(1+n)} px + \tau_x \frac{(2+n)}{(1+n)} p x_{\tau_x} \right] = 0, \tag{A5.1c}$$

where $V_T^* = -[u'(c_{1t}) + \phi u'(c_{2t+1})]$, $V_{\tau_b}^* = pb V_T^*$, and $V_{\tau_x}^* = (1 + R^{-1}) px V_T^*$.

Applying Slutsky decompositions, we can re-express (A5.1b) and (A5.1c) as

$$\Omega_{\tau_b} = pb\Omega_T + \lambda \left[\tau_b p b_{\tau_b}^c + \tau_x \frac{(2+n)}{(1+n)} p x_{\tau_b}^c \right] = 0 \tag{A5.2a}$$

$$\Omega_{\tau_x} = (1 + R^{-1}) px\Omega_T + \lambda \left[\tau_b p b_{\tau_x}^c + \tau_x \frac{(2+n)}{(1+n)} p x_{\tau_x}^c + p\varepsilon \right] = 0, \tag{A5.2b}$$

where $pb\Omega_T = pb \left[V_T^* + \lambda(1 + \tau_b p b_T + \tau_x \frac{(2+n)}{(1+n)} p x_T) \right]$,

$$(1 + R^{-1}) px\Omega_T = (1 + R^{-1}) px \left[V_T^* + \lambda(1 + \tau_b p b_T + \tau_x \frac{(2+n)}{(1+n)} p x_T) \right] \quad \text{and}$$

$$\varepsilon = \left(\frac{1}{1+n} - \frac{1}{1+r} \right) x.$$

If the lump-sum tax is set at the optimal level, then the first RHS terms are zero in (A5.2a) and (A5.2b) because $pb\Omega_T = 0$ and $(1 + R^{-1}) px\Omega_T = 0$. Hence we obtain the following expressions:

$$\Omega_{\tau_b|T=T^*} = \tau_b p b^c_{\tau_b} + \tau_x \frac{(2+n)}{(1+n)} p x^c_{\tau_b} = 0 \tag{A5.3a}$$

$$\Omega_{\tau_x|T=T^*} = \tau_b p b^c_{\tau_x} + \tau_x \frac{(2+n)}{(1+n)} p x^c_{\tau_x} - \frac{(2+n)}{(1+n)} - (1+R^{-1}) = 0. \tag{A5.3b}$$

Equations (A5.3a) and (A5.3b) define a system of two equations with optimal tax rates as endogenous variables, and text equations (5.30a) and (5.30b) can be solved with this equation system.

B. Amenity Services as Public Goods

The Lagrangian to the government's maximization problem is now

$$\max_{\{T, \tau_b, \tau_x\}} \Omega^A = V^*(T, \tau_b, \tau_b) + (n+2)(q-1)[v(k_{1t}) + \beta v(k_{2t+1}) + \phi v(k_{1t+1}) + \beta v(k_{2t+1})]$$

$$- \lambda \left[G^0 - \tau_b p b - T - \tau_x \frac{(2+n)}{(1+n)} p x \right]. \tag{A5.4}$$

For convenience, define $D = (\phi + \beta)v'(k_{2t})(1+g)$ and $D^0 = \beta v'(k_{2t+1})$, giving:

$$\Omega^A_T = V^*_T + (n+2)(q-1)(Dx_T - D^0 b_T) + \lambda \left[1 + \tau_b p b_T + \tau_x \frac{(2+n)}{(1+n)} p x_T \right] = 0 \tag{A5.5a}$$

$$\Omega^A_{\tau_b} = V^*_{\tau_b} + (n+2)(q-1)(Dx_{\tau_b} - D^0 b_{\tau_b}) + \lambda \left[p b + \tau_b p b_{\tau_b} + \tau_x \frac{(2+n)}{(1+n)} p x_{\tau_b} \right] = 0 \tag{A5.5b}$$

$$\Omega^A_{\tau_x} = V^*_{\tau_x} + (n+2)(q-1)(Dx_{\tau_x} - D^0 b_{\tau_x})$$

$$+ \lambda \left[\tau_b p b_{\tau_x} + \frac{(2+n)}{(1+n)} p x + \tau_x \frac{(2+n)}{(1+n)} p x_{\tau_b} \right] = 0. \tag{A5.5c}$$

Applying Slutsky decompositions, we can now re-express (A5.5b) and (A5.5c) as

$$\Omega^A_{\tau_b} = p b \Omega^A_T + (n+2)(q-1)(D^0 b^c_{\tau_b} - Dx^c_{\tau_b}) + \lambda \left[\tau_b p b^c_{\tau_b} + \tau_x \frac{(2+n)}{(1+n)} p x^c_{\tau_b} \right] = 0 \tag{A5.6a}$$

$$\Omega^A_{\tau_b} = (1+R^{-1}) p x \Omega^A_T + (n+2)(q-1)(D^0 b^c_{\tau_x} - Dx^c_{\tau_x})$$

$$+ \lambda \left[\tau_b p b^c_{\tau_x} + \tau_x \frac{(2+n)}{(1+n)} p x^c_{\tau_x} + p \varepsilon \right] = 0. \tag{A5.6b}$$

Hence when the lump-sum tax has been set at the optimal level, we have

$$\Omega^A_{\tau_b|T=T^{*A}} = (n+2)(q-1)(D^0 b^c_{\tau_b} - Dx^c_{\tau_b}) + \lambda \left[\tau_b pb^c_{\tau_b} + \tau_x \frac{(2+n)}{(1+n)} px^c_{\tau_b} \right] = 0 \quad \text{(A5.7a)}$$

$$\Omega^A_{\tau_b|T=T^{*A}} = (n+2)(q-1)(D^0 b^c_{\tau_x} - Dx^c_{\tau_x}) + \lambda \left[\tau_b pb^c_{\tau_x} + x \frac{(2+n)}{(1+n)} px^c_{\tau_x} + \varepsilon \right] = 0.$$
$$\text{(A5.7b)}$$

These optimality conditions (A5.7a) and (A5.7b) define the optimal harvest and bequest tax rates as the two-equation system. The optimal tax equations in the text can now be derived by solving (A5.7a) and (A5.7b).

6 Deforestation: Models and Policy Instruments

The permanent clearing of forest cover was the rule for temperate forests in the industrial world through the 1940s. Vast areas of Europe and North America were cleared as industrial expansion and development of infrastructure proceeded. Deforestation today occurs largely in the tropical native forests of Africa, Asia, and Latin America. The figures are alarming. General estimates suggest that during the 1980s and 1990s, the annual rate of deforestation has averaged 1% worldwide, with deforestation in tropical rain forests averaging 0.6% (FAO 2001). While these rates may not seem especially large, consider that between 1990 and 2000, more than 14.2 million hectares of tropical rain forests disappeared. What has captured the attention of economists is that these forests are not only responsible for more than 80% of the world's biodiversity, but they are also thought to play important roles in maintaining global climate conditions.

In public discussions stressing the environmental role of tropical forests, deforestation is usually taken to mean decreases in native forestland area. This definition, however, ignores the impact of secondary changes in forestland area that follow from harvesting, such as selective taking of trees, the establishment of plantations, or growth of secondary forests on idle cutover land left underutilized. In response to these issues, the UNs' Food and Agricultural Organization (FAO 1996) more recently proposed an updated definition of deforestation, stating "Deforestation occurs...when the forest canopy cover is reduced to 10% or lower of its natural level."

This chapter is devoted to policy instruments aimed at slowing deforestation. Any discussion of policy instruments opens many important and challenging research questions, such as the most effective instruments in changing private and government behavior toward sustainable forestry. Although these ideas sound like simple extensions of the material in chapter 5, there are many important differences. Both empirical evidence and the literature suggest that any policy design targeting deforestation must account for many institutional and economic factors prevailing in underdeveloped tropical countries. Imperfect credit and labor markets bias agents' decision making toward short-term capture of rent from land use. Insecure property rights imply that illegal logging is rampant, although there are increasing though possibly suboptimal investments in plantation forestry. Poor and inefficient governments in these countries often sell timber through a concessions process, raising money by using royalty fee systems applied to the sales. The literature has pointed out numerous cases where royalty systems are improperly designed, and government enforcement of these payments is lacking, with low enforcement effort and penalties. Perhaps most important, corruption among public officials and bribery are common in harvesting of public forests in tropical areas. Combined with expanding populations and increasing access to frontier areas, tropical forests therefore face constant pressure.

We begin by providing a synthesis of the literature concerning the basic forms and causes of deforestation. In the second section we extend material from chapter 5 to examine a public forest concession model with and without corruption. In the third section we consider deforestation within the context of land use, extending material in chapter 3 to include insecure property rights and migration to frontier forests. Finally, in the last section, we offer a summary for policy design. We will focus on static models to explain deforestation in this chapter. Dynamic approaches are discussed in chapter 12.

6.1 Basic Forms of Deforestation

The literature focusing on deforestation is immense. There have been studies covering market-based and household-based problems, micro- and macroeconomic aspects, and regional and global aspects. Kaimowitz and Angelsen (1998), Angelsen and Kaimowitz (1999), Angelsen (1999), and Barbier and Burgess (2001a) collectively survey theory and econometric studies on the key factors affecting deforestation. The unanimous conclusion from these efforts is that the causes of deforestation remain an important and fairly open question. While institutional factors are obviously important, clearing land for crops and pasture, illegal logging, and poorly designed concessions are nearly always implicated.

The literature in economics has universally studied deforestation through four forms of forest clearing: conversion of land use toward crops and livestock grazing, expansion of commercial logging in and out of concessions, illegal logging and other illicit activities, and in some arid countries, pressure on forests from fuel collection by local communities. Often these uses become important only because of government actions, such as building roads and encouraging access to previously inaccessible public forests, poor enforcement, or providing subsidies for clearing land.

6.1.1   Conversion to Agricultural Land

Conversion of native forest land to crop or grazing uses typically follows classic "slash-and-burn" methods, with trees burned to make way for agriculture. While some resulting agriculture may be intensive, most is extensive, being based on shifting cultivation, that is, the production of crops that deplete soil resources, reducing the productive nature of the land over time and requiring new land to be cleared for crop areas. Shifting cultivation is perhaps more responsible for permanent deforestation than slash and burn. Pasturing is also a common cause of land clearing because cattle provide higher levels of wealth and typically result in larger farms.

Insecure property rights are likely to promote unsustainable agriculture over sustainable forest uses. In many countries, a land user's property rights are made secure simply by clearing land because this is an indication to the government that the land exists in a "beneficial use." Alston et al. (1999) have also argued that clearing reduces the risks of future government expropriation of the land.

Recent reports, however, suggest that more than 1 million hectares in the Amazon have been reforested into plantations since 1990 (FAO 2001). As the International Tropical Timber Organization (ITTO) in its 2005 status report states, planted forests are now playing a much more significant role in the supply of timber in tropical countries. The area of planted tropical forests has expanded considerably in the past 15–20 years, with some tropical countries becoming increasingly reliant on planted forests for their domestic timber supply. Planted forests will likely reduce pressure on native forests to satisfy demand for wood, and forest plantation holders practice a form of repeated rotation forestry that some consider a good alternative to clearing land for slash-and-burn agriculture; however, the debate about whether to deduct regeneration of plantation forest cover from deforestation rates continues. Also, plantation users face many difficulties in protecting their investments from illegal trespass when prices and costs make this activity profitable. This is simply because forest cover makes it more difficult for landowners to monitor these lands.

6.1.2   Commercial Harvesting through Concessions

A large share of commercial logging in tropical forests takes place through concessions. These are contracts between a forest owner (typically the government) and a forest user that permit harvesting specified volumes for a specified forest-land area. Concessions may cover small or large areas and are common in fringe native tropical forests with low or expanding infrastructure. They may be short term and awarded annually, or last for many years. In tropical developing countries, concession contracts typically define allowable volume or area to be harvested and more recently they specify environmentally sensitive logging methods that harvesters must use (for a comprehensive discussion of these contracts, see Gray 2000). Examples of the latter include harvesting only specified species, diameter limits for felling, and the use of skidding and road-building techniques that control soil erosion, protect species diversity, and limit damage to residual trees. Even when harvesters are given the right to harvest any trees on a site, there are at least some environmentally sensitive logging stipulations in the contracts. In most cases, harvesters bidding for the concession rights are large firms with enough capital to undertake the concession management obligations.

Concessions for public forests are deliberate acts by governments, suggesting that the opportunity costs for standing tropical native forests are still high. Concessions are also an undeniable and quick source of government revenues. The presence of binding government revenue constraints is manifest in the frequent inability (and even unwillingness) of governments to enforce the stringent rules of concession contracts. Lack of enforcement makes illegal logging easier and concessions have been heavily criticized as contributing indirectly to greater deforestation than intended.

6.1.3 Illegal Logging

Illegal logging is an undeniable threat to the future health of native forests worldwide. Although the role of illegal logging in promoting deforestation varies over countries, a recent study of trade flows by Guertin (2003) finds that more than 80% of the volume harvested in Latin American tropical forests may be undeclared, lending credence to the fact that harvesters cut beyond the bounds specified by the government. A World Bank cross-country study by Contreras-Hermosilla (2002) finds evidence of illegal logging nearly everywhere in the world, and especially in tropical settings. In Indonesia, illegal logging has been well recognized both in the applied literature and in recent research studies (Smith et al. 2003, Contreras-Hermosilla 2002).

Illegal logging typically occurs in one of two forms. The first is by harvesters who break concession agreements by harvesting more than is allocated by the contract; practicing unsustainable and unallowed forms of harvesting, such as high grading, where only the selected and best species are taken; or using improper logging that damages residual trees and soil resources (Tacconi et al. 2003). The second form is practiced by harvesters who trespass on private forest land and remove valuable trees when prices and costs make these activities profitable. Both forms of illegal logging arise largely from insufficient enforcement by the government and by private landowners.

6.1.4   Fuel Collection

We should briefly note another form of pressure on forests. This comes from collection of fuel by villagers in developing countries. This is mainly a problem in arid countries where fuelwood is a limiting constraint in household production and consumption decisions. Fuelwood collection pressures on open-access forest land in these situations result from poor households in nearby villages who collect fuelwood from unprotected forests. Property rights problems and difficulty enforcing these rights also exist here, but instead of modeling this as land-clearing behavior, arid-country models are based on understanding household labor decisions as they relate to on-farm and off-farm activities. These problems are often studied using a different approach than that taken in this book, with fuelwood collection examined in a household production framework. Often, the complexity of labor decisions when markets are incomplete is a main focus, and econometric approaches have been used to link resource quality to household decisions and pressures on forest exploitation. Interested readers can turn to Arnold et al. (2006), Hyde and Amacher (2001), and Sills et al. (2003) for reviews of these types of forest-use models in arid and tropical contexts.

6.2   Causes of Deforestation

The study of deforestation has focused on both causes and approaches to explain the mechanism of forest loss and degradation. The study by Deacon (1994) is important in that he provides one of the most interesting and early empirical tests of the causes of deforestation, using panel data from numerous countries. He examined population, property rights and political stability, and national income factors as potential drivers of forest clearing. He finds statistical support that deforestation is higher in countries with positive population growth, negative income growth, and poorly defined property rights systems with a lack of enforcement.

Barbier and Burgess (2001a) also provide an important synthesis that goes further in suggesting there have been two "waves" of deforestation literature. The first wave relied on empirical analyses to establish important driving factors of tropical deforestation. The second wave has proposed models of economic behavior for households and firms to investigate how deforestation is affected by microeconomic choices. Barbier and Burgess propose and then apply an econometric synthesis (first wave type) model to evaluate important deforestation factors using panel data for countries judged important to world forest loss by the World Bank. For their sample with all countries combined, they find that predictions about deforestation factors will not be correct if indices for corruption, security of property rights, and political stability are not included as potential factors. Once these variables are included, for all countries combined, they find that among other things, the most significant predictors of deforestation include positive effects from agricultural export pressures, population growth, a high corruption index, and a political instability index. They do not find secure property rights to be important, but they argue that the effects of corruption, political stability, and property rights can vary dramatically across countries and could all be important in specific cases. Their review of second-wave studies leads to the conclusion that agricultural and forest sector policies might be more important in predicting deforestation pressures for this level of analysis. They also caution that many of the predictors of deforestation are likely to be determined locally and to vary widely across areas even within the same country.

A large part of the literature on policy instruments surrounding tropical forest concessions has blamed the improper application of royalties for illegal logging within the forests (see for example, Barr 2001, Richards 1999, Gray 2000, and Hardner and Rice 2000).[1] Some important studies of illegal logging incentives are Boscolo and Vincent (2000), who analyze the impact of royalties on the use of minimum site impact (i.e., environmentally sensitive) logging practices by loggers, Clarke et al. (1993), who examine the role of penalty schemes and optimal dynamic enforcement expenditures in preventing exploitation of open-access forests, and Walker and Smith (1993), who model noncompliance of loggers facing a given concession contract. Barbier and Burgess (2001b) analyze costly enforcement efforts expended by landowners to prevent illegal logging in plantation forests and, finally, Amacher et al. (2007) examine reform in royalties that can reduce illegal logging in the form of harvesting beyond the limits of the concession area.

Notwithstanding illegal logging, there is another even more insidious feature that must be mentioned in any discussion of deforestation. Corruption certainly exists among government officials, meaning that the potential exists for loggers to pay bribes for illegally logging in government-managed forests. In many cases, bribes have been singled out as an inherent part of harvesting in tropical countries, confounding any central government's attempt to control deforestation or promulgate forest policy (see e.g., Palmer 2000, 2005).[2] Smith et al. (2003) specifically review and study forest corruption in Indonesia, with its long history of

---

1. The applied literature has recommended several changes in royalty systems. The most common suggestion has been to raise royalty rates as a way of increasing government collection of rents and reducing excessive harvesting (Gray 2000, Vincent 1990, Merry et al. 2004, Palmer 2000). Others have called for a shift toward using area-based lump-sum royalties (Barr 2001, Richards 1999, Gray 2000, Hardner and Rice 2000). Another common recommendation is to increase enforcement effort (Hardner and Rice 2000).

2. Excellent surveys of corruption and policy instruments from the general public finance literature include those by Bardhan (1997), Adit (2003), Svensson (2005), and Schleifer and Vishny (1993).

concessions, defining corruption as either collusive or noncollusive. The latter is defined as a situation where government officials demand bribes for approval of an unauthorized activity, such as logging beyond a concession boundary or failing to use environmentally sensitive logging methods. The former occurs when resource users and government officials work together to steal rents that the government could capture. As Smith argues, collusive corruption is more difficult to detect and penalize because the incentives for the briber and the government official align.

Contreras-Hermosilla (2002) proposes several reasons for corruption in forestry settings, including underpaid logging inspectors, complex regulations for harvesting, numerous bureaucratic steps required to obtain permits, low penalties for illegal activities and the unwillingness of governments to enforce penalties, and the distant, open-access nature of native forests. Not surprisingly, Contreras-Hermosilla's examples include virtually every tropical developing country that still contains the bulk of the world's remaining native forests.

Forest economists have not yet had much to say about policy instruments in a climate of corruption. Exceptions at the time this chapter was written amount only to Delacote (2005) and Barbier et al. (2005). Delacote finds that corruption may induce a government policy maker to designate large areas of forest resources for harvesting, or set less stringent regulations targeting forest use. In Delacote's problem, harvesters act as a group and negotiate bribes directly with the central government to obtain more favorable concession terms, i.e., he considers collusive corruption. Later in this chapter we will take a more decentralized approach and focus on bargaining for bribes between individual harvesters and local government inspectors responsible for detecting and penalizing illegal logging. Barbier et al. focus on an open-economy model of corruption, studying both theoretically and empirically how lobbying by groups involved in exploiting forest resources influences corrupt resource-rich governments and ultimately affect deforestation rates and trade parameters. They find that corruption can lead to greater deforestation; however, terms of trade can play an important role in mitigating corruption effects.

Despite the information that these studies give us, there is still no clear agreement on what are the "true" causes of deforestation. The complexity of the problem means that the causes probably include combinations of factors, such as changes in access to native forests (road paving); macroeconomic conditions and economy-wide policies; government incentives that make agricultural production relatively more profitable than forest production; poverty and population pressures that encourage migration into forested areas; incomplete property rights or the inability of the government to enforce property rights on distant, open forests; poorly designed government policies; and rapid technological change in agricultural industries. The precise nature of how these stimulate deforestation and how they are collectively important is quite complicated and not fully resolved, but clearly, insecure property rights and illegal logging are themes that run through all of these causes.

6.3   Forest Concessions

We first turn to forest concessions, focusing on perhaps the most common form of illegal logging, in which a harvester takes more volume than is allowed by contract. We will consider concession design under no corruption and under corruption of government officials. As we noted earlier, the harvester pays a royalty to the government for a concession right. Royalty rates are levied on either the volume or the value of harvested timber. Alternatively, a lump-sum royalty is sometimes recommended, with the harvester paying a given sum that reflects the estimated value of timber in the concession irrespective of the amount actually harvested (Richards 1999, Hardner and Rice 2000). Fines for illegal logging, when applied, are typically lump sum in nature and include confiscation of harvests or a fine, although there has been discussion of basing fines on the extent of illegal activity (Hardner and Rice 2000). We will look at fines based on this suggestion in this section. In what follows, all parties will be assumed risk neutral.

6.3.1   Optimal Policy Design—Absence of Corruption

Concessions provide governments with revenue, but illegal harvesting within them generates externalities and social costs. Since enforcement to prevent these acts is costly and governments are poor, the question is how to design a royalty and concession system that balances these opposing forces. Consider a harvester who receives permission to harvest a specified volume, $Q = \bar{Q}$. The size of the concession is chosen by the government, who also charges a value-based royalty against permitted harvesting, $tq\bar{Q}$, where q is the timber price and t is the royalty rate. If the harvester does not engage in illegal activities, then the exact amount of volume specified in the concession is logged and the harvester earns the following rents:

$$\pi = q\bar{Q}(1 - t) - c(\bar{Q}), \tag{6.1}$$

where $c(Q)$ is a convex cost of harvesting valued at $\bar{Q}$, with $c'(Q), c''(\bar{Q}) > 0$. The harvester will not necessarily be honest, however, if the marginal rent for harvesting is positive at $\bar{Q}$, i.e., if $q(1 - t) - c'(\bar{Q}) > 0$. Define the volume of timber illegally logged as "excessive" harvesting, expressed using the difference between actual logging Q and permitted logging $\bar{Q}$, i.e., $X \equiv (Q - \bar{Q})$. Let p denote the probability that this illegal logging is detected when the government monitors the forest and $(1 - p)$ be the probability that it is not detected. Greater effort on the government's part implies a higher probability of detection. If the harvester is caught, a penalty is applied on the extent of the illegal activity, so that the penalty is fqX, where f is an exogenous penalty assessed and $f > 1$. Such a fine is similar to confiscation of the harvest by the government. Using these assumptions, the harvester's profits when he is and is not detected, respectively, are

$$Y = [qQ - tq\bar{Q} - c(Q)], \tag{6.2}$$

and

$$Z = [Y - fqX]. \tag{6.3}$$

Thus the harvester maximizes expected profits:

$$\max_{Q} \ E\pi = qQ - c(Q) - tq\bar{Q} - pfqX \tag{6.4a}$$

which, given the following first-order condition for harvesting,

$$E\pi_Q = q - c'(Q) - pfq = 0. \tag{6.4b}$$

The second-order condition holds: $E\pi_{QQ} = -c''(Q) < 0$. Equation (6.4a) implies that the volume harvested, Q, is chosen to equate marginal revenue (q) and expected marginal cost, the latter of which consists of harvest cost plus payment of expected fines. Obviously, changes in the detection probability and size of the concession all affect the harvester's choice. Using the standard comparative statics applied to (6.4b), and the definition of illegal logging ($X \equiv Q - \bar{Q}$), these effects are summarized as $X = X(\underset{0}{t}, \ \underset{-}{p}, \bar{Q})$. The royalty affects profits in a lump-sum manner and does not affect illegal logging, but a larger concession size reduces the incentives to cheat by increasing potential harvested volume. As expected, a higher probability of detection reduces illegal activity.

The government's optimal concession policy problem takes the behavior described here as given. The government must be sensitive to the external costs associated with illegal logging and the concession choice itself. These are captured in the following function: $v(\bar{Q}, X)$, with $v_{\bar{Q}} < 0$, $v_X < 0$, $v_{\bar{Q}\bar{Q}}, v_{XX} > 0$. This assumes that illegal logging is ecologically more harmful than the government designating a larger concession. As in chapter 5, social welfare is defined as the profits of the harvester net of negative externalities, $SW = E\pi^* + v(\bar{Q}, X)$, where $E\pi^*$ is the harvester's expected indirect profit function.

The government's cost of auditing is assumed to be a convex function of the probability of detection, $\phi(p)$, such that $\phi'(p) > 0$, $\phi''(p) > 0$, and $\phi'(0) = 0$. Expected government revenues are then defined as the sum of royalty payments collected, (exogenous) fines collected, and the cost of auditing:

$$E(R) = tq\bar{Q} + pfqX - \phi(p). \tag{6.5}$$

The government now chooses a concession policy $(t, \bar{Q}, p)$ to maximize SW subject to (6.5). The first-order condition for the royalty t, using the envelope theorem for $E\pi^*$, yields

$$\Omega_t = -q\bar{Q} + \lambda q\bar{Q} = 0. \tag{6.6}$$

Condition (6.6) indicates that $\lambda = 1$, meaning that revenues can be collected without any efficiency losses (see chapter 5). Using $\lambda = 1$, the first-order conditions for concession size and the detection probability are

$$\Omega_{\bar{Q}} = v_{\bar{Q}} - v_X = 0 \tag{6.7a}$$

$$\Omega_p = v_X X_p + pfqX_p - \phi'(p) = 0. \tag{6.7b}$$

From (6.7a) we find $v_{\bar{Q}} = v_X$, so that the size of the concession is expanded up to the point where the marginal negative externality that is due to the choice of concession size (LHS) equals the reduced marginal externality from lower illegal logging (RHS). Thus the government can increase the concession size in order to reduce incentives for ecologically more harmful illegal logging. Expressing (6.7b) in the form $\phi'(p) = (v_X + pf)X_p$ shows that resources are devoted to auditing up to a point where the marginal audit cost equals the difference (net benefit) of reduced negative externalities and reduced income from fines. These findings are collected in proposition 6.1.

Proposition 6.1   The optimal concession policy without corruption   For risk-neutral harvesters, the optimal design of policy instruments consists of the following:

(a) a concession size that equates the negative externality from the concession with the reduction in externality that is due to lower illegal logging,

(b) a royalty rate used to meet the government's budget constraint, and

(c) an auditing probability equal to the marginal benefit from decreased illegal logging net of tax revenue lost from lower fines collected.

Policy design is this case is very clear and simple. It is clearly optimal to increase the size of the concession to reduce some incentives for illegal logging if the external costs are high enough.

6.3.2   Optimal Policy Design—Presence of Corruption

As Smith and Contreras-Hermosilla point out in their examples, corruption most typically takes the form of bribes made by illegal harvesters to government-employed inspectors. Thus the important issue is how to include bribery in the model of section 6.3.1. Recall that the detection effort of the government is modeled using a relatively anonymous cost function. Now, we must allow for a field logging inspector, paid a wage w by the government, who checks whether harvesters are following their concession contract. Without corruption, it was not necessary to include the role of this inspector in the government's enforcement possibilities because he was implicitly assumed to act as one with the government. With corruption, the inspector's auditing effort is a positive function of his wage, following what the multitude of nonforestry studies have contended and shown. Thus a higher wage for inspectors implies an increased audit probability, and government enforcement costs now become in their simplest form the wage paid to the inspector. We should further point out that should the inspector detect illegal logging and not take a bribe, then a fine is imposed on this activity by the government and we have a case similar to that of no corruption in proposition 6.1.

Of course, incentives may exist for the inspector to take a bribe from the harvester and subsequently overlook illegal logging activity. Taking Smith's non-collusive corruption as our modus operandi by assuming that the central government will punish bribery, there is some positive probability that the government will detect the bribe. The timing of inspection and bribery must therefore be made explicit. There are two options. First, the logger may try to illegally log, paying a bribe only if he is detected by the inspector with probability p. Second, a logger may act more aggressively and first pay a bribe to the inspector for an agreed amount of illegal harvesting before he begins harvesting. Here we assume that the logger tries his luck and harvests first, which makes sense as a baseline case. The resulting policy problem is solved using backward induction. First, we solve the negotiation problem between the harvester and inspector over the bribe, given any level of illegal logging. Second, we then solve for the level of illegal logging taking the optimal bribe as given. Finally, we solve the government's concession design problem, taking all bribery and illegal logging responses as given.

The most convenient way to solve for bribes is to use Nash bargaining between harvester and inspector.[3] Recall that illegal harvesting is defined by the term $X = Q - \bar{Q}$. Let the penalty when illegal harvesting is detected continue to be $f > 1$. Let b denote the size of a bribe between the harvester and inspector. Let $0 < m < 1$ be the probability that the government detects the inspector taking a bribe. If the government detects bribery, then the inspector and harvester each

---

3. Nash bargaining is popular for bribery when corruption is studied in the general economics literature, such as with Acconcia et al. (2003), Sanyal et al. (2000), Basu et al. (1992), Chander and Wilde (1992), Mookherjee and Png (1995), and Cahuc and Zylberberg (2004, chapter 7).

must pay a penalty equal to ϕb, where $0 < \phi < 1$ is the bribery fine rate, and the harvester in addition must pay the fine for illegal logging, fqX.

We need to characterize the expected costs to the harvester from illegally logging when the harvester has been detected by the inspector ($p = 1$). The expected costs to the harvester are

$$[\phi b + fqX]m + b(1 - m). \tag{6.8}$$

The first term is the expected penalty for bribery and illegal logging when these are detected by the government with probability m, while the second term is the expected bribe paid to the inspector when government does not detect bribery with a probability $1 - m$. The harvester has an incentive to pay a bribe whenever the expected cost in (6.8) is less than what the harvester would have to pay in the absence of corruption if caught, i.e.,

$$[\phi b + fqX]m + b(1 - m) < fqX \Leftrightarrow b < \frac{fqX(1 - m)}{1 - m(1 - \phi)}. \tag{6.9}$$

Turning to the inspector, he will accept a bribe if his expected gain is positive. Assuming that the inspector is paid the same wage rate, w, before and after being caught and that the same penalty rate on the bribe is applied to both harvester and inspector, the expected revenue to the inspector from bribery includes the revenue under the cases where the government detects it and where it does not detect it, $(w + b)(1 - m) + (w - \phi b)m = w + b(1 - m) - \phi bm$. The difference of the two latter terms must be positive for the inspector to take a bribe:

$$b(1 - m) > \phi bm \Leftrightarrow \phi < (1 - m)/m. \tag{6.10}$$

The inequality in (6.10) holds for any size of bribe b, as $\phi < (1 - m)/m$. Thus a bribe is agreed upon between inspector and harvester whenever b is positive but less than the size indicated by condition (6.9).

The next task is to determine the exact size of the bribe within the range set by (6.9). The following Nash bargaining problem defines this bribe as maximizing the product of expected gains to harvester and inspector from (6.9)–(6.10),

$$\underset{(b)}{\max}\ \Omega = EH^\beta EK^{1-\beta}, \tag{6.11}$$

where $EK = w + b[1 - m(1 + \phi)]$, and $EH = fqX(1 - m) - b[1 - m(1 - \phi)]$. The parameters β, $1 - \beta$ measure the relative bargaining power of the harvester and inspector, respectively. Bargaining is of course only meaningful if both parties have some bargaining power, $0 < \beta < 1$. The first-order condition of (6.11) is, after some rearranging,

$$\Omega_b = 0 \Leftrightarrow \beta \frac{EH_b}{EH} + (1 - \beta) \frac{EK_b}{EK} = 0, \tag{6.12}$$

where $EH_b = -[1 - m(1 - \phi)] < 0$ and $EK_b = 1 - m(1 + \phi) > 0$. The first-order condition can be solved explicitly for the optimal bribe level:

$$b^N = (1 - \beta) \frac{fqX(1 - m)}{[1 - m(1 - \phi)]} - \beta \frac{w}{[1 - m(1 + \phi)]}. \tag{6.13}$$

The optimal bribe depends on relative bargaining powers, the wage of the inspector, the penalty rate for illegal logging and bribery, the timber price, the degree of cheating by the harvester, and the probabilities of detection. The single-equation comparative statics of this optimal bribe are straightforward and can be used to show how the bribe depends on these factors:

$$b^N = b(\underset{-}{\beta}, \underset{+}{f}, \underset{-}{m}, \underset{+}{q}, \underset{0}{t}, \underset{+}{X}, \underset{-}{\bar{Q}}, \underset{-}{w}, \underset{-}{\phi}). \tag{6.14}$$

The most interesting result is that the royalty used here does not affect the bribe. This may seem as if this type of royalty will avoid corruption. Unfortunately, it does not because all other concession design parameters affect incentives for bribery and the size of the bribe. It is interesting that a lower concession size $\bar{Q}$ simply decreases the optimal bribe as the incentive to illegally log beyond the concession border increases. Other parameters have expected effects, including the fact that a higher wage reduces the bribe (see Di Tella and Schargrodsky 2003).

In the second stage, the harvester takes the optimal bribe in (6.14) as given and chooses harvested volume Q. Using b^N in place of b, the expected profits of the harvester when facing probabilities of detection p and m and a fine for bribery ϕ are

$$E\pi = qQ - tq\bar{Q} - c(Q) - p\{mfqX + b^N[1 - m(1 - \phi)]\}. \tag{6.15}$$

Substituting from (6.13) into (6.15) and rearranging gives

$$E\pi = qQ - tq\bar{Q} - c(Q) - p\{fqX[1 - \beta(1 - m)] - \beta w\xi\}, \tag{6.16}$$

where $\xi = [1 - m(1 - \phi)]/[1 - m(1 + \phi)] > 0$. Comparing (6.16) with expected profits under no corruption, a harvester who pays a bribe has additional factors affecting his decisions, such as his bargaining power, the detection probabilities for both illegal harvesting and bribery, and the wage of the inspector as this factors into the optimal bribe. At the very least, this shows that predicting a harvester's illegal behavior in a concession is much more complicated when corruption is present.

Before examining the harvester's choices under corruption, we need to add one more wrinkle to the problem that reflects what happens in practice, and which makes a link between other corruption work that has not focused on forestry. Cross-country corruption studies of tax collectors have shown statistically less corruption, however defined, when inspectors are higher paid. There is no reason not to expect this for forest concessions, as suggested by Contreras-Hermosilla (2002), and so we assume that the probability that the inspector detects the harvester depends endogenously on the size of his wage, or $p \equiv p(w)$ in (6.16), with $p'(w) > 0$. We will utilize this assumption now when we examine how the inspector's wage affects illegal logging.

The first-order condition of the harvester's choice of volume harvested is from (6.16)

$$E\pi_Q = q - c'(Q) - p(w)fq[1 - \beta(1 - m)] = 0. \tag{6.17}$$

The second-order condition holds, as $c''(Q) < 0$. Comparing this with the first-order condition in the absence of corruption, there is an extra term in (6.17) equal to $p(w)fq[\beta(1 - m)]$. This represents an additional marginal benefit of avoiding getting caught engaging in bribery if the harvester cheats and is detected by the inspector but the bribe is not detected by the government.

Denote the optimal choice of harvesting satisfying (6.17) as Q^c. As we assumed earlier, if the harvester's cost function is convex, then one can show that the harvesting choice is determined by parameters in the following manner: $Q^c(\underset{-}{w}, \underset{0}{t}, \underset{0}{\bar{Q}}, \underset{+}{q}\underset{-}{f}, \underset{-}{m}, \underset{+}{\beta})$. Recalling that $X = Q - \bar{Q}$, a harvester with greater bargaining power harvests further beyond concession limits, while a higher probability of bribery being detected by the government affects this choice in the opposite way, i.e., there is less cheating. Since the audit probability increases with the wage, $p'(w) > 0$, a higher wage paid to the inspector implies less cheating. Other effects are similar to the noncorruption case.

The concession policy set chosen by the government now includes the inspector's wage (which, recall, uniquely determines the inspector's detection probability), the royalty level, and the size of the concession. The social welfare function for the problem under corruption is $SW = E\pi^* + EK^* + v(\bar{Q}, X)$, with the harvester's choice of Q^c defining $E\pi^*$ from (6.16). Using the definition of the optimal bribe in (6.16), the inspector's expected income is $EK^* = w + b^N[1 - m(1 + \phi)]$. Expected government revenues can now be defined as a function of the inspector's wage, royalty payments collected, and fines collected:

$$ER = tq\bar{Q} + p(w)fqXm + b^N m[1 + p(w)]\phi - w. \tag{6.18}$$

The first term is the royalty collected from the concession contract, while the remaining terms measure expected collections from fines for bribery and illegal harvesting when they are detected. It is interesting that in terms of revenue generation, the existence of bribery and corruption might net the government greater revenues if it can detect it, leading to a potentially perverse situation that we will not explore here.

Facing a revenue target of $\bar{R}$, the government's policy choice problem is to choose instruments that will maximize the following Lagrangian function: $\Omega^c = E\pi^* + EK^* + v(\bar{Q}, X) + \lambda[ER - \bar{R}]$. The first-order condition for the royalty rate is now

$$\Omega_t^c = -q\bar{Q} + \lambda q\bar{Q} = 0. \tag{6.19a}$$

As with no corruption, we again find that $\lambda = 1$ at the optimal royalty rate because it is not distortional. However, the other instrument choices diverge from the no-corruption case. Using $\lambda = 1$, the necessary conditions for the optimal concession size $\bar{Q}$ are after some rearranging,

$$\Omega_{\bar{Q}}^c = [v_{\bar{Q}}(.) - v_X(.)] + b_{\bar{Q}}^N[1 - p(w)](1 - m) = \Omega_{\bar{Q}}$$

$$+ b_{\bar{Q}}^N[1 - p(w)](1 - m) = 0, \tag{6.19b}$$

where $b_{\bar{Q}}^N = -\dfrac{(1 - \beta)fq(1 - m)}{[1 - m(1 - \phi)]} < 0.$

Since $p(w)$ and m are both less than one, the second term in each equation is nonpositive. $\Omega_{\bar{Q}}$ is the government's first-order condition under no corruption (6.7a). In the first RHS expression, the first term in parentheses is the effect of the choice of concession size on amenities, which is the same as in the no-corruption case. The second (new) term is an additional negative marginal change in government revenues from bribery that undermines detection and collection of the penalty. We therefore find, even under noncollusive corruption, that the concession size is smaller under corruption, with the divergence being larger when the bargaining power of the inspector is higher, the inspector's wage is lower, or the timber price is higher.

Finally, the wage of the inspector is set to balance the marginal benefits and costs of enforcement, but now the benefits include reductions in bribes:

$$\Omega_w^c = v_X(.)X_p p'(w) + b^N[p'(w)(m - 1)] + b_w^N[1 - p(w)](1 - m), \tag{6.19c}$$

where $b_w^N = -\beta/[1 - m(1 + \phi)] < 0$. This is a more complicated condition than under no corruption (6.7b) and includes additional terms that are due to changes in bribery expected from the government's choice of w. The third term is a reduction in inspector income that is due to a decrease in the bribe at the margin (a marginal cost), while the second term is a change in the optimal bribe that represents a marginal benefit. The first two amenity terms are identical to the condition before and represent the marginal benefit through greater detection and lower cheating that comes from increasing the inspector's wage. The precise choice of w and the resulting probability of auditing depend on the relative probabilities of detection as well as bargaining power of the inspector. One can expect higher enforcement costs under corruption than under no corruption. We leave it to the simulation in box 6.1 to show the precise differences between the corruption and no-corruption cases. We collect our findings in proposition 6.2.

Proposition 6.2 Optimal concession policy with corruption Corruption between risk-neutral harvesters and inspectors leads to an optimal concession policy design that differs from the corruption case in that there is pressure to reduce the concession size to a level smaller than it waved be in the absence of bribery, and enforcement costs of the government are higher.

The presence of bribery clearly requires modification of any optimal concession policy regime in important ways, offering an interesting extension to the ideas we discussed in chapter 5. Simply expanding the size of the concession can no longer be used efficiently to reduce illegal logging, as it could in the noncorruption case. Furthermore, the size of the auditing probability, chosen indirectly through the wages paid to inspectors, amends the earlier results and generally will not be equal, as box 6.1 certainly hints. It is interesting that the lump-sum nature of the royalty we examine here again means that the government can raise revenue to support its enforcement costs, even with bribery. Thus in the corruption case, royalties must still be set high enough to raise revenues. While this may appear to be an unexpected result, it is fully consistent with calls in the literature outside of corruption to raise royalties of the form we have discussed in this chapter.

6.4 Competing Land Uses and Deforestation

Insecure property rights may have more profound effects on deforestation than illegal harvesting in concessions because they matter for all land uses in developing, and especially tropical, countries. Indeed, the large body of empirical research in tropical fringe areas suggests that population growth (through migration), infrastructure changes, and insecure property rights for land ownership are the most critical factors associated with forest decline and agricultural expansion worldwide (Wibowo and Byron 1999, Barbier and Burgess 2001a). Bohn and Deacon (2000) find in a cross-country study that insecure property rights are a highly significant predictor of deforestation. However, when explaining more specific features of population growth and insecure property rights, empirical work has

Box 6.1
An Example of Illegal Logging and Corruption

This box presents a numerical analysis of concession design drawing on a parametric model found in Amacher et al (2008), calibrated for forest concessions in the Amazon. From the text, an important calibration assumption is for the cost function of the harvester. This is assumed to equal $c(Q) = 0.34(5 - 07 * Q + 0.0045 * Q^2)$, where Q is cubic meters per hectare of wood harvested and processed. Given an estimate of roughly 40 cubic meters per hectare, a 400 hectare concession would contain about 16,000 total cubic meters of wood. Let the timber price per cubic meter equal $q = \$43$. Define the lost public goods from illegal logging beyond the concession boundary equal $v(\bar{Q}, X) = -0.04(0.006\bar{Q}^2 + 0.03(Q - \bar{Q})^2)$. This follows the theory in the text, as lost public goods are increasing in the concession size and in illegal logging. In the function here, it is assumed that illegal logging yields higher damage to public goods at the margin than does the government designating a larger concession.

Let the probability of inspector detection of illegal logging equal the following function of the wage, $p(w) = 0.0008 * w$, which gives a detection rate of 0.8% for $w = \$100$. Government enforcement costs are assumed given by $c(w) = 150 + (0.175 + 0.00345w)w$. Note that this equation is applied differently in the corruption and noncorruption cases. In the corruption case, the wage choice determines both enforcement costs and the probability of government detection at the margin. However, for the corruption case, because the logging inspector is dishonest, the wage has no impact on the detection probability for illegal logging by the government. The probability that the government detects bribery by the inspector and the harvester, and the costs associated with this detection, are exogenous. Other assumptions are that the harvester's bargaining power is 40% ($\beta = 0.40$), so that the inspector has more (60%) of the power. The exogenous fine value f and the fine for bribery ϕ are set equal to $30 and $3 respectively. Finally, a binding tax revenue requirement (net of enforcement costs) is assumed to be $120,000; this represents the revenue that the government must maintain through its concession design choices.

The benchmark case is to consider harvester behavior in the absence of any enforcement policy. Solving the harvester's expected income maximization problem under no enforcement results in the harvester removing roughly 13,200 cubic meters of wood from the concession and receiving $109,000 in net stumpage revenues.

The results for solving the socially optimal policy instrument design in the no corruption and corruption cases are presented in table 6.1. Reported in the table are the optimal instruments, total enforcement costs, and the volume of timber actually harvested under the optimal instruments. We also show the impact of the government applying the socially optimal enforcement policy to harvested volume; this is defined as the difference between wood actually harvested under zero enforcement and wood harvested under the optimal policy instrument solution.

In line with the theory, the results in table 6.1 underscore the need to consider concession instruments as a system and not individually like most of the literature's focus on royalty rate choices. This shows up clearly when comparing the cases of corruption and non corruption. Referring to the table, the presence of corruption

Box 6.1
(continued)

Table 6.1
Socially Optimal Concession Design under Corruption and No Corruption

Optimal instruments	No corruption	Corruption
Concession size	9,758 m^3	7,017 m^3
Royalty rate	27%	36%
Inspector wage	$80	$173
Detection probability (rate)	0.07	0.14
Enforcement costs	$186	$284
Actual harvested volume	11,710 m^3	10,772 m^3
Impact of enforcement on the actually harvested volume	−1,483 m^3	−2,473 m^3
Harvester profits	$148,700	$119,600
Bribe	—	$15,700

ultimately means smaller concessions but much higher area royalty rates, along with increased enforcement costs due to higher wages that must be paid to inspectors to reduce their incentives for taking bribes. In fact, the government needs to charge a relatively high royalty to finance not only its budget requirement but also to pay higher wages to inspectors in order to strengthen enforcement at the margin. The government's policy choices in the corruption case have high effects at the margin, leading to significantly reduced illegal logging (−2473 m^3) when compared to the noncorruption case.

The auditing effort employed by the government is much higher (double) under corruption. This is because there are now two sources of illegal behavior for this case (illegal logging and bribery). Furthermore, although not presented in the table, Amacher et al. (2008) show that illegal logging has critical impacts on public goods provision from native forests. In the absence of corruption, amenity benefit losses due to illegal logging are $14,174, while amenity losses due to the concession size itself are $22,857. If illegal logging were zero, then this latter number gives the loss that society would bear in order to obtain rents from harvesting native tropical forests in this example. It is important to note how these respective figures differ in the presence of corruption. The losses in amenities due to illegal logging are larger under corruption ($16,478), because as we know from the model that the harvester has increased opportunities to cheat through bribes. Interestingly, though, the losses from designating the concession of a certain size are lower at $11,818. This results from the smaller concession that is chosen in the absence of corruption.

In the case of no corruption, the harvester will choose to remove 11,710 cubic meters of volume, which amounts to cheating of about 20% of the optimal concession size of 9800 cubic meters. The relative amount of cheating is much higher under corruption, because of the added potential for the harvester to bribe an inspector and avoid paying fines. Here, for example, the harvester removes 10,772 cubic meters, which implies he removes more than 40% beyond what is allowed in the concession.

Box 6.1
(continued)

We can draw two conclusions from the results in this box. First, clearly, corruption leads to greater illegal logging and should be an important focus of policy making regardless of whether the government thinks it can be completely controlled or not. Second, the solution to concession design and capturing rent for the government is not simply to raise royalty rates blindly. In fact, as corruption is controlled, larger concessions can be chosen and lower royalty rates become optimal. In many countries with very large concessions, such as the Congo and recently Brazil, the potential presence of corruption makes these allocations problematic.

been inconclusive. On the other hand, Barbier and Burgess (2001b), Shively (2002), and Shively and Pagiola (2004) all find evidence that higher wages can slow deforestation. None of these studies considered costly site protection paid by landholders that is made necessary by migration and pressure from the resulting population growth.

Using these empirical findings as motivation, Amacher et al. (2008) develop an analytical land-use framework to examine how migration pressure, insecure property rights, and costly public and private enforcement jointly affect incentives for and the profitability of important land uses, and ultimately deforestation, in tropical frontier forest areas. Three land-use classes are considered in this section: agriculture, plantation forestry, and public native fringe forest or idle forested areas. Two concepts of deforestation are used, one defined as a change in any type of forest cover and the other defined as a change in only native forest cover.

Some land-use models have been developed to study deforestation. At the heart of these models is an assumption that net returns, or rents, to each land use vary across land quality or distance from markets in a classic von Thunen sense. In the simplest form, distance increases transportation costs and reduces access to the resource. This decreases rents from any productive land use, making some frontier native forests inaccessible and unprofitable to log. Parks et al. (1998) and Barbier and Burgess (1997) specify land-use margins between agriculture and plantation forest uses. Parks et al. (1998) and Hardie and Parks (1997) further develop a framework combining land quality and a von Thunen distance approach (for an application of a pure von Thunen model, see Chomitz and Gray 1996). Barbier and Burgess (1997) use a more simple cost–benefit ratio to define land use between agriculture and forestry. Recently, Alix-Garcia (2007) developed a rent-based model of native forest land clearing for agriculture and pasture uses in Mexico where rents can vary over both land quality and distance to markets, the latter of which affects transactions costs of production. Deforestation in this model depends on the proportion of a given land parcel cleared, and this is driven by demand changes over time.

This section builds upon all of this work and also links to another line of literature dealing with the effects of insecure property rights on deforestation. This literature has sorted through many issues relevant here, covering risks of land expropriation and illegal logging, and some authors look at the role of migration in enhancing these risks. In most cases, native tropical forests are considered as a sole land-use form and other land uses are not modeled. Expropriation risk in agriculture and forestry have been discussed separately in many articles, the most important of which for our purposes are Armsberg (1998), Alston et al. (2000), Contreras-Hermosilla (2000), Blaser and Douglas (2000), and van Kooten et al. (1999). Alston et al. (2000) demonstrate that insecure property rights and expropriation risk provide incentives for both migrant settlers and private landowners to accelerate deforestation. They discuss examples where migrants (squatters) are granted legal rights by the government to occupy designated open-access land areas (public or private) as long as they convert the land to a "beneficial" use. Land not deemed to be in a beneficial use is at risk of being expropriated by the government. Alston et al. argue that the only way private landowners can avoid expropriation is to keep the land cleared of forest for at least 5 years.

Concerning the Brazilian Amazon, Contreras-Hermosilla (2002) provides examples where establishing forest plantations and clearing forests are two responses to this risk. Furthermore, Blaser and Douglas (2000) argue that expropriation of forests will be a critical impediment to future management for tropical forests. Wibowo and Byron (1999) point out that there are "probabilities" that land users will be evicted by the government or other land users if they establish long-term forest investments such as plantations, and they point out that this is an important feature of insecure property rights in tropical countries that retards the establishment of forest plantations. Given all of this evidence and discussion, it is therefore not difficult to imagine that both native forests and an increasing number of forest plantations will come under expropriation pressures as population growth continues in fringe areas.

Costly private enforcement is discussed in many papers. Hotte (2005) and Clarke et al. (1993) examine agricultural land users' costly private enforcement efforts, such as building fences or obtaining formal title. Hotte (2005) goes further to allow for a case where the landowner may either devote more efforts to protect his property, or purposely overexploit it to reduce returns that could be lost from illegal activities. Miceli et al. (2002) also analyze the relative merits of two title systems in resolving conflicting claims on land.

Illegal logging and expropriation risk for both planted and native forest land uses have been studied in a forest rotation framework by Mendelsohn (1994),

Barbier and Burgess (2001b), and Zhang (2001), and in the context of productive forest investments by Bohn and Deacon (2000). Mendelsohn (1994) was the first to examine a problem in which forests are subject to a constant confiscation risk unless the owner expands costly efforts to defend against this risk. Barbier and Burgess (2001b) develop Mendelsohn's analysis further to show that under a high level of tenure insecurity, it may not be worthwhile to invest in long-term productive uses of land where efforts of landowners to reduce confiscation risk are a lump-sum payment that eliminates the risk completely. Zhang (2001) introduced a known probability of losing plantation forest to expropriation by the government, examining the impact of this on rotation age but not land use, while Bohn and Deacon (2000) show how risk of capital confiscation, through expropriation or illegal logging, reduces investment in forest plantations.

The push of migration has been linked to deforestation in several empirical studies (see e.g., Barbier and Burgess 2001a, Angelson and Kaimowitz 1999, Shively 2002, and Shively and Pagiola 2004). Higher population growth in fringe areas certainly increases the possibility of illegal logging, owing to the number of people, but also from expropriation as the government faces pressure to create formal settlements and develop the frontier. Besides these direct push effects, migration brings more available labor and possibly lower wages in frontier areas. This lowers the transaction costs of finding and employing laborers, tending to improve the profitability of crop production and promoting clearing and deforestation of land. Barbier and Burgess (2001b) confirm this empirically, finding that higher wages tend to reduce deforestation by making other forms of land use profitable.

Finally, native forests are open-access lands, meaning that illegal logging is ever present. As a landowner, the government can devote resources to monitoring and punishing illegal loggers, although the probabilities of being fined and caught are both low in developing countries (Gray 2000).

6.4.1   Insecure Property Rights

Suppose that landowners face three types of risks: expropriation, illegal logging on private and public lands, and unapproved harvesting on public lands; all are violations of existing law. Expropriation risk is defined as the possibility that the government will take part of the private land base from agriculture or plantation forestry for its own use or for future settlement by migrants. Illegal logging risk in plantation forestry is defined as the possibility that forest rents from private plantation forests will be taken by a sudden illegal trespasser who harvests the trees. Finally, we use the notion of illegal logging risk in native fringe forest areas to describe unapproved harvesting by trespassers in publicly owned native forests within the fringe area.

Native forests typically have dual benefits: they provide nontimber benefits to local and indigenous people and revenue when harvested. Plantation forestry is present in many tropical countries and, as noted earlier, is increasing in importance.[4] When one considers abandoned cleared land, plantation forestry provides a potential source of sustainable timber supply, and this could prove useful in sustaining native forest benefits and reducing deforestation.

The nature of the model is illustrated in figure 6.1. The approach is to link migration pressure and insecure property rights to land use on a continuum of land quality; this ultimately determines how much land is subject to deforestation through clearing and illegal harvesting. Migration pressure and insecure property rights affect the profitability of agricultural production, plantation forestry, and the use of native public forests, and this affects input choices and ultimately rents for land uses on the continuum. Land is allocated among these land-use forms according to their relative profitability in a recursive land use model already familiar from chapters 2 and 3.

Three models in the literature serve as our starting point. Agricultural production (6.20a) with land conversion is described by a conventional and well-known agricultural input choice model. Plantation forestry (6.20b) is described by a rotation model under risk of loss by trespass (Reed 1984), which we will take up in more detail in chapter 10. Finally, illegal logging in native tropical fringe areas (6.20c) is similar to our concession-based model in the previous section. In the presence of heterogeneous land qualities, the per-parcel present-value returns for agriculture, plantation timber production, and illegal logging on native fringe forests are given by the following three "rent" functions, respectively:

$$\pi = \int_0^\infty [p_a f(l;q) - wl]e^{-rt}\, dt - \psi, \tag{6.20a}$$

$$J = (r+\lambda)e^{-(r+\lambda)T}[p_f F(T;q) - \bar{c}][r(1 - e^{-(r+\lambda)T})]^{-1}, \tag{6.20b}$$

and

---

4. The ITTO (2005) report also presents tables showing that there are nearly 45 million hectares of rotation-based plantation forests in tropical countries, which include Asia and the Pacific (38 million hectares), Latin America (5.6 million hectares), and Africa (825,000 hectares). Tropical plantation forestry is managed to different extents in many Asian, African, and South American countries. Whereas many plantation forests are publicly owned in Africa and Brazil, one can find private plantations and programs promoting them in India, Bangladesh, Indonesia, and Brazil (Albers et al. 1996, Hyde et al. 1996). Furthermore, plantation forestry has also been cited as a potentially important source of timber products and other services, and it has been argued as a means for developing secondary forestry on previously grazed and abandoned lands (FAO, 2005). Malaysia even provides an incentive for establishing forest plantations to relieve pressure on natural forests.

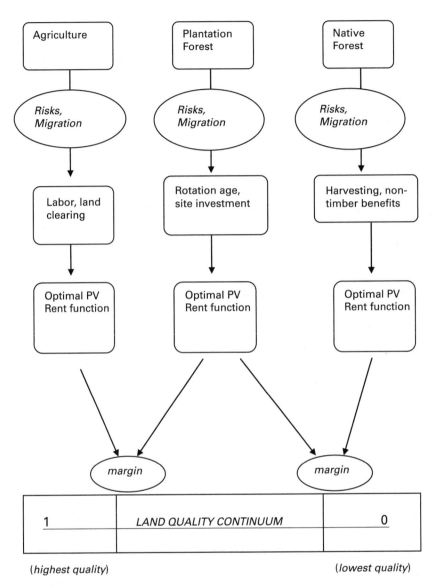

Figure 6.1
Allocation of private land as a function of the market.

$$V = \Omega + (1 - \rho\tau)P_f Q - \phi(q)c(Q), \tag{6.20c}$$

where q is the quality of the land parcel, p_a and p_f are the crop and timber prices net of harvesting costs, r is the interest rate, ψ is a constant cost of land conversion (such as clearing and burning), $\bar{c}$ is forest planting and site preparation costs, and λ is the probability of a catastrophic loss for the forest plantation owner (see chapter 10), which is interpreted here as illegal logging risk (i.e., timber trespass). The agricultural production function is denoted $f(l;q)$, while $F(T;q)$ is the plantation forest yield at rotation age T. The term Ω in (6.20c) is harvest revenue received from concession sales in native forests. Because there can be illegal logging in concessions, Q, ρ, and τ denote, respectively, the volume of timber illegally logged, the probability that illegal logging will be detected and fined, and the penalty for illegal logging in native public forests following our earlier discussion. Like in chapters 2 and 3, rents increase with improved land quality, $f_q > 0$, $F_q > 0$ and $\phi_q c(Q) < 0$.

We begin by adding migration pressure to the frontier by having population n be a factor in all return functions. There are two features of migration that are noteworthy. First, migration affects the cost structure of production within each land use by changing the availability of labor and second, migration serves as a "push" factor for deforestation by increasing the likelihood of agricultural production-related land clearing and trespass-related harvesting in plantation forests and native fringe forests. The first feature of migration is introduced into (6.20a)–(6.20c) through the following assumptions:

$$l = l(n), \quad \text{with } l'(n) > 0 \quad \text{and} \quad \psi = \psi(n) \quad \text{with } \psi' < 0, \tag{6.21a}$$

and

$$\bar{c} = c(n) \quad \text{with } c'(n) < 0 \quad \text{and} \quad C = \phi(q, n)c(Q) \quad \text{with } \phi_n c(Q) < 0. \tag{6.21b}$$

These show that migration reduces the costs of land conversion, planting, and logging by increasing labor supply. The fact that $l'(n) > 0$ reflects an assumption that migration improves input efficiency in agriculture (i.e., land clearing) because agricultural production in tropical fringe areas is typically hindered by insufficient input supplies and credit rationing, and these are somewhat alleviated by migration.[5]

The second feature of migration is introduced by recognizing that migration increases the risks to property rights in all land-use forms. These risks exist in

---

5. Many studies suggest that lack of labor is a central limiting factor in agricultural production, but underdeveloped money and credit systems are also important (Angelson 1999, Kaimowitz 1996). Migration combined with developing markets gradually helps to mitigate these problems.

the two earlier-defined forms: as a risk of land expropriation and a risk of illegal logging in plantation and native forests. Landowners can protect their lands by employing costly private enforcement. Examples include building fences when converting land to agriculture and applying for and obtaining formal ownership title. In forest plantations, site investment includes planting, monitoring, and protection costs. Therefore denote site investment cost as

$$c(n) = \bar{c}(n) + \int_0^T c(n)e^{-rx}\,dx.$$

Let α define the probability that a percent share δ of agricultural land area is expropriated by the government. For agricultural and plantation forestry use, we then modify (6.20a) and (6.20b) through the following assumptions about risk:

$$\alpha = \alpha[n, \theta(n)] \quad \text{with } \alpha_n > 0, \text{ and } \alpha_\theta < 0;\ \theta = \psi, c(n) \tag{6.22a}$$

$$\lambda = \lambda[n, c(n)], \quad \text{with } \lambda_c < 0 \text{ and } \lambda_n > 0. \tag{6.22b}$$

Differentiating (6.22a) with respect to migration yields $d\alpha/dn = (\alpha_n + \alpha_\theta\theta')\,dn > 0$, so that higher migration increases the expropriation risk directly through the push factor (first term) and indirectly through labor availability by decreasing land-clearing and site investment costs (second term). Owing to lower costs, protection efforts can be increased while keeping outlays at their previous level. This increased enforcement effect may be a dominant factor during early stages of migration, but eventually the direct risk effect determines ex post migration risks. This interpretation will have important implications for deforestation in what follows.

A second extension is to include illegal logging in native forests; this affects sustainable production of nontimber benefits. Assume that public enforcement is exogenously given in native forest areas and that these areas exhibit two overlapping land uses: sustainable production of nontimber benefits (B^*) and exploitation of forest stocks through illegal logging. Illegal logging decreases the flow of nontimber benefits to local people, who have no means of safeguarding against the logging. Depending on the actual tree species in a given site, illegal logging in native forests occurs in two forms: either by selective logging (also known as "high-grading") of only the most valuable trees, or through excessive harvesting of timber volume beyond the boundaries of the concession. In what follows we use ω to denote the share of high-valued species in a native forestland unit, P_f to denote the price of the most valuable trees, and a small p_f to refer to the price of the least valuable trees.

Given the assumptions of insecure property rights, migration pressure, and high-grading, the objective functions for present value return per parcel in (6.20a)–(6.20c) must be re-expressed as follows:

$$\pi = \int_0^\infty [1 - \delta\alpha(n, \psi)][p_a f(l; q) - wl]e^{-rt}\, dt - \psi, \tag{6.23a}$$

$$\hat{J} = [1 - \delta\alpha(n, c)]J, \tag{6.23b}$$

and

$$V = \Omega + (1 - \rho\tau)[P_f \omega Q + p_f(1 - \omega)Q] - c(Q). \tag{6.23c}$$

Equations (6.23a) and (6.23c) present the new models we will work with in what follows. The optimal choice of the agricultural input, rotation age, and illegally logged volume defines the indirect net present value rent functions that determine the pattern of land allocation in our model (see figure 6.1).

Previous literature has already established some results for how input choices depend on parameters. For agricultural labor, rotation age, and illegal logging choices, the literature cited earlier collectively finds the following results: $l^* = l(p_a, w, q,)$, $T = T(p_f, \lambda, \bar{c}, r)$ and $Q = Q(p_f \rho, \tau)$. Proposition 6.3 provides a new insight concerning the effects of migration and insecure property rights on these choices.

Proposition 6.3 Expropriation and illegal logging risk   When insecure property rights exist in the form of land expropriation and illegal logging and landowners are risk neutral, then:

(a) For agriculture, expropriation risk and private enforcement do not affect the optimal choice of inputs, but increased migration increases input intensity.

(b) For plantation forestry, expropriation risk has no effect on the optimal rotation age; an increase in migration decreases rotation age, but increased private enforcement increases it.

(c) For native fringe forests, public enforcement via detection and penalties reduces the amount of illegally logged volume but increases high-grading, while migration increases illegally logged volume.

Proof   To prove parts (a) and (b), differentiate (6.23a)–(6.23b) to obtain $\pi_l = [1 - \delta\alpha(n, \psi)](p_a f_l - w) = 0$ and $\hat{J}_T = (1 - \delta\alpha)J_T = 0$; thus both conditions are independent of expropriation risks at the margin. Migration effects in parts (a)–(c) are obtained by conventional comparative statics. To prove part (c) of proposition 6.3, define the expected indirect revenue for illegal logging beyond the concession through "pure" high-grading and through harvesting all kinds of trees as $V^{h*} = \Omega + (1 - \rho\tau)P_f\omega Q - c(\omega Q)$ and $V^* = \Omega + (1 - \rho\tau)[P_f\omega Q^* + p_f(1 - \omega)Q^*] - C(Q^*)$, respectively. Taking the difference yields $V^* - V^{h*} = p_f(1 - \omega)Q^*(1 - \rho\tau) - C(Q^*) + C(\omega Q) \geq (<) 0$. Thus the higher ρ and τ are, the smaller is the advantage from harvesting all trees.

Expropriation risk does not affect the optimal intensity of input use in agriculture and forestry (i.e., for crop labor and rotation age decisions). The reason for this is that costly private enforcement efforts, such as obtaining formal title, are activities that are independent of input intensities. However, (6.23a) and (6.23b) show that land values in these uses will clearly be reduced by costly enforcement, and expropriation risk will clearly lead to a change in land use. Also, the finding that a higher penalty and detection probability increase high-grading of only the best species is interesting. Restricting illegal logging of highly valuable trees clearly requires imposing greater penalties specifically for high-grading.

6.4.2   Land Allocation by the Private Market

Denote by G the entire set of potentially usable forest land outside of concessions. Within G, land quality differs according to site factors. All land can be divided into separate parcels having a uniform quality. Following Lichtenberg (1989), one can rank land quality by the scalar measure q, with a scale chosen so that minimal land quality is zero and maximal land quality is equal to one, i.e., $0 \leq q \leq 1$. $G(q)$ is then the cumulative distribution of q, i.e., the set of parcels having at most a quality level of q. Let $g(q)$ be the density function for $G(q)$, i.e., $g(q) = G'(q)$. The bottom of figure 6.1 describes this continuum of qualities and is useful in helping to follow the discussion here.

The total amount of forest land is given by

$$G = \int_0^1 g(q)\, dq. \tag{6.24}$$

Land area G will be allocated to agriculture and the two forestry uses, or will remain idle. The profits for all land-use activities increase with improved land quality, $\pi_q^* > 0$, $\hat{J}_q^* > 0$ and $V_q^* > 0$, so any type of production is more profitable on better land. We now need another assumption concerning the relative profitability of land uses in terms of land quality. This is needed to ensure unique interior solutions for land uses.

Assumption 6.1   For land qualities and land uses, the following relationships hold:

6.1a. $V^* < \hat{J}^* < \pi^*$ for $q = 1$,

6.1b. $V^* > \hat{J}^* > \pi^*$ for $q = 0$,

and

6.1c. $V^* = B^*$ for $q = \bar{q} \geq 0$.

6.1a and 6.1b define the relative profitability of different land uses. From 6.1a, at the best land quality level, agricultural production is most profitable and plantation forestry is more profitable than illegal logging. Assumption 6.1b indicates that the order is reversed on the lowest-quality land. Assumptions 6.1a and 6.1b, together with $\pi_q^* > 0$, $\hat{J}_q^* > 0$ and $V_q^* > 0$, ensure that agriculture performs best when it is practiced on the highest-quality land and native public forests remain on the lowest-quality land. This allows a partition of our land quality continuum (figure 6.1) that is consistent with what has been observed in tropical countries (Parks et al., 1998). Finally, assumption 6.1c allows the possibility that some land will remain intact as native forests. In this form of land use, nontimber benefits (B^*) are produced in a sustainable fashion; we assume B^* is exogenous. Illegal logging in native forest areas reduces the possibilities of utilizing nontimber benefits. The margin for native forest use is then defined by the equality between marginal net returns in illegal logging activities and nontimber benefits.

Denote the share of land devoted to agriculture and combined public native forests and idle land by h_a and h_{m+i}, respectively. The share of land area devoted to plantation forestry is therefore given by $h_f = 1 - h_{m+i} - h_a$. The share of land area subject to illegal logging, h_m, is defined by a zero-profit condition. The market allocation of land in the economy solves the following problem:

$$\max_{h_a, h_{m+i}} \ PV = \int_0^1 [V^* h_{m+i} + \hat{J}^*(1 - h_{m+i} - h_a) + \pi^* h_a] g(q) \, dq. \tag{6.25}$$

The necessary conditions for interior solutions of h_a and h_{m+i} are

$$\frac{\partial PV}{\partial h_a} = \pi^* - \hat{J}^* = 0, \quad \frac{\partial PV}{\partial h_{m+i}} = V^* - \hat{J}^* = 0 \quad \text{and} \quad V^* = B^*. \tag{6.26}$$

The first condition defines the upper intensive margin of land quality between agriculture and plantation forestry uses, which we denote as q_1. The second condition defines the lower intensive margin between plantation forestry and illegal logging, which is denoted as q_2. Finally, the zero-profit (third) condition defines the extensive margin between illegal logging and idle unexploited land, denoted by $\bar{q}$. The amount of land devoted to agriculture and the two forestry uses is determined by these three margins (see appendix 6.1).

Figure 6.2 shows land allocations, margins between land uses, and present value rent curves as a function of quality (see also the lower part of figure 6.1).

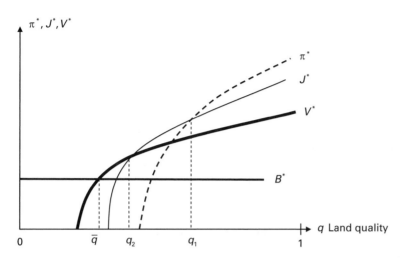

Figure 6.2
Land allocation as a function of land quality and rents to agriculture (π^*), plantation forestry (J^*), illegal logging in native forests (V^*), and nontimber benefits (B^*).

Optimal land use is determined by the intersection of rent lines as given by the first order conditions (6.26) and the zero-profit condition of illegal logging on native forests. The dashed line in the figure is the optimal rent function for agriculture π^*, in that optimal choices for the labor input have been substituted into the rent function; the thin solid line is the corresponding optimal rent function for plantation forestry $\hat{j}^*$; the thick solid line is the optimal rent function for native forest exploitation V^*. The horizontal line is the rent function for non-timber benefits. The first order conditions determine the margins, which in the figure are the points where the optimal rent functions intersect. The distribution function of land quality determines the amount of land devoted to each land use. For example, $G(q_2) - G(\bar{q})$ is the length of the land quality continuum on the X axis between the two margins q_2 and $\bar{q}$ and gives the amount of land between these margins L_m, $G(q_1) - G(q_2)$ is the length of the land quality continuum on the X axis between the two margins q_1 and q_2 and gives the amount of land between these margins L_f, and $G(1) - G(q_1)$ is the length between q_1 and 1 and gives the amount of land here L_a (see appendix 6.1). The shape of figure 6.2 reflects assumption A; that is, the rent curve for agricultural production when graphed over land qualities is everywhere steeper than the rent gradient for plantation forestry, and the rent gradient for plantation forestry use is steeper than the rent gradient for illegal logging.

6.4.3 Land-Use Results

The following two definitions are helpful for our discussion.

Definition 1   Deforestation takes place whenever overall forest cover decreases owing to clearing for agricultural land or illegal logging.

Definition 1 gives equal emphasis to exploitation of native and plantation forests, i.e., it simply concerns any removal of forest cover. By this definition, reforestation reduces deforestation when trees are planted to establish a plantation forest on previously cleared land, or when forest cover appears on cleared land through natural regeneration. This definition has been heavily criticized by environmentalists and nongovernmental organizations for giving too much emphasis to plantation forests, because it implies that clearing native forests and establishing an equivalent area in forest plantations are perfect substitutes and do not constitute deforestation per se. Of course, it is evident that the biodiversity provided by plantation forests differs critically from that provided by native forests in terms of all forms of nontimber benefits and ecosystem services. We therefore also apply the following stricter definition of deforestation.

Definition 2   Deforestation takes place when native forest land is cleared either for agricultural land, plantation forests, or by illegal logging.

Clearly, definition 2 ignores the role to plantation forestry to forest cover. In fact, whenever native forest land is cleared for plantation establishment, it would amount to deforestation. Also, by this definition restoration of deforested land can only take place when natural regeneration or nature-mimicking regeneration is used in a way that eventually leads to the original native old-growth forest containing a large variety of tree species and nontimber benefits.

Migration Pressure   The variables L_a, L_f, L_m, and L_B refer to the amount of land allocated to agriculture, plantation forestry, native forests in illegal logging, and native forests producing nontimber benefits, respectively. The details of all derivations here are given in appendix 6.1.

Migration pressure affects the profitability of all forms of land use by increasing expropriation and illegal logging risks and by decreasing the costs of illegal logging:

$$\frac{dL_a}{dn} > 0, \quad \frac{dL_f}{dn} < 0, \quad \frac{dL_m}{dn} > 0, \quad \frac{dL_B}{dn} < 0. \tag{6.27}$$

As appendix 6.1 shows, migration affects rents through multiple channels associated either with expropriation and illegal logging risks, or reduced costs owing to migration. Risks increase with migration more rapidly than costs decrease.

Hence the effect of migration on rents is plausibly negative. Despite this, the signs of changes in land area between agriculture and plantation forestry are still ambiguous through the term $(\pi_n^* - \hat{J}_n^*)$ because migration reduces the returns in both activities. Thus, referring to figure 6.2, the relative shift of both rent curves for plantation and agricultural use is indeterminate. A plausible assumption that $(\pi_n^* - \hat{J}_n^*) > 0$. This reflects the observation that at the upper intensive margin q_1, increasing migration pressures probably raise expropriation risk more for plantation forestry than for agriculture because expropriation has been noted as a special risk for forests (Blaser and Douglas 2000, p. 11). It also follows the beneficial-use and expropriation issues for forests discussed in Alston et al. (2000) and Armsberg (1998). This effect is further reinforced by an increased risk of illegal logging for plantation forestry that follows from population growth. At the lower intensive margin q_2, the effect of migration pressure on plantation forestry and illegal logging is clear. Illegal logging benefits from reduced logging costs as a result of migration, and thus the lower intensive margin shifts to the right in figure 6.2 in the direction of higher land qualities, while the extensive margin shifts to the left in the direction of lower land qualities that become profitable to exploit. Land area under both plantation forestry and native forests decreases with migration pressure.

One can conclude that migration promotes deforestation according to definitions 1 and 2 through two channels: by expanding agriculture and increasing illegal logging activity. This is not the entire story though. Recall our discussion of the two opposing effects that migration has on expropriation and illegal logging risks. Migration directly increases these risks through a larger population, but the direct risk effect is offset somewhat because private enforcement is made less costly by greater availability of labor and decreasing labor costs. From equations (6.21a) and (6.21b), we employed the assumption that ultimately the direct risk effect dominates the private enforcement cost effect as migration pressure increases. However, it is conceivable that in early stages of migration the indirect protection cost effect could dominate the risk effect. In this case, referring to figure 6.2, the ultimate location of the upper and lower intensive margin may change as migration progresses. These results provide theoretical insight into specific empirical findings discussed in a recent survey article by Barbier and Burgess (2001b). Using various model specifications, they found an ambiguous effect of population growth on agricultural land expansion. This ambiguity is noted also in other empirical deforestation studies (see e.g., Contraras-Hermosilla 2002).

Expropriation and Illegal Logging Risks   The effects of expropriation and illegal logging risks on land area are given by

$$\frac{dL_a}{d\alpha} < 0, \quad \frac{dL_f}{d\alpha} = ?, \quad \frac{dL_m}{d\alpha} > 0, \quad \frac{dL_B}{d\alpha} = 0 \tag{6.28a}$$

$$\frac{dL_a}{d\lambda} > 0 \quad \frac{dL_f}{d\lambda} < 0 \quad \frac{dL_m}{d\lambda} > 0, \quad \frac{dL_B}{d\lambda} = 0. \tag{6.28b}$$

We start our interpretation with expropriation risk in (6.28a). Referring to appendix 6.1, the effect of a higher expropriation risk on the upper intensive margin q_1 is generally ambiguous and depends on the sign of the term $(\pi_\alpha^* - \hat{J}_\alpha^*)$. However, given that agriculture takes place on better-quality lands, it is plausible to assume that the expropriation risk effect has a larger effect on expected returns to agriculture than to plantation forestry (this would also be true if expropriation risk for forest plantations was zero). In this case, expropriation risk shifts the upper intensive margin to the right in figure 6.2, and less land is devoted to agriculture as plantation forestry expands into higher-quality land. Despite this, rents in plantation forestry are reduced, so that illegal logging becomes more profitable on lower-quality lands that are devoted to plantation forestry. This serves to expand the lower intensive margin, q_2, to the right. The overall effect on land allocated to plantation forestry is ambiguous.

The illegal logging risk in equation (6.28b) has clearer effects since it reduces the present value rents from forest plantations. In this case, illegal logging in native forests expands as the extensive margin $\bar{q}$ moves to the left in figure 6.2. Illegal logging risk would also imply that the upper intensive margin q_1 moves to the left and agricultural use expands to lower-quality lands currently in plantation forestry.

When expropriation risk alone is considered, and when it is relatively more important in reducing incentives for agriculture, the total effect on deforestation depends, as evidenced by our earlier discussion of margins, on the sign of $dL_f/d\alpha + dL_m/d\alpha$. Applying definition 2 then implies that there is no change in deforestation because land in native forests is unchanged, $dL_B/d\alpha = 0$.

This is not the case if illegal logging risk increases. Then deforestation always occurs, based on definition 1. Such a result is evidence of a double mechanism of deforestation and provides new theoretical support for calls in the applied policy literature to implement subsidies and reduce the costs of sustainable forest practices, such as reduced-impact logging and the establishment of plantations (Winkler 1997, Barreto et al. 1998). It is interesting that by definition 2 we would conclude that no deforestation takes place because of expropriation risk.

This result is revealing for the deforestation debate in many ways. It places empirical studies in context, showing expropriation and illegal logging risks as important variables explaining deforestation (Barbier and Burgess 2001a,b), the recognized increasing importance of forest plantations in tropical countries, and

discussions about illegal logging. It also provides specific reasons why the empirical literature suggests that migration pressure may have both increasing and
decreasing effects on deforestation (see Angelson and Kaimowitz 1999).

In most cases, insecure property rights lead to agricultural expansion and
increased illegal logging. Both expropriation and illegal logging risks reduce the
establishment of plantation forests and the stock of native public forests. If plantations are a desirable land component, then a double benefit could be obtained
by increased enforcement in fringe areas to decrease illegal logging, because it
promotes allocation of land to plantation forestry and production of nontimber
benefits on native forest land.

Public and Private Enforcement   The separate effects of public enforcement by
the government and investment-based enforcement by private landowners are
interesting because the latter has the effect of reducing expropriation risks by serving as a signal of beneficial economic use. The former has the effect of a higher
probability of detection and higher penalty rate. The private effects are

$$\frac{dL_a}{d\psi} \lessgtr 0, \quad \frac{dL_f}{d\psi} \gtrless 0, \quad \frac{dL_m}{d\psi} = 0, \quad \frac{dL_B}{d\psi} = 0 \tag{6.29a}$$

$$\frac{dL_a}{dc(n)} \lessgtr 0 \quad \frac{dL_f}{dc(n)} \gtrless 0 \quad \frac{dL_m}{dc(n)} \gtrless 0, \quad \frac{dL_B}{dc(n)} = 0. \tag{6.29b}$$

The signs for agriculture and plantation forestry in (6.29a) and (6.29b) are generally ambiguous, as appendix 6.1 shows. As in the case of migration, both risk
and cost changes affect rents in agriculture and plantation forestry. Higher site investment costs directly decrease the profitability of plantation forestry, but they
also indirectly decrease illegal logging and expropriation risks, which increases
profitability. These two countervailing effects clearly open up many interesting
interpretations not yet considered in the theoretical literature despite the empirical debate that has surrounded site protection.

We can trace out two alternative outcomes concerning site investments that
depend on the degree to which property rights are insecure. The more secure
property rights are, the more likely it is that the direct cost effect will dominate
land-use changes. In other words, higher protection costs for agriculture and
forest plantation uses will decrease the profitability of both land uses, leaving
unclear what happens to the upper intensive margin. Land would shift to the use
with the lower relative protection costs.

By definition 1, deforestation decreases (or increases) when the direct effect of
land-clearing costs (or site investment) dominates. In neither case would we find
deforestation according to definition 2.

Public enforcement parameters are qualitatively identical, and so we report them using a general index, $k = \rho, \tau$:

$$\frac{dL_a}{dk} = 0, \quad \frac{dL_f}{dk} > 0, \quad \frac{dL_m}{dk} < 0, \quad \frac{dL_B}{dk} > 0. \tag{6.30}$$

A higher probability of detection (or a higher enforced fine) will decrease the profitability of illegal logging activities at both the extensive and lower intensive margins, $\bar{q}$ and q_2, in figure 6.2. Plantation forestry then expands to additional lower-quality lands as q_2 moves to the left. Illegal logging becomes less profitable at the extensive margin $\bar{q}$. The land area under illegal logging unambiguously decreases, implying that larger areas of native forests remain intact for production of nontimber goods. Deforestation decreases according to both definitions this time, although high-grading may increase. Public enforcement always increases the area of native forest land even though private enforcement has an ambiguous effect.

This is important for deforestation policy. First, it implies that governments must find ways to enforce restrictions on illegal logging and collect stiff fines for such behavior if the goal is to reduce the area subject to this use, and thus to ultimately reduce deforestation no matter how it is defined. Doing so decreases the profit from illegal logging and reduces deforestation by shifting the extensive margin. Second, migration pressure must be reduced to make private enforcement successful.

These results can also be used to shed light on some puzzles found in empirical work. For example, Bohn and Deacon (2000) obtained a statistically significant effect for investment in site protection in their empirical cross-country deforestation equation, while Barbier and Burgess (2001b) found that it was not statistically significant. Our results provide some context for this by showing that the outcome depends on whether or not the management cost effect dominates the risk-reducing effect.

Timber Price and Value of Nontimber Benefits The development of timber markets is often said to be crucial for reducing deforestation and establishing more secure land-use practices. Consider the effects of increased timber price and increased value of nontimber benefits, denoted by B. Timber price directly affects the profitability of plantation forestry and illegal land-use activities:

$$\frac{dL_a}{dp_f} \gtreqless 0, \quad \frac{dL_f}{dp_f} \gtreqless 0, \quad \frac{dL_m}{dp_f} \gtreqless 0, \quad \frac{dL_B}{dp_f} < 0 \tag{6.31a}$$

$$\frac{dL_a}{dB} = 0, \quad \frac{dL_f}{dB} = 0, \quad \frac{dL_m}{dB} \leq 0, \quad \frac{dL_B}{dB} \geq 0. \tag{6.31b}$$

The timber price effect in (6.31a) is generally ambiguous and depends on which land-use forms have the highest marginal benefits from a price increase. However, a higher timber price does not necessarily promote plantation forestry. While higher timber prices push plantation forestry onto land currently occupied by an agricultural use, illegal logging also becomes more profitable at the lower intensive and extensive margins. Only if plantation forestry rents increase more than rents from illegal logging will the total amount of land in plantation forests increase.

This aspect this has been a point of debate within the empirical literature. Angelson (1999) and Angelson and Kaimowitz (1999) point out that previous models provide conflicting results concerning the effects of tenure insecurity and timber prices, while others have argued that increases in prices will lead to greater incentives to establish sustainable forest plantations (see e.g., Barbier et al. 1995). Our model provides a context for thinking through both sides of this story.

Equation (6.31b) describes some possibilities of reducing illegal logging by increasing the value of nontimber benefits in native forests.[6] The outcome is conditional on how increased B affects the incentives of illegal loggers. Provided illegal loggers are local people, a higher B naturally reduces such logging because they derive higher benefits from unharvested forest. However, if illegal loggers come from outside the fringe area, one cannot reduce illegal logging by increasing B. Hence the identification of illegal logger types is vital for policies aiming at reducing this logging by increasing the value of nontimber benefits. In terms of deforestation, we obtain either decreased deforestation according to both definitions or no change at all when nontimber benefits increase.

Finally, narrowing our analysis to the idea of deforestation only as it applies to the removal of native forests, the following changes to the findings here would be necessary: Provided that the effects of higher migration, or higher risks from expropriation, and illegal logging are strong enough, or plantation forestry is not well established, native forestland area will shrink as these property risks increase.

6.5   Summary

We have examined two mechanisms for deforestation: land-use decisions in fringe forests and corruption in concessions. Illegal activities are a common thread in both mechanisms. Concession design must accommodate bribery. Most concession parameters change the nature of bribes between government officials and illegal loggers, and the government is less able to control concession harvesting through the size of the concession than is possible without corruption and

6. An example would be through development of markets or education programs.

bribery. Bargaining power between the parties here is critical, and harvesters with greater power harvest further beyond concession limits. The only way of controlling these effects is a higher probability of punishment, and this means higher wages for inspectors. Still, higher timber prices and low penalty rates will always lead to greater cheating and bribery.

The presence of bribery modifies a revenue-constrained government's design of concessions. Most of the literature mentioned earlier suggests that royalties are too low. We show indeed that royalties, at least if they are lump sum, should be increased to levels that allow governments to capture rents adequate to cover budget requirements. However, what a government really needs to do to keep corruption in check is to have smaller concessions. As corruption is controlled, concessions can be larger and royalties can actually be decreased due to lower enforcement costs. This is not normally the case in practice, where concessions span thousands of hectares in some tropical countries, and large firms with considerable bargaining power over logging inspectors are the rule.

The land-use mechanism reveals other interesting means for controlling deforestation. Public enforcement of illegal logging effectively reduce deforestation, as expected, but the success of private enforcement depends on the intensity of migration. Higher timber prices also have an ambiguous effect on deforestation because besides improving profits from plantation forestry, they increase incentives for illegal logging. Policies targeting migration pressures are thus one of the key parameters in reducing agricultural land clearing and illegal logging in fringe tropical forest areas. However, the success of private enforcement needs to be weighed when evaluating any migration-based policy.

Plantation forests also raise an important policy issue. Key factors that promote the use of plantation forest land are lower illegal logging and expropriation risks that are jointly present with higher timber prices and lower costs of site investment. If these factors were promoted in a way that did not allow plantation forestry to expand and crowd out native forestland areas, then conditions could be created for sustainable forest management and reforestation of areas abandoned by agriculture or grazing. One topic for further work concerns evaluating how uncertainty in prices and costs affect land uses in our type of model. Albers et al. (1996) have an interesting application of this.

One final comment concerns public enforcement and high-grading of forest stocks. We have shown here that public enforcement plays a critical role in reducing the destruction of native forests. It can only be successful, however, if detection activities (i.e., the probability of detection) are expanded to an adequate level. That said, it will always be difficult to detect and finish high-grading of native forests, and new satellite technologies help less here than in monitoring a larger-scale activity such as illegal clear-cutting. Higher enforced fines are probably the only means a government has to affect high-grading.

Appendix 6.1   Comparative Statics of Land Allocation

The land areas in agriculture and the two forestry uses are defined as follows:

$$L_a = \int_{q_1}^{1} g(q)\, dq = G(1) - G(q_1), \tag{A6.1a}$$

$$L_f = \int_{q_2}^{q_1} g(q)\, dq = G(q_1) - G(q_2) \tag{A6.1b}$$

$$L_m = \int_{\bar{q}}^{q_2} g(q)\, dq = G(q_2) - G(\bar{q}). \tag{A6.1c}$$

Define a parameter vector $\theta = [p_a, p_f, c, r, \lambda, w, \rho, \tau]$. By differentiating (A6.1a)–(A6.1c), we obtain for the changes in the land use

$$\frac{dL_a}{d\theta} = -g(q_1)\frac{\partial q_1}{\partial \theta}, \quad \frac{dL_f}{d\theta} = g(q_1)\frac{\partial q_1}{\partial \theta} - g(q_2)\frac{\partial q_2}{\partial \theta}, \quad \frac{dL_m}{d\theta} = g(q_2)\frac{\partial q_2}{\partial \theta} - g(\bar{q})\frac{\partial \bar{q}}{\partial \theta}. \tag{A6.2}$$

In (A6.2) the change of intensive and extensive margins, $\partial q_1/\partial \theta$, $\partial q_2/\partial \theta$, and $\partial \bar{q}/\partial \theta$, can be defined by differentiating the land allocation conditions in (6.26). Differentiating equation (6.26) with respect to land quality yields $D = (\pi_q^* - \hat{J}_q^*)$, $\Delta = (\hat{J}_q^* - V_q^*)$, which are both positive because of assumption 6.1 in the text. The comparative statics are

Migration pressure

$$\frac{dL_a}{dn} = \frac{[\pi_n^* - \hat{J}_n^*]}{D} g(q_1) > 0$$

$$\frac{dL_f}{dn} = -\frac{[\pi_n^* - \hat{J}_n^*]}{D} g(q_1) + \frac{[\hat{J}_n^* - V_n^*]}{\Delta} g(q_2) = ?$$

$$\frac{dL_m}{dn} = -\frac{[\hat{J}_n^* - V_n^*]}{\Delta} g(q_2) + g(\bar{q})\frac{V_n^*}{V_q^*} g(\bar{q}) = ?$$

Expropriation area

$$\frac{dL_a}{d\delta} = \frac{[\pi_\delta^* - \hat{J}_\delta]}{D} g(q_1) < 0$$

$$\frac{dL_f}{d\delta} = -\frac{[\pi_\delta^* - \hat{J}_\delta]}{D} g(q_1) + \frac{\hat{J}_\delta}{\Delta} g(q_2) = ?$$

$$\frac{dL_m}{d\delta} = -\frac{\hat{J}_\delta}{\Delta} g(q_2) > 0,$$

where

$$\pi_n^* = -\int_0^\infty \delta(\alpha_n + \alpha_\psi \psi_n)[p_a f(l) - wl]e^{-rt}\, dt - \psi_n = ?,$$

$$J_n^* = -([[(r + \lambda)\delta\alpha_n + (\lambda_n + \lambda_c c_n)\psi][p_f F(T;q) - c] + c_n(r + \lambda))[r(1 - e^{-(r+\lambda)T})]^{-1} = ?$$

$$V_n^* = -\phi_n c(Q) > 0; \quad \pi_\delta^* = -\alpha \int_0^\infty [p_a f(l) - wl]e^{-rt}\, dt < 0; \quad J_\delta^* = -\alpha J < 0.$$

Expropriation risk

$$\frac{dL_a}{d\alpha} = \frac{[\pi_\alpha^* - \varpi_\alpha J]}{D} g(q_1) = ?$$

$$\frac{dL_f}{d\alpha} = -\frac{[\pi_\alpha^* - \hat{J}_\alpha]}{D} g(q_1) + \frac{\hat{J}_\alpha}{\Delta} g(q_2) = ?$$

$$\frac{dL_m}{d\alpha} = -\frac{\hat{J}_\alpha}{\Delta} g(q_2) > 0$$

Illegal logging risk

$$\frac{dL_a}{d\lambda} = -\frac{\hat{J}_\lambda^*}{D} g(q_1) > 0$$

$$\frac{dL_f}{d\lambda} = \frac{\hat{J}_\lambda^*}{D} g(q_1) - \frac{\hat{J}_\lambda^*}{\Delta} g(q_2) < 0$$

$$\frac{dL_m}{d\lambda} = -\frac{\hat{J}_\lambda^*}{\Delta} g(q_2) > 0,$$

where $\pi_\alpha^* = -\delta \int_0^\infty [p_a f(l) - wl]e^{-rt}\, dt < 0;$ $\hat{J}_\alpha^* = -\delta J < 0,$ and $\hat{J}_\lambda^* = -(1 - \delta\alpha) \times [(r + \lambda)T(e^{-(r+\lambda)T}(1 - e^{-(r+\lambda)T})^{-1} - 1][p_f F(T;q) - c][r(1 - e^{-(r+\lambda)T})]^{-1} < 0.$

Land-clearing cost

$$\frac{dL_a}{d\psi} = \frac{\pi_\psi^*}{D} g(q_1) = ?$$

$$\frac{dL_f}{d\psi} = -\frac{\pi_\psi^*}{D} g(q_1) = ?$$

$$\frac{dL_m}{d\psi} = 0$$

Site investment

$$\frac{dL_a}{dc(n)} = -\frac{\hat{J}_c^*}{D} g(q_1) = ?$$

$$\frac{dL_f}{dc(n)} = \frac{\hat{J}_c^*}{D} g(q_1) + \frac{\hat{J}_c^*}{\Delta} g(q_1) = ?$$

$$\frac{dL_m}{dc(n)} = -\frac{\hat{J}_c^*}{\Delta} g(q_1) = ?$$

where $\pi_\psi^* = -\delta\alpha_\psi \int_0^\infty [p_a f(l) - wl]e^{-rt}\, dt - 1 = ?$ and $J_c^* = [(r + \lambda) + (r + \lambda)\delta\alpha_c\lambda_c\xi \times [p_f F(T;q) - c][r(1 - e^{-(r+\lambda)T})]^{-1} = ?,$

Public enforcement

$$\frac{dL_a}{d\tau} = 0 \qquad\qquad \frac{dL_a}{d\rho} = 0$$

$$\frac{dL_f}{d\tau} = -\frac{V_\tau^*}{\Delta} g(q_1) > 0 \qquad \frac{dL_f}{d\rho} = -\frac{V_\rho^*}{\Delta} g(q_1) > 0$$

$$\frac{dL_m}{d\tau} = \frac{V_\tau^*}{\Delta} g(q_1) < 0 \qquad \frac{dL_m}{d\rho} = \frac{V_\rho^*}{\Delta} g(q_2) < 0,$$

where $V_\tau^* < 0$ and $V_\rho^* < 0.$

Timber price

$$\frac{dL_a}{dp_f} = -\frac{\hat{J}^*_{p_f}}{D}g(q_1) < 0$$

$$\frac{dL_f}{dp_f} = \frac{\hat{J}^*_{p_f}}{D}g(q_1) + \frac{[\hat{J}^*_{p_f} - V^*_{p_f}]}{\Delta}g(q_2) = \,?$$

$$\frac{dL_m}{dp_f} = -\frac{[\hat{J}^*_{p_f} - V^*_{p_f}]}{\Delta}g(q_2) + g(\bar{q})\frac{V^*_n}{V^*_q}g(\bar{q}) = \,?$$

Value of nontimber benefits

$$\frac{dL_a}{dB} = 0$$

$$\frac{dL_f}{dB} = 0$$

$$\frac{dL_m}{dB} = -\frac{V^*_B}{V^*_q}g(\bar{q}) < 0,$$

7 Conservation of Biodiversity in Boreal and Temperate Forests

As indicated earlier, the 1992 Convention on Biological Diversity was perhaps most responsible for introducing the concept of biodiversity to the general public. Biological diversity is multifaceted and important to life in all possible forms on Earth. It applies to all living organisms in terrestrial and aquatic ecosystems, as well as organisms living in human-managed environments. To make the concept operative, biologists typically distinguish among diversity of species, diversity of habitats, and genetic diversity with regard to number, composition, and relationships among organisms (Sprangenberg 2007, Armsworth et al. 2004).

These distinctions are interesting from an economics perspective. Biodiversity in any of these forms can be thought of as a public good. From chapter 5 we know that provision of public goods cannot be trusted solely to private markets. The socially optimal level of biodiversity conservation for forest land depends on the social benefits and costs of providing the biodiversity. The aim of policies is to internalize any social costs generated by individual landowners who do not manage specifically for biodiversity. This type of problem will no doubt be an important policy area for forest resources in the decades to come. A clear example arose in chapter 6 with rent-driven deforestation of open-access native tropical forests. These forests contain roughly 80% of all known species on the Earth, and so preserving their native state is crucial in maintaining biodiversity for the planet. It should be obvious that continued deforestation is one of the greatest threats to biodiversity in the world, and thus the policies we presented in chapter 6 to control deforestation in a sense also promote global biodiversity.

Many challenges also lie ahead for conserving biodiversity in boreal and temperate forests. These forests are no longer subject to the deforestation found in developing tropical countries, and in fact temperate and boreal forestland has actually been increasing over time in many developed countries. Still, there are threats to biodiversity here in many forms. Most important, management decisions by individual forest landowners may not be consistent with practices designed to maintain or increase biodiversity.[1] Intensive forest management practices, such as planting single-species stands, improving timber stands, suppressing fires, and harvesting have all replaced the natural disturbance dynamics that have driven forest renewal and helped to maintain the diversity of habitat for centuries.

Work in forest biology and ecology comprises the largest volume of research in boreal-temperate biodiversity. One important aspect of this work has been the design of conservation networks (arrangements of habitat), and the development of cooperative forestry management techniques for creating these networks through collective and cooperative silvicultural management. At the same time, some social-based research has targeted encouraging the adoption of conservation network-based management practices among commercial forest landowners.

From the perspective of forest economics, these two paths of research immediately bring to mind three interesting issues. The first concerns how much ecologically valuable forestland areas should be strictly conserved (i.e., placed in a biodiversity reserve), with limited or no commercial practices allowed. The second concerns how current management practices might be modified to maintain biodiversity in commercial forests. Third, it is of interest to consider how policy instruments can be used to modify landowner behavior. Given that in many countries, including large areas in North America and Scandinavia, forest land is owned by private nonindustrial landowners, the scope for policy instruments is considerable. While all these problems are clearly pressing, it may surprise the reader to learn that rigorous research in these areas is still in its infancy.

This chapter is organized into three parts. We first review the ideas of designing forest habitat networks based on basic forest ecology and popular site selection models. Following the spirit of this book, we then examine two key policy questions in biodiversity conservation. First, we investigate how a green market-based auction approach can be used to promote voluntary participation of private landowners in building biodiversity reserves for forest land. Second, we examine the design of policies that can promote maintenance of biodiversity in regular commercial forests; our special focus here is on green tree retention as a means of increasing the dead and decaying wood needed for threatened old-growth forest habitat. Finally, we discuss some other aspects of biodiversity and forests, including genetic diversity and invasive foreign species.

---

1. Excellent examples are the U.S. spotted and horned owl conservation policies (e.g., see Nalle et al. 2004).

7.1 Conservation Networks

Ecological models of forests generally advocate biodiversity conservation and maintenance at three levels: the stand, the landscape, and the larger forest region (Hunter 1990, 1999). The most popular ecological frameworks used in describing the hierarchical structure of forests on these scales are the corridor-patch-matrix model (Forman 1995) and the landscape continuum model (McIntyre and Hobbs 1999). The corridor-patch-matrix model constructs the forest landscape in three basic forms: habitat patches, corridors (transition areas) connecting them, and surrounding areas that are unsuitable for the species in question. This model draws on the theory of island biogeography (MacArthur and Wilson 1967) but has also found support in metapopulation theory (Hanski 1999). The landscape continuum model views the forest landscape with the help of environmental factors, such as forest cover, which changes smoothly in terms of its structure. This model applies well to forests where habitat patches are not distinct and is particularly relevant for natural forest areas (Wiens 1997).

Empirical research in boreal and temperate forests has revealed the many ways in which modern forestry practices can threaten biodiversity. Most importantly and obviously, clear-cutting of large forest areas and subsequent artificial regeneration of forest has replaced disturbances (such as fire) needed for natural forest renewal (Franklin 1993, Holling 2001, Bergeron et al. 2002, Kuuluvainen 2002). Although these actions improve stand growth and profitability, the resulting even-aged management regime leads to a uniform forest structure. Such a structure cannot sustain the variety of species associated with forests in natural states and is sometimes referred to as a biological desert.

The continuum of decaying and dead wood has decreased dramatically in timber-producing regions as old-growth natural forests have been replaced by even-aged managed plantation forests. Efficient elimination of forest fires has reinforced this impact by keeping forests in the same successional state over large areas. Some empirical studies have found evidence that commercial forestry has decreased the volume of dead wood to a tenth of the amount present in virgin forests. In boreal forests, this represents a critical biodiversity loss because roughly 25% of all species are dependent on dead wood for propagation (Siitonen 2001, Similä et al. 2003).

From the perspective of the forest as a hierarchical and structurally varied landscape, the (even aged) single stand–based management regime cannot lead to a connected network of habitat types (Lindenmayer and Franklin 2002). At the landscape level, the appropriate network must consist of different types of forests and stands. An entirely preserved forest area is the core of any biodiversity conservation network. Around this core should be built a pattern of ecological corridors, stepping stones (transitions), buffer zone forest areas, and commercial forests in which biodiversity conservation is actively taken into account. All parts of the network are linked to each other in order to ensure the interconnection and continuum of forest landscapes over time and space (Franklin and Forman 1987).

If society is interested in transforming forests into biodiversity networks, then there are several important policy implications to be gleaned from the discussion so far. First, greater areas of ecologically valuable forest land should be strictly protected to ensure enough reserves for endangered species. Large unified areas are regarded as the best means for preserving these species. Second, according to metapopulation theory, in order to improve the viability of scattered and threatened populations, the connectiveness of stands at the landscape level must be increased using ecological buffers, corridors, and ecological transition zones.[2] While buffer areas protect ecologically valuable core areas from edge effects, corridors and transition (or stepping stone) zones connect scattered but ecologically valuable core areas and promote both species viability and colonization of new areas.

Finally, as the last and largest part of the conservation network, maintenance of biodiversity in commercial even-aged forests must be strengthened by providing incentives for landowners to adopt uneven-aged forest structures through changes in their decision making. Such structures require preserving small but ecologically highly valuable habitat patches, called "key biotopes," such as ditches, rocky patches, and various types of wetlands. Another important challenge in commercial forests is to artificially create a continuum of decaying and dead wood by leaving tall trees standing at final harvests. The trees left standing are called "retention" trees. When these die and decay, they provide habitats for many kinds of species. Retention trees are the basis of several silvicultural practices, such as shelterwood and seed tree harvesting methods, but in those practices the trees left from the first harvest are selected specifically to provide a seed bank for regeneration and are eventually harvested later. Conservation of biodiversity requires allowing retention trees to die and decay naturally, and trees are selected based on their potential to harbor species.

Site selection models have been used by ecologists and some economists to choose the "best" stands in a conservation network. The best stands are typically chosen according to their representativeness, that is, according to the extent to which they produce key natural features needed for biodiversity. A practical ecological concept in promoting representativeness is complementarity (Faith and

---

2. Swallow et al. (1997) provide an early intuitive analysis of connectivity between ecological core areas. Jacobson considers the use of forest cooperatives in wildlife-sensitive corridors (Jacobson 2002). However, more precise analytical treatments and applications in economics are still missing.

Walker 1996, May 1990, Vane-Wright at al. 1991). This measures the contribution of an area, or a set of areas, to an existing network of reserves in terms of unrepresented natural features (Margules and Pressey 2000). Most often, site selection aims specifically at promoting species representativeness, measured by the number of species included in the reserve network. The complementarity principle argues that new sites should be chosen so that the increase in the number of species in the conservation area network is large enough to compensate for the opportunity costs of reserving the new sites. Species abundance and various spatial constraints can then be included to stress the role of endangered species and interdependence between stands in sustaining species.

Most of the site selection studies are purely ecological in scope. Ando et al. (1998), Balmford et al. (2000), and Polasky et al. (2001) were the first to include a conservation budget in these models in order to undertake an integrated ecological-economic analysis of site selection. While they applied their resulting models to large land areas, Stokland (1997) and Juutinen et al. (2004) focus on forest management and site selection at the stand level, that is, stands are the operative units for site selection in their approaches. Siitonen (2001), and Juutinen et al. (2008) further extend this analysis to include spatial constraints. Nalle et al. (2004) is an example of work where the focus is on the viability of a given species, in a dynamic and spatial setting.

All of these economic applications rely on the assumption that society can simply add any forest site it wishes to conserved land areas in order to satisfy the conservation budget. This may not be feasible. For example, if land is owned by private landowners, then land takings for strict protection can create negative consequences as landowners undertake avoidance behavior; these spillover effects can ultimately reduce biodiversity and thereby prevent land takings (Innes et al. 1998). Zhang (2004) and Mehmood and Zhang (2005) further show that forest harvesting increases in regions where endangered (or red-listed) species are found, because landowners have strong incentives to effectively remove habitat before they must adhere to strict harvesting laws that discovery of these types of species ultimately lead to (see also Lueck and Michael 2003).

The problems inherent in having the government effectively take away landowners' rights to make management decisions about their own land imply that there is potentially great scope for voluntary programs to promote biodiversity. Voluntary conservation measures may lead to higher biodiversity outcomes, and they can be used to complement mandatory rules for forest management. Temporary voluntary conservation can also help in the transition to permanent takings for reserves. It is less expensive than strict protection and ensures better targeting of conservation, given that many red-listed species are those living in a given patch of land only during some phase of forest succession.

7.2   Auctions for Biodiversity Conservation

We now turn to voluntary biodiversity conservation measures for private forest-land. We first outline biodiversity auctions as a conservation instrument and then examine how policy instruments can be used to promote voluntary conservation in commercial forests.

A well-known market-based means suitable for voluntary conservation of forests is a "green auction" mechanism. These are common in agricultural sectors, and examples can be found throughout the world, such as the Conservation Reserve Program in the United States, the Trading and Nature Values Program in Finland, and the Bush Tender Trial and Landscape Recovery program in Australia. In these programs, farmers agree to remove land from crop production for a specified period of time in return for an annual payment received from the government (Latacz-Lohman and Schillizi 2005 provide an overview of many of these types of programs). The Trading in Nature Values Program (TNV) in Finland is a voluntary program specifically designed to provide biodiversity. Forest landowners offer bids (suggested prices) to the government, which then, if it is willing to pay for the land offered, accepts the bid and begins transfers of subsidy payments to the landowners. Other similar programs include forest banks in the United States. There, landowners voluntarily give up their rights to forest management by enrolling their land in a cooperative through an auction system. The bank managers, who are often environmental interest groups, then make forest management decisions to promote biodiversity and other goals, with the enrolled forest landowners receiving a share of annual payments from whatever harvesting is carried out by the managers (Sullivan et al. 2005).

In this section, we will apply the general theory of green auctions outlined by Latacz-Lohman and van der Hamsvoort (1997) to biodiversity conservation in forestry. We follow Juutinen and Ollikainen (2008), who provide a green auction-based analysis of a system in Finland that has immediate relevance for policy instruments.

7.2.1   Basic Framework

Landowners who can potentially participate in any voluntary conservation program must own stands that can exhibit different ecological qualities. They participate in the conservation program by supplying one or several stands, with the idea that they will strictly protect the land offered. The landowners and the government sign a contract defining the conservation payment to the landowner for a fixed period of time. Prior to receiving bids, a government regulator announces the desired ecological characteristics for stands to be supplied and defines the weights the government gives to each characteristic. Moreover, it explains how the ecological features of forest land are related to the size of the bid that will be accepted.

To formalize this process, let the number of the desired ecological characteristics set by the regulator be n, and let $\alpha_1, \ldots \alpha_n$ denote weights associated with each characteristic such that the sum of weights is equal to one. Each stand supplied to the conservation program is assigned a score value b that is related to its level of desired ecological characteristics. Let E represent the maximum ecological value that can be obtained under the given weights. The ratio of the biodiversity score of each stand to the maximum biodiversity score, b/E, then measures how close each stand comes to the desired natural forest state. For any biodiversity score value, we have $0 < b/E \leq 1$.

Let σ denote the bid of a landowner for a stand having a biodiversity score of b. Let the regulator's maximum willingness to pay for this stand be $R = R(b)$, with $R'(b) > 0$, indicating that the regulator will pay more for a stand with a higher biodiversity score. Let, ω_b and ω_r denote weights given to the biodiversity score value of the stand and to the associated bid, respectively, with $\omega_b + \omega_r = 1$. Using this notation, the regulator computes a single overall score value I for any stand offered, as follows:

$$I = \left[\omega_b \frac{b}{E} + \omega_\sigma \left(1 - \frac{\sigma}{R} \right) \right] \bar{I}, \tag{7.1}$$

where $\bar{I}$ denotes the maximum obtainable score value. Equation (7.1) defines the overall score value of the supplied stands as the share of the maximum obtainable score value. The overall score value depends positively on the weight given to the biodiversity score and negatively on the payment required by the government to secure the bid. All stands are ranked using the overall score in equation (7.1).

Acceptance into the program requires a landowner's score value I of the supplied stand to be above an endogenously determined cutoff value, denoted by I^c. The representative landowner's bidding strategy will therefore be guided by expectations concerning where his stand falls relative to this cutoff value. Let $\underline{I}$ denote the overall score value below which the bidder's expectation of being accepted to the program is zero. Then the probability of being accepted into the program is defined by:

$$P(I > I^c) = \int_{\underline{I}}^{I} f(I)\, dI = F(I). \tag{7.2}$$

We assume that the landowner is risk neutral. He will therefore submit a bid if the expected return from forest management under participation exceeds the return from not participating.

Denote by A the age of the stand at the point of time when the conservation program starts, and let p and r denote the stumpage price and real interest rate. Let $Q(b;A)$ be the forest volume function of an initial stand of age A that exhibits biodiversity level b. The size of the biodiversity score b could depend on actions that a landowner takes, such as increasing the volume of deadwood or restoring water conditions in a stand. Let the privately optimal rotation age be given by T^*. The difference $T^* - A$ indicates how long it would be optimal for the landowner to wait for the first harvest if he were making his own decisions in the absence of a biodiversity conservation program.

In the absence of voluntary conservation (denoted by subscript 0), the fact that b is a nonmarket good implies that landowners maximize the present value of harvest revenue. For a given biodiversity level b, forest rent from optimal management is then a function only of the optimal rotation age choice,

$$V_0 = pe^{-r(T^*-A)}Q(b; T^* - A) + e^{-r(T^*-A)}V^*. \tag{7.3}$$

In equation (7.3) V^* refers to the maximum net present value of returns from all future (privately optimal) rotations.

To examine rents once the landowner has a bid accepted and is enrolled in a conservation reserve program, let γ denote the length of the contract. To allow for the possibility that $T^* - A$ is longer than the length of the contract, we denote the optimal rotation age under the conservation contract by $A + \delta$. Hence, if $\gamma > T^* - A$, then we have $\delta = \gamma$. However, if the number of years to the optimal rotation age exceeds the contract length, then we have $\delta = T^* - A$. Under this notation, the net return from forest management not including the conservation payment under the conservation program (denoted by subscript 1) is given by:

$$V_1 = pe^{-r\delta}Q(b; A + \delta) + e^{-r\delta}V^*. \tag{7.4}$$

It makes a great difference whether $\gamma > T^* - A$, or $\gamma < T^* - A$, that is, whether the length of the contract exceeds or falls short of the optimal time for the private landowner to harvest. In the latter case ($\delta = T^* - A$) the landowner does not need to change his privately optimal rotation age to participate in the program. This is not true for the former case, because the landowner must adopt a new rotation age that was not originally optimal for him.

The expected net return (i.e., net of opportunity and other costs) to the landowner from participating in the program can now be expressed using (7.4) and (7.3) as

$$E\pi = [V_1 - V_0 + \sigma - h(b)]F(I). \tag{7.5}$$

The costs of providing a higher level of biodiversity through the conservation program b is denoted by $h(b)$, with $h'(b) \geq 0$. These costs could equal those required to change the structure of the forest stand, or they could represent trans-action costs incurred by the landowner for participation. Equation (7.5) exhibits two interesting features. First, although the ecological character of the supplied stands is common knowledge, there are some economic parameters that only the landowner knows. The actual management costs, forest growth, net timber price, and thereby the private optimal rotation age, may be imperfectly known to the regulator. The landowner can use this private information to collect information rents, that is, to obtain additional rents by setting the bid, σ, at a level that exceeds his actual private costs from participation. The regulator cannot completely elimi-nate this asymmetric information in the bidding process.

7.2.2   Optimal Bidding

The economic problem of the landowner is to choose σ and b for a given stand to maximize the expected profit from participating (7.5) subject to (7.1) and the obvious constraints $b \leq E$ and $\sigma \leq R$. The Lagrangian for this problem is

$$L = [V_1 - V_0 + \sigma - h(b)]F(I) + \lambda_R(R - \sigma) + \lambda_E(E - b), \qquad (7.6)$$

where λ_R and λ_E are multipliers associated with each constraint. The interior solu-tion for (7.6) is given by

$$\frac{\partial L}{\partial b} = -h'(b)F(I) + [V_1 - V_0 + \sigma - h(b)]F'(I)\left[\omega_b + \omega_\sigma \frac{\sigma R'(b)}{R^2}\right]\bar{I} = 0, \qquad (7.7a)$$

$$\frac{\partial L}{\partial \sigma} = F(I) - [V_1 - V_0 + \sigma - h(b)]F'(I)\frac{w_\sigma}{R}\bar{I} = 0. \qquad (7.7b)$$

From equation (7.7a), the landowner supplies a stand with the optimal bio-diversity level obtained by equating the marginal costs of providing it (the first term) to the expected marginal net return from rents generated from the stand in the conservation program (the second term). The bid in (7.7b) is chosen to equate the probability of acceptance with the expected marginal net return from participation.

Further insight concerning optimal decisions can be obtained by shifting the second terms to the right-hand side of equations (7.7a) and (7.7b), and then divid-ing equation (7.7a) by equation (7.7b) to obtain the ratio of marginal biodiversity management costs to the size of the bid:

$$\frac{h'(b^*)}{1} = R(b^*)\frac{\omega_{\sigma r}}{\omega_b} + \frac{\sigma^i R'(b^*)}{R}. \qquad (7.8)$$

This shows that the landowner makes decisions to set the ratio of marginal biodiversity conservation costs for the bid $(h'(b)/1)$ equal to the ratio of the weights times the maximum rental payment plus the relative change in the maximum rental payment times the bid. Note that if the maximum rental payment is independent of ecological characteristics, then the latter term is zero and we have simply $h'(b^*)/1 = (\omega_\sigma/\omega_b)R$.

If there is no need or chance to improve the ecological characteristics of the stands submitted to the conservation program, then $h(b) = 0$. Under this simplification, the optimal bid is

$$\sigma = \frac{F(I)}{F'(I)\dfrac{w_\sigma}{R}\bar{I}} - [V_1 - V_0]. \tag{7.9}$$

Clearly, the size of the bid depends on the costs of conservation defined by the difference in forest rents lost from participating in the program $[V_1 - V_0]$ and the marginal effect of the bid on the acceptance probability.

Let us finally assess how some key exogenous parameters affect the optimal combination of the bid and the stand's ecological performance. For a change in the weights and the timber price, we can apply Cramer's rule (see the mathematical review at the end of this book) to equations (7.7a)–(7.7b) to obtain

$$\sigma = \sigma(\omega_b, \omega_\sigma, \ p \ ); \quad b = b(\omega_b, \omega_\sigma, \ p \ ). \tag{7.10}$$
$$+ \quad - \quad +/- + \quad ? \quad +/-$$

Hence, a higher weight attached by the regulator to the ecological performance standard for a stand increases both the value of the biodiversity score and the bid level. Increasing the regulator's weight on the bid decreases the optimal bid because the landowner increases rents at the margin from participation by reducing his bid. This weight, however, has an ambiguous effect on the value of the biodiversity score. The effect of a higher timber price depends on the relationship between the optimal rotation age and the contract period.

7.2.3 Parametric Solution

Using data from the Finnish Trading in Natural Values Program, Juutinen and Ollikainen (2008) develop simulation data for 400 stands. They solve the model here using a budget level (€200,000) that results in roughly the same number of enrolled stands as in the actual TNV data (67 forest areas). Table 7.1 condenses the basic results of the biodiversity auction. Reflecting the range of weights in the actual data (0.7–0.5), the weights for the biodiversity score and the bid used were chosen as $\omega_e = 0.7$ and $\omega_\sigma = 0.3$. The outcome of the auction is expressed for the

Table 7.1
Biodiversity Auction: The Number of Enrolled Stands and Average Characteristics by Forest Types

Forest type and age class (years)	Enrolled stands	Diversity score[b]	Bid (€/ha)	Conservation cost[a] (€/ha)	Rent (€/ha)
Herb-rich (100)	7	0.80	3,183	1,629	1,554
Herb-rich (60)	7	0.73	2,914	576	2,338
Herb-rich (40)	5	0.70	2,804	−326	3,129
Mesic (120)	15	0.76	3,027	1,501	1,526
Mesic (70)	17	0.75	3,009	440	2,569
Mesic (55)	7	0.71	2,831	−44	2,874
Dry (140)	6	0.72	2,866	662	2,204
Dry (85)	3	0.69	2,779	129	2,651
Dry (60)	0	—	—	—	—
Average (all stands in sample)	77	0.74	2,964	714	2,250

[a] A negative value indicates that there are stands that would not be harvested during the fixed contract period.
[b] Biodiversity score is the ratio of a stand's ecological index value (b) to the maximum obtainable index value (E), b/E.

key forest types and the three stand classes within each type. Table 7.1 reports the number of enrolled stands and the averages of the biodiversity scores, bids, conservation costs, and rents to the private landowners in each stand class. The last row of this table reports the total number of enrolled stands and averages for the entire sample.

Table 7.1 is revealing in many ways. The number and characteristics of enrolled stands differ considerably across forest types and stand ages. Herb-rich forests have the highest timber volumes. Therefore only a few of the oldest stands (over 100 years) are enrolled for this forest type. The conserved stands in herb-rich forests include even the youngest class of stands; these are relatively cheap in terms of lost discounted rents. The same is true for mesic forests, except in this case stands of 70 years are the dominant conserved class. Not surprisingly, stands in the dry forest type have the least representation among enrolled stands because their biodiversity values are the lowest.

Conservation funds for payments made to landowners are limited. Therefore it is interesting to discuss the bids, costs, and rents paid separately. Without exception, bids are much higher than the cost of conserving forest land, implying that landowners' rents from participation can be quite high. This reflects the generous and uniform government rental payment function. If the government were to decrease the maximum compensation level, then payments to landowners would also decrease. Still, positive rents are inevitable for a uniform payment function. In the theory of green auctions, this type of rent is known as an information rent, in that it arises because the government cannot entirely eliminate asymmetric information by setting the rental payment in a uniform manner.[3]

Another feature of the results in table 7.1 is the fact that the government will pay for stands that would not have been cut during the contract period. For herb-rich and mesic forest types, the youngest stands enrolled have negative costs. This indicates that even young stands may exhibit structural features or exist in a location that makes them worth enrolling. The desirable structural features are typically thought to be high deadwood volumes or standing large, broad-leaved trees in a young stand, both of which are beneficial to biodiversity given the earlier discussion of ecology. Location is also a desirable characteristic if the young stand borders another highly biodiversity-valuable stand. The uniform rental payment function makes the rents for these stands high, and the landowner receives income without any forgone costs associated with giving up harvest rights. However, this is a case where it makes sense for the government to pay for these stands and thereby secure their biodiversity benefits. Finally, sensitivity analysis shows that both the government's willingness to pay for biodiversity benefits and the weights given to the biodiversity score and the bid both are crucial in affecting the outcomes of the biodiversity auction. Box 7.1 compares the simulation results with actual enrollment in the TNV program.

7.3   Green Tree Retention

Another important policy dimension for biodiversity provision concerns management of stands in more commercial forest settings. In these problems, the landowner essentially chooses the level of b given here, not for strict protection, but for maintaining biodiversity along with profitable timber production. Obviously, an important issue here is how, in the spirit of chapter 5, private landowners make decisions that may not be socially optimal.

An important example of this problem concerns a component of biodiversity called "green retention trees." Green retention trees refer to tall or older trees that are left permanently unharvested to die and decay in the new replanted stand (they are sometimes also called "leave" trees for this reason). These trees are not left to assist regeneration through seed banks, but rather to provide dead and decaying wood for old-growth forest species. In Finland and Sweden, leaving retention trees of 5–10 stems per hectare is recommended by national forest laws. Moreover, forest certification systems (Forest Stewardship Council, FSC, and the

---

3. A similar problem exists in principal–agent models with hidden actions or moral hazard, i.e., the agent has private information not known by the principal. Thus, the principal may not be able to achieve a first-best outcome (where the agent behaves in a way that maximizes rents to the principal) if a uniform payment schedule over the agent's actions is assumed (Hey and Lambert 1987).

Box 7.1
Comparison of Simulation with the Actual TNV Outcomes

Table 7.2
Number of Enrolled Stands and Their Average Characteristics in the Finnish Trading in Natural Values Program, 2003 and 2004

Forest type and age class (years)	Enrolled stands (number)	Stand age (years)	Diversity score	Bids (€/ha)	Costs (€/ha)	Rents (€/ha)
Herb-rich (87–160)	17	111	0.48	2,125	1,893	232
Herb-rich (56–80)	13	68	0.48	1,838	937	901
Herb-rich (41–50)	5	45	0.53	1,620	60	1,560
Mesic (102–170)	12	123	0.42	1,908	1,673	235
Mesic (70–95)	15	83	0.40	1,654	943	712
Mesic (50–61)	3	57	0.24	1,177	248	929
Dry (150–178)	4	165	0.42	1,355	1,039	316
Dry (98–110)	3	103	0.25	567	611	−44
Average (all stands in sample)	72	95	0.43	1,757	1,189	568

The results of the theoretical biodiversity auction model can be compared with actual empirical results of the TNV in the South-West Finland Forestry Center. For a general discussion, see Juutinen et al. (2008). Table 7.2 condenses the basic characteristics of the Finnish TNV program in 2003 and 2004. It shows the number of enrolled stands, the average properties for all stands in the sanple (age and biodiversity score), and landowners' bids for each forest type. The last row in the table gives the total number of enrolled stands and the averages of the overall sample.

In general, the enrolled stands are fairly old on average. Relative to table 7.1, the share of stands conserved in the oldest forest age class of herb-rich forest types is now higher. These stands account for 41% of all conserved stands, while this share is 21% in table 7.1. However, the rents for these stands are low. In fact, with one exception (dry, 98–110 years), these stands yield the lowest rents.

The high number of enrolled old stands on the best forest habitat types indicates strong conservation motives by Finnish landowners in the region covered by the program. These landowners choose rotation ages longer than the typical commercial rotation because they value amenity benefits from their stands. A landowner of this type implies that the bids are reduced for the conservation reserve program. Indeed, table 7.2 shows that the average bid is €1,757/ha which implies it is €1,207 lower than those reported in table 7.1. This is not the whole story of the difference though. The rent calculated in table 7.2 is biased because the costs of conservation are defined in terms of efficient timber production, assuming application of the commercial rotation period for each stand type. Thus, for conservation-motivated landowners, the costs here clearly overestimate the landowner-specific costs of enrolling forest stands. This implies that the actual rents are higher than those imputed in table 7.2. Despite undervaluation of forestry rents, table 7.2 suggests a wider use of voluntary conservation instruments. Green auction instruments appear useful in channelling private conservation motives for public provision of biodiversity.

Finnish Forest Certification System, FFCS) in these countries require the same rule concerning green tree retention (GTR) at harvest time. Some provinces in Canada have similar requirements, as do some states in the United States (for a survey of recommendations and laws across countries, see Vanha-Majamaa and Jalonen 2001). In this section we examine the socially optimal rotation age and retention tree volume for forests managed under either Faustmann or Hartman assumptions, i.e., under the case of no amenities and the case where amenities are present. The presentation follows Koskela et al. (2007a and 2007b). Other recent studies of green retention trees using different frameworks include those by Arnott and Beese (1997), Ranius et al. (2005), Jonsson et al. (2006), and Wikström and Eriksson (2000).

7.3.1 Commercial Forests

Green tree retention has two main objectives according to Franklin et al. (1996). First, by creating uneven-aged forests and increasing the volume of deadwood present in the long run, these trees increase structural variation in the forest stand. Second, retention trees enhance the connectivity of forest stands of different ages across a landscape. There are generally two basic ways of implementing GTR whenever a stand is harvested: either by dispersing unharvested trees throughout the stand more or less uniformly, or by aggregating unharvested trees into groups. Aggregate retention is typically recommended in the boreal forests of Finland, Sweden, and Norway. The most frequently saved trees are aspen, birch, and other related hardwood tree species, all of which support a higher number of old-growth wildlife species than softwood tree species, and many of which are the only tree types that harbor several protected and endangered species. In the United States, retention trees have been recommended for several species, perhaps most important for protected woodpeckers that nest in decaying softwood stands throughout the Southeast. In many cases, the uniform method is recommended here because it tends to protect woodpeckers from predation, enhances mating opportunities, and provides an accessible and continuous food source throughout a large forested area.

As this discussion shows, the main biodiversity benefit from retention trees is habitat support from a steady flow of old, decaying trees and deadwood (Ehnström 2001). Even one individual retention tree per hectare can increase the amount of snags and downed wood logs, all of which are important habitats for many species (Jalonen and Vanha-Majamaa 2001). Recent ecological studies in boreal forests indicate that forests managed with retention trees tend to have as many beetles and fungi on average as those present in unharvested old-growth forests with the same amount of dead and decaying wood present (see e.g., Jonsell et al. 2004, Junninen et al. 2006).

Green tree retention can be also promote understory vegetation, especially with regard to nonwoody vascular plants, although their relative abundance may change when compared with true old-growth stands (see e.g., Jalonen and Vanha-Majamaa 2001 and Koivula 2002). There is also evidence that GTR promotes sustainable ecological processes that are beneficial to species that are particularly sensitive to forest management operations (this is known to biologists as "lifeboating"). For instance, some lichens can continue to live in an actively harvested stand with retention trees over the long term (see e.g., Hazell and Gustafsson 1999 and Hilmo 2002).

Relative to GTR, prolonged rotation ages produce dead and decaying wood through the natural mortality of trees. However, to produce larger amounts of dead and decaying wood, the landowner would need to implement rather long rotation ages. These could be costlier (in terms of lost rents and management costs) than techniques designed to create deadwood artificially (see Ranius et al. 2005). Notwithstanding this observation, longer rotation ages provide other services besides simply generating deadwood. Prolonged rotation ages promote other habitats by creating more shade and suitable microclimates that promote an abundance of old-growth species.

A natural way for economists to model the biodiversity benefits of forest management is to modify the felicity function of the basic Hartman model in chapter 3 to incorporate these features. To do this, we assume that biodiversity benefits can be expressed as a sum of the benefits accruing from the age of the stand becoming harvested and from the retention trees left after harvesting. Furthermore, we assume that retention trees reach their biological maturity and decay during the next rotation period after harvesting. The additive form assumed for biodiversity benefits reflects the fact that prolonged rotation ages and artificial creation of deadwood serve two different purposes; that is, they increase habitats for two different sets of old-growth species.

Denote the rotation age and retention tree volume to be chosen in the current rotation period by T and G, respectively. Let $\bar{G}$ indicate the volume of retention trees left from the previous rotation period, and let $\bar{T}$ denote the rotation age that is chosen in the next rotation period. Using this notation, the biodiversity benefits, BB, from retention trees can be expressed as

$$BB = a(T) + v(T, G) = \int_0^T F(x)e^{-rx}\, dx + \int_0^T B(\bar{G}, x)e^{-rx}\, dx + e^{-rT} \int_0^{\bar{T}} B(x, G)e^{-rx}\, dx.$$

$$(7.11)$$

The first term, $a(T) = \int_0^T F(x)e^{-rx}\, dx$, is a conventional amenity valuation function studied in chapter 3, but here it is applied to biodiversity produced throughout the rotation. It makes sense to assume that older stands yield higher biodiversity benefits, so that $F'(T) > 0$. Recall that this reflects the earlier-mentioned fact that some old-growth species (such as lichens) require live old trees. The latter term in (7.11), $v(T, G)$, is a sum of two terms. The first term describes the biodiversity benefits (accruing in the current rotation period) from retention trees left from the previous harvest. The second term measures discounted benefits accruing during the next rotation period from the retention trees left at the end of the current rotation period.

We make the following derivative sign assumptions concerning biodiversity benefits in (7.11):

$$v_T = \hat{B}(T, G, r) \equiv e^{-rT}\left[B(\bar{G}, T) - r\int_0^{\bar{T}} B(x, G)e^{-rx}\,dx\right] > 0 \tag{7.12a}$$

$$v_{TT} = \hat{B}_T(T, G, r) = -rv_T + e^{-rT}B_T(T, \bar{G}) < 0 \tag{7.12b}$$

$$v_G = e^{-rT}\int_0^T B_G(x, G)e^{-rx}\,dx > 0; \quad v_{GG} = e^{-rT}\int_0^T B_{GG}(x, G)e^{-rx}\,dx < 0 \tag{7.12c}$$

$$v_{TG} = v_{GT} = -re^{-rT}\int_0^{\bar{T}} B_G(x, G)e^{-rx}\,dx < 0. \tag{7.12d}$$

Thus the marginal biodiversity benefit in (7.12a) is a positive difference of the biodiversity benefits from green retention between the beginning and the end of the second rotation period. According to (7.12b), marginal biodiversity benefits decrease with the age of retention trees. The same holds true for the marginal benefits from the volume of green tree retention, G by (7.12c). Finally, equation (7.12d) indicates that the cross-derivative between T and G is negative, owing to the fact that retention trees decay and die over time.

7.3.2   Socially Optimal Biodiversity Management

Consider a social planner who replants bare land and decides how many retention trees to leave in a steady state at the final felling.[4] The bare land contains a given amount of retention trees $\bar{G}$ from the just-previous harvest. The new stand to be planted grows according to a growth function, $f(T, \bar{G})$, which has the exogenous inherited (left from the previous harvest) volume of retention trees as an additional argument. The growth function exhibits conventional properties in rotation age; $f_T > 0$ and $f_{TT} < 0$ over the relevant range of the rotation age.

---

4. Our assumption implies that the "initial" stand, from which the first green tree retention originates, has already been harvested. The analysis of the initial stand case is qualitatively similar to the steady-state analysis made later, and so it is omitted.

In addition to rotation age, the planner decides the volume of retention trees to leave at the end of the rotation period. Therefore the planner must know how retention trees affect conventional even-aged forest management. Beyond producing biodiversity benefits, retention trees affect the profitability of forestry in two ways. First, by increasing shade, they decrease the growth of the new stand planted at the beginning of the next rotation period in increasing fashion, so that $f_G < 0$ and $f_{GG} < 0$.[5] Second, the regeneration costs of the new stand are affected by the volume of retention trees. However, to simplify the presentation, we assume here that this effect is small and can be neglected.

Under a constant timber price p and real interest rate r, the planner's economic problem is to maximize social welfare by choosing the optimal rotation age and optimal retention tree volume, with social welfare for a single stand given by

$$SW = \left[pe^{-rT}[f(T,\bar{G}) - G] - c + \int_0^T F(x)e^{-rx}\,dx + \int_0^T B(\bar{G},x)e^{-rx}\,dx \right](1 - e^{-rT})^{-1}$$

$$+ e^{-rT}\left[\int_0^{\bar{T}} B(x,G)e^{-rx}\,dx + pf(\bar{T},G)e^{-r\bar{T}} \right](1 - e^{-rT})^{-1}. \tag{7.13}$$

To make the link between the current and next rotation period more transparent in equation (7.13), we have collected the effects of retention trees extending to the next rotation period into the second bracketed term. These effects include biodiversity and reduced growth of the new stand. The first-order conditions for this problem can be expressed as

$$SW_G = -p + \int_0^{\bar{T}} B_G(x,G)e^{-rx}\,dx + pf_G e^{-r\bar{T}} = 0 \tag{7.14a}$$

$$SW_T = pf_T - rp[f(T,\bar{G}) - G] + F(T) + B(\bar{G},T)$$

$$- r\left[\int_0^{\bar{T}} B(x,G)e^{-rx}\,dx + pf(\bar{T},G)e^{-r\bar{T}} \right] - rSW = 0. \tag{7.14b}$$

From equation (7.14a), the optimal volume of retention trees is chosen to equate the present value of the sum of marginal utility from retention trees over their whole process of decay to the sum of the marginal loss of harvest revenue from leaving these trees, in addition to the value of the decreased growth of future stands. According to (7.14b), the optimal rotation age is chosen so that the

---

5. Using data for Scotch Pine, Valkonen et al. (2002) find that retention trees decrease the growth of a new stand by a 1–2% overall.

marginal return of delaying harvest by 1 year equals the opportunity cost of delaying harvesting year. While the former is defined by the sum of the harvest revenue and biodiversity benefits during the first and the second rotation periods, the latter is different from what we found in chapter 3 and is defined as the interest lost on both the standing timber and the land (as before), as well as the interest lost through the lower value of future growth as a result of tree retention.

Using the notation for marginal biodiversity benefits from green retention adopted in equation (7.12a), the second-order conditions can be expressed as

$$SW_{GG} = \int_0^T B_{GG}(x, G)e^{-rx}\, dx + pf_{GG}e^{-rT} < 0 \tag{7.15a}$$

$$SW_{TT} = pf_{TT} - rpf_T + F'(T) + \hat{B}_T < 0 \tag{7.15b}$$

$$D = SW_{GG}SW_{TT} - SW_{TG}^2 > 0, \tag{7.15c}$$

where $SW_{TG} = SW_{GT} = 0$. Given our assumptions, condition (7.15a) holds automatically. Condition (7.15b) holds under the conventional Hartman model requirement that biodiversity benefits should not be too large relative to other terms of (7.15b).

The comparative statics will differ partly from the results in chapter 3, because now we have a 2×2 equation system of first-order conditions. However, given that the cross-derivative between the rotation age T and the volume of retention trees G is equal to zero, the comparative statics of the rotation age T remains conventional. The results for the retention tree volume are new. These results are

$$T = T(\underset{-}{p}, \underset{-}{r}, \underset{+}{c}), \quad G = G(\underset{-}{p}, \underset{-}{r}, \underset{0}{c}). \tag{7.16}$$

From (7.16), a higher timber price leads to a shorter rotation age and decreased level of retention tree volume. The former effect is well known from the basic Hartman model. The latter effect results from the fact that a higher timber price increases the profitability of harvesting, making it more costly to leave retention trees for biodiversity purposes. A higher interest rate works similarly by decreasing the opportunity cost of both delaying harvesting and leaving retention trees. Finally, while a higher regeneration cost lengthens the rotation age in a conventional way, it has no impact on the optimal retention volume, owing to our simplifying assumption that retention trees do not affect regeneration costs.

7.3.3 First-Best Policy Instruments

We already know from chapters 2–5 that we cannot expect biodiversity to enter the objective functions of private landowners in the right way if this public good is important to nonlandowners. This means that the socially optimal and privately optimal rotation age and retention volume will likely differ, and the socially optimal level of biodiversity will not generally be provided by the market. This calls for policy instruments that remove the difference in incentives between society and private landowners to provide biodiversity.

We now outline the policy design problem needed to promote maintenance of biodiversity in commercial forests, with a special focus on conservation of threatened old-growth species through increasing dead and decaying wood at the stand level. We do this assuming that the landowner behaves according to either a conventional Faustmann or Hartman model. The Faustmann-based landowner maximizes only the present value of harvesting rents, ignoring both biodiversity benefits during a rotation and retention tree benefits at harvest time. The Hartman-based landowner ignores only retention tree benefits. Although the landowner's amenity values may also not be consistent with social biodiversity values, we ignore this case and assume that the landowner and society have identical preferences for amenity benefits accruing in the stand during any rotation; thus, all differences in rotation age and retention volume result from the retention tree aspect of the problem.

The goal of any biodiversity policy should be to internalize the externality inherent in private landowners ignoring biodiversity when making decisions. There are two potential targets for policy in this problem: the rotation age and the volume of retention trees left after harvesting. Generally, the government should punish landowners for choosing a private rotation age that is too short, and it should reward landowners who leave retention trees. This is a typical carrot-and-stick type of Pigouvian policy. Thus a subsidy is needed for retention trees, whereas forest taxes can be used to affect the choice of rotation age. Chapters 2–5 show that there are many taxation alternatives if the goal is to lengthen the rotation age. The government can use harvest (yield or unit) taxes, where we previously showed that $T_x^F > 0$, $T_x^H > 0$, as $F'(T) > 0$ for $x = \tau_1, u$. It can also use a timber tax because we showed that $T_\alpha^F < 0$, $T_\alpha^H < 0$ in both Faustmann and Hartman models. Finally, the government can apply a site value tax since we know that $T_\beta^H > 0$, when $F'(T) > 0$ for the Hartman model.

More specifically, three instrument combinations are feasible: a retention tree subsidy can be used jointly with (1) the yield (or unit) tax, (2) a timber tax subsidy, or (3) a site value tax. The optimal combination can be developed using the same approaches applied in chapter 5 when deriving first-best taxes. In particular, we introduce the instruments into the landowner's target function, equate the resulting private first-order conditions with the socially optimal choices, and then solve for the implicit socially optimal instrument rates that equate these first-order conditions defining rotation age and green retention tree choices.

Consider first a combination of a retention tree subsidy and a yield tax. The general tax structures are fairly similar for both Faustmann (F) and Hartman (H) cases:

$$s'(G)_i^* = (1 - \tau_i^*) \int_0^{\bar{T}} B_G(x, G)e^{-rx}\, dx, \quad \text{with } i = F, H \tag{7.17a}$$

$$\tau_i^* = -\frac{\Phi + E_i}{\Omega}, \tag{7.17b}$$

where $E_F = [\Phi + [F(T) - rE] + [B(T, \bar{G}) - rH]]$　　$E_H = [\Phi + [B(T, \bar{G}) - rH]]$,

$\Omega = p[f_T - r\eta([f(T, \bar{G}) - G] + e^{-rT}f(\bar{T}, G))]$　　and

$$\Phi = r\eta \left[s(G) - \int_0^{\bar{T}} B(x, G)e^{rx}\, dx \right],$$

$$\eta = (1 - e^{-rT})^{-1}, \quad E = (1 - e^{-rT})^{-1} \int_0^T F(x)e^{-rx}\, dx,$$

$$H = (1 - e^{-rT})^{-1}e^{-rT} \int_0^{\bar{T}} B(x, G)e^{-rx}\, dx.$$

In (7.17a), the optimal yield tax rate is positive but less than one, and the marginal subsidy paid for retention trees reflects decreasing social marginal biodiversity benefits (recall assumption 7.12c). Because the yield tax lengthens rotation age, the marginal subsidy is adjusted accordingly to the optimal yield tax. The terms in equation (7.17b) in turn indicate that the optimal yield tax in the Hartman model is smaller than in the Faustmann case. In contrast, the marginal subsidy for retention trees is higher than in the Faustmann model because the after-tax timber price and opportunity costs of retention trees are higher in the Hartman model.

Suppose the government employs a timber tax α. Now the optimal instrument mixes in the Faustmann and Hartman cases are as follows:

$$s'(G)_i^* = \int_0^{\bar{T}} B_G(x, G)e^{-rx}\, dx, \quad i = F, H \tag{7.18a}$$

$$\alpha_i^* = -\frac{A_i}{pf(T, \bar{G}) - rU}, \tag{7.18b}$$

where $A_F = [[F(T) - rE] + [B(T, \bar{G}) - rH]]$, $A_H = [[B(T, \bar{G}) - rH]]$,

$$U = (1 - e^{-rT})^{-1} \int_0^T pf(s, \bar{G}) e^{-rs} \, ds \quad \text{and} \quad [pf(T, \bar{G}) - rU] > 0.$$

As earlier, the marginal subsidy for retention trees decreases in G for both Faustmann and Hartman models. It is independent of the timber tax, which does not distort the timber price and works only through the opportunity cost of leaving retention trees. Therefore the subsidy is the same in both the Faustmann and Hartman cases. For the timber tax in (7.18b), both numerators and denominators are positive, so that the optimal instrument turns out to be a timber subsidy rather than a tax. This makes sense because we already know that the timber tax shortens the rotation age (a subsidy will lengthen it). The level of the optimal timber subsidy depends on the present value of the retention tree subsidy. Its optimal size reflects the ratio of the net marginal biodiversity benefits (adjusted for opportunity costs) and the effect of the timber subsidy on timber production rents. Finally, we see that the optimal timber subsidy is smaller in the Hartman case than in the Faustmann case because the externality in terms of the rotation age is smaller.

Finally, suppose a combination of a tree retention subsidy and site value tax, β, is used. This combination can only be applied for the Hartman case. The optimal instrument mix is

$$s'(G)^* = \int_0^{\bar{T}} B_G(x, G) e^{-rx} \, dx;$$

$$\beta^* = -\frac{[(B(T, \bar{G}) - rH + \Omega]}{pf_T - r\eta(p[f(T, \bar{G}) - G] + pe^{-r\bar{T}}f(\bar{T}, G) + s(G))} \tag{7.19}$$

The marginal retention tree subsidy is now similar to that of (7.18a) and independent of the level of the site value tax. The optimal site value tax is positive because the numerator is negative given the first-order conditions; the denominator is positive. The optimal site value tax reflects the net marginal biodiversity benefits relative to net harvest revenue.

7.3.4   Retention Tree Volumes and Instruments—Example

Drawing on the theoretical framework outlined here, we now report on some simulation results concerning rotation ages, retention volumes, and the levels of policy instruments. The simulation model is described in detail in Koskela et al.

Table 7.3
Privately and Socially Optimal Rotation Ages: Faustmann, Hartman, and Biodiversity Models

	Faustmann private	Hartman private	BIODIV social
Rotation age (years)	60	66	67
Retention volume (m^3/ha)	0	0	7.9
Mean annual harvest (m^3/ha)	4.20	4.42	4.08
Mean annual net income (€/ha)	152	171	159
Timber benefit (€/ha)	1,013	1,026	873
NPV of amenity benefit (€/ha)	0	518	524
Corresponds to annual flow (€/ha)	0	15.5	15.7
NPV of biodiversity benefit (€/ha)	0	0	320
Corresponds to annual flow (€/ha)	0	0	9.6
Total benefit (soil expectation value) (€/ha)	1,013	1,544	1,717

(2007b). The results are considered for pine under typical growth conditions in southern Finland without thinning. Table 7.3 reports optimal rotation ages, retention volumes, and economic benefits of the Faustmann and Hartman models for private landowners using the BIODIV model in which society values amenities during a rotation as well as biodiversity. In all models the real interest rate is 3%.

From Table 7.3, the rotation ages range from 60 to 67 years and are oldest for the socially optimal biodiversity solution. Mean annual harvest volumes are 4.08–4.42 m^3/ha. Thus conservation of biodiversity does not imply a major decrease in timber supply per hectare. The volume of retention trees is zero in the Faustmann and Hartman models, but equals about 8 m^3/ha (3% of standing volume) for the BIODIV model. Assuming that the tree diameter at breast height (dbh) is about 24 cm (which was the mean diameter of pines at final felling), the solution is consistent with leaving about 18 retention trees per hectare. The same volume would be obtained by leaving 10 large trees (dbh 30 cm). However, a closer look at the optimization results reveals that the retention of 8 m^3/ha consists of naturally regenerated spruces and birches, the diameter of which is only about 17 cm. This corresponds to as many as 50 retention trees per hectare. The optimization result is logical because small-sized spruces and birches have lower opportunity costs (lower harvesting rents), are more expensive to harvest, and contribute more than large pines to the biodiversity of forest stands.

Table 7.4 shows the five possible sets of instrument combinations for biodiversity policy where s denotes the retention tree subsidy, τ the yield tax, α the timber tax, and β the site value tax. As the theory suggests, the Faustmann and Hartman results differ for the (s, τ) instrument combination. The theoretical

Table 7.4
Optimal First-Best Total Subsidy Payments (s) and Tax Instrument Rates (τ, α, β)

Combination	Faustmann		Hartman	
s, τ	$s = 1{,}000$ (€)	$\tau = 40–65\%$	$s = 750$	$\tau = 20–40\%$
s, α	$s = 1{,}500–2{,}500$ (€)	$\alpha = -1\%$	$s = 1{,}900$	$\alpha = -0.5\%$
s, β	N/A	N/A	$s = 1{,}700$	$\beta = 0.01$

framework suggests that the yield tax rate should be higher in the Faustman model than in the Harman model, and the results of table 7.4 verify this. Also, the retention tree subsidy is higher for the Faustmann model. It is interesting that for the Hartman model, a harvest tax rate range is obtained that actually fits the currently applied tax rate in Finland of 29%, while the tax rate exceeds this for the Faustmann model. The theory also suggested that retention tree subsidies should be equal in Faustmann and Hartman models when combined with the timber subsidy [the combination (s, α) in table 7.4]. The minus marks for α in the third and fifth columns demonstrate that indeed it is optimal to have a timber subsidy and not a tax. The second column in turn shows the range of a retention tree subsidy in the Faustmann model. This range has a mean almost identical to the optimal retention tree subsidy in the Hartman model, just as the theoretical analysis suggested. The final instrument combination, a retention tree subsidy with the site value tax, is reported for the Hartman model in the last row. The site value tax was modeled by charging it in the initial planting year. Calculating this sum on an annual basis yields the rate given in column five. Recall that the theoretical model predicts the same retention tree subsidy as in the previous case. Owing to discontinuities of the simulation model, the retention tree subsidy is, however, slightly lower than in the previous case.

It is interesting to compare how our instrument combinations affect government budgets. Even though we assume that the instrument combinations are first-best, it is always important to evaluate the budget burden of alternative instrument combinations to aid the choice of policy, especially when, as in the biodiversity case, instrument combinations are needed. These effects are collected in table 7.5 for the optimal policies in table 7.4. In the table, the labels "F" and "H" refer to the Faustmann and Hartman models.

Using the optimal computed rates of subsidies and taxes we derived here, we report in the fourth and fifth columns the present values of revenues for instruments applied over one rotation. The government's net budget effect is shown in the sixth column. When the forestry budget exhibits a surplus, it is indicated by $(+)$ and a budget deficit is denoted by $(-)$. Only when a yield tax is applied will the government budget yield a surplus. The budget deficit is largest with the

Table 7.5
Budget Effects of Optimal Instrument Combinations

Model and instruments (F, Faustmann; H, Hartman)	Size of subsidy per 10 m^3	Tax rate	Received subsidies (NPV/ha) (€)	Tax payment (NPV/ha) (€)	Budget burden
F: subsidy (s) and yield tax (τ)	€1,000	60	916	8,644	7,728 (+)
F: subsidy (s) and timber tax (α)	€2,000	−1	1,866	−496	2,362 (−)
H: subsidy (s) and yield tax (τ)	€750	30	689	4,468	4,298 (+)
H: subsidy (s) and timber tax (α)	€1,900	−0.5	1,786	−248	1,538 (−)
H: subsidy (s) and site value tax (β)	€1,700	0	1,490	0	1,490 (−)

combination (s, α) for the Faustmann model and smallest for the combination (s, β) for the Hartman model.

There are some important lessons here for using policy instruments to ensure the optimal provision of biodiversity in forest regions. First, different instrument combinations have different budget impacts. The net effects depend on the characteristics of available forest taxes and the assumption concerning whether or not private landowners value amenities during a rotation. This means that the government needs to carefully consider the makeup of forestland ownership before employing the optimal biodiversity policy if budget effects are important. Second, if the landowners behave as the Hartman model predicts, then budget burdens for biodiversity policies will be lower than if landowners follow Faustmann model-based decision making.

7.4   Modifications

This chapter has focused on the simplest and most typical site selection and policy instrument examples relevant for private forest landowners. We omitted a large body of general biodiversity literature that does not have a specific forestry focus, such as work by Weitzman (1992, 1993), Polasky and Solon (1995), Li and Lofgren (1998), Brock and Xepapadeas (2003), Kassar and Lasserre (2004), and Hamaide et al. (2006). In much of this work, the focus is on determining the socially optimal set of species to preserve over time over all ecosystem types (water, forest, etc) when complementarities among species may not be perfectly known. Nonetheless, the reader should be aware of these studies, because they certainly apply tangentially to forest ecosystems. Another branch of literature that may have implications for biodiversity policy choices are those studies concerning the economic and ecological performance of ecosystem management in forestry, such as Haight (1995), Bevers and Hof (1999), and Calkin et al. (2002). Furthermore, Boscolo and Vincent (2003), and Potts and Vincent (2008) show how to integrate timber production and species viability aspects in forest management decisions, where use (or overuse) of one species may be a detriment to other dependent species. All of this work has potential to add to the selection models discussed earlier.

One glaring omission in this chapter is a rigorous discussion concerning the new and increasingly important problem of invasive foreign species. Invasive species are non-native species that have been introduced, often by accident, into the forest landscape. Examples in North America are the gypsy moth, hemlock wooly adelgid, oak-ash borer, and cogongrass. These species in some cases quickly overcome existing species and are a well-recognized threat to biodiversity worldwide. Often, invasive plants grow faster than domestic species and may be resistant to weather changes, leading to rapid changes in the structure of forest landscapes once they appear. Furthermore, their speed of spread can make monitoring early appearance a difficult detection problem. As a result, the losses that are due to invasive species, both commercial and related to socially desirable biodiversity, have the potential to be quite large. For instance, Pimentel et al. (2005) estimate that the annual cost of invasive species in the United States is $120 billion.

One of the most basic economic questions concerning invasive species is whether society should apply preemptive or reactive controls (Perrings 2005). Preemptive control refers to (possibly costly) actions that retain the original landscape infrastructure while maintaining vigilance concerning possible invasion events. Of course, if invasive species are detected, then in principle one should totally eradicate the species using available resources. Thus preemptive control reduces the probability that invasive species will become established in the area in the first place. Reactive control, on the other hand, refers to letting an invasion take place and then using control measures that balance the damage imposed by invasive species with the costs of control. In this approach, the costs of preemptive controls are not spent and society plays a wait-and-see approach. The main purpose of reactive control is to reduce the extent and magnitude of damage by invasive species if they arrive.

The problem of preemptive versus reactive control is well known in economics outside of the invasive species problem (see e.g., Polinsky and Shavel 1994), but the question is a new issue in invasive species problems. Applications to forestry are almost nonexistent. Most papers so far examine invasive pests that affect agriculture, another industry, or households by assessing the relative net costs of the two alternative control approaches. Krcmar-Nozic (2000), Finnoff et al. (2006), and Kim et al. (2006) examine optimal policy design problems in which resources are allocated to monitoring and subsequent control subject to budget constraints, threshold effects, and other factors. Barbier and Shogren (2004) have an interesting analysis of policy instruments under conditions where a household can take costly actions to protect against income-reducing invasions.

In forests, when invasive species become present on a large scale, they will nearly always lead to a reduction in the number and quality of all forest species through competition for growing space. There are a few relevant articles that study these effects. Matthews et al. (2002) examine biodiversity policies matched with carbon sequestration in forests. McAusland and Costello (2004) show conditions under which trade policies should be revised according to infection rates of invasive species among trading partners. Mehta et al. (2007) develop a model to assess the importance of uncertainties in detecting invasive species for eventual policy targeting. Using a dynamic search problem of minimizing detection and control costs, they find that the difficulties of detecting invasive populations are a key factor in developing a strategy for agency monitoring. This has clear applicability to forests, where remote detection of invasive species with low population densities but potentially high spread factors is not yet fully developed. Finally, Prestemon et al. (2006) consider how arrival of an invasive species (gypsy moth) is likely to affect global trade in forest products.

At present the debate concerning invasive species in forests should probably focus on two key but unstudied issues among economists. The first is whether forest management techniques can affect the presence of invasive species and the conditions under which they spread or are detected, and, if so, how this affects the application of any policies at a landscape scale that seek to minimize damage by invasive species. The second issue is how the government can target policy instruments toward invasive species that are hard to detect and for which biological information is still being gathered. We have no doubt that invasive species will play a more important role in forest economics in the years to come, but this area remains too much of an open issue for discussion here.

7.5   Summary

This chapter examined the conservation of biodiversity, an urgent problem in temperate and boreal forest policy circles. The overall concept of biodiversity conservation networks from ecology research is an important starting point. This work suggests that forest biodiversity is enhanced when three types of habitat structures exist: strictly protected core areas surrounded by buffer zones, transition corridors, and commercial forests. With this as a backdrop, we formulated associated economic and policy problems with a special focus on the design of biodiversity policy instruments. This included command-and-control regulations as well as voluntary provision of biodiversity, the selection of biodiversity reserves through a green auction system, and provision of biodiversity through retention trees.

Biodiversity conservation is a relatively new topic for forest resource economics. We end this chapter with a discussion of what we view as a critical issue for future work. Clearly, developing new and efficient voluntary instruments for forest landowners will be one basic research direction for the future. Invasive species provide yet another fruitful topic because they remain a growing and critical threat to forest biodiversity, and forest economics has barely touched it. The third issue in our view is the challenge of analyzing the simultaneous tradeoffs among the following three topics: conservation of biodiversity, carbon sequestration, and invasive alien species. Until we do this, we will not have the full picture of biodiversity and forests within our grasp.

8 Forest Age Class Models

One challenge in forest economics for decades has been the management of forest-land areas composed of land units with different-aged trees. The main questions have surrounded what land units should be harvested in a first cutting, then the optimal units to cut and the rate of cutting over time. These questions can only be answered once we define an optimal long-run structure for a forest consisting of many age classes, as any long-run structure can in theory be achieved through a series of harvests over time. This problem is specific to even-aged forest management where a stand is the unit of management and the forest is composed of many individual stands. This contrasts with work in uneven-aged forest management. In uneven-aged management, the problem is to determine the optimal diameter class distribution of the forest that should exist in the long run, where trees of different ages are present on any given land unit. Uneven-aged management is a very difficult problem to study mathematically, and most applications in forest economics have been either simulation- or operations research-based.

In previous chapters we assumed that the landowner made decisions concerning only a single stand. In this chapter we assume that the landowner (either a private agent or society) has command over a forest area large enough for our research questions to be relevant. The long-run management problem of the landowner is now to harvest over time in a way that divides this area into units that each support an even-aged stand of a different age. At the steady-state solution, each stand is managed on the same rotation age and once it is harvested, a new stand is established by replanting the land unit. We are interested in examining the optimal steady-state age class structure of the forest in this problem. While some interesting issues exist concerning the dynamics by which stand structure converges to various steady states, these types of analyses require numerical methods that go beyond the scope of this book.

The oldest suggestion for an optimal steady-state age class structure is something called a normal forest. This follows the notion of achieving an even flow of timber volume harvested in each period. In the history of forest policy discussions, even-flow solutions have been sought in order to meet the needs of forest industries supplying mills or to ensure the stability of local communities dependent on wood-based revenues or tax collections. Even-flow solutions imply there should be one stand available for harvest in each period. If each stand is economically mature at an age of T years, then a normal steady-state age class structure will have T units of land (the given land area divided in $1/T$ number of units), which collectively have stands of ages from 0 to T years. In a normal forest, exactly one age class becomes mature in each year, and this oldest age class it harvested and the bare land is replanted to become the youngest age class in the forest. Although a normal forest clearly leads to even flow, it may not be an optimal long-run structure from an economics point of view.

The formal examination of optimal age class structures in forest economics is of relatively recent origin, dating from Kemp and Moore (1979), Heaps and Neher (1979), and Heaps (1984). These authors treated both time and land areas as continuous variables, but none provided a proof of whether the normal forest is an optimal long-run state. Lyon and Sedjo (1983) were the first to use an empirical age class model with time measured in discrete intervals but land treated as a continuous variable. This discrete-time–continuous-land (or space) approach has also been adopted in subsequent modeling.[1] Mitra and Wan (1985, 1986) were the first to provide formal proofs concerning the long-run steady-state optimum for forest structure. Using an assumption of zero discounting and a strictly concave forest landowner utility function, they showed in a discrete-time–continuous-space model that the normal forest is an asymptotically stable equilibrium given any initial forest structure. However, if utility is discounted, Mitra and Wan (1985) also show numerically that equilibrium forest stocks can oscillate optimally around the normal forest structure. Their analysis has been carried further by Salo and Tahvonen in several articles (2002a, 2002b, 2003, 2004), who consider models combining nonlinear utility functions of consumption and multiple forest vintages in discrete time.

Apart from theoretical research on this subject, there are also many numerical applications of age class models. Lyon and Sedjo (1990) used the model to examine timber supply, while Uusivuori and Kuuluvainen (2005) introduced amenity benefits and analyzed the resulting solutions both theoretically and numerically. Uusivuori and Laturi (2007) recently have examined how policies targeting carbon sequestration differ when applied under the assumption that a landowner

---

1. An argument in favor of this specification is that it frequently is used as the basis for analyzing forest policy (see e.g., Sedjo and Lyon 1990). Moreover, to the extent that use of forest resources is marked by some seasonality, this model can accommodate a discrete solution where harvesting is concentrated in some specific period of time (Getz and Haight 1989 argue for this case using numerous examples of discrete-time models in forestry and wildlife problems).

holds multiple even-aged stands of different ages rather than a single even-aged stand.

In this chapter, we provide a relatively brief introduction to the theory behind age class models with emphasis on those specifications that might be used to examine forest policy questions. We first discuss a general structure of modeling that is followed by most of the articles forming this literature. While the mathematical detail needed to solve for various cyclical steady states in these models goes beyond what is reasonable for a survey text such as ours, we will provide a flavor for the types of results that have been obtained in the different models that have recently comprised this literature. We pay attention to both cases where a competing land use is and is not available, as these along with the specification of utility of consumption have important implications for the optimal long run forest structure. Next, we discuss the currently important world issue of carbon policies to consider whether policy design tailored for one stand differs from an optimal design made under the assumption that a landowner manages multiple-age classes when making decisions. We end this chapter by going beyond even-aged management of multiple stands to consider uneven-aged models.

8.1   Basic Structure

In this section we take for granted that a forest can be divided into units of land each having an age class unique to its unit. Our goal is to develop and describe the basic features of an age class model and present the harvesting choices of the landowner. Our presentation in this section relies on a version of Uusivuori and Kuuluvainen (2005) that omits amenities, although the structure of their basic model also has many elements in common with the other previously noted articles.

Consider a landowner who manages a forest with n stands representing n distinct age classes. Denote the land unit with all stands by x. Let the initial time zero age class distribution for all stands be given by the vector $x_0 = (x_{01}, x_{02}, \ldots x_{0n})$, ordered from the youngest (1) to the oldest (n) stands [there is a similarly defined age class distribution at any time t given by $x_t = (x_{t1}, x_{t2}, \ldots x_{tn})$]. Let aggregate growth be defined through a vector $q = (q_1, \ldots q_n)$ describing the rates of change in timber volumes (per hectare) in each age class. In a given period t, the landowner harvests a share a_{ti} of each stand ($i = 1, \ldots n$). Each share must be between 0 and 1; if it equals zero, then this means that age class i is not harvested at time t, whereas if it equals one, then the age class is cleared entirely of occupying trees. Hence the volume of timber (Q) in the entire forest in period t is given by the product of growth and age class vectors, $Q = q x_t$. The age classes at time $t+1$ then follow from those existing at time t and from harvesting decisions made at time t. Define $x_{t+1} = A_t x_t$, where each element of matrix A_t describes periodic harvesting shares and determines the age class dynamics of the model. Thus, A_t is as follows:

$$
A_t =
\begin{bmatrix}
a_{t1} & a_{t2} & \cdot & \cdot & a_{tn} \\
1 - a_{t1} & 0 & \cdot & \cdot & \cdot \\
0 & 1 - a_{t2} & 0 & \cdot & 0 \\
\cdot & \cdot & \cdot & \cdot & \cdot \\
0 & 0 & 1 - a_{tn-1} & 1 - a_{tn}
\end{bmatrix}.
$$

The first row gives the periodic harvest shares of each age class and the following rows give the unharvested shares of subsequent age classes (Uusivuori and Kuuluvainen 2003).

The landowner is assumed to maximize utility from consumption in each time period. Let the preferences of the landowner be defined by a concave utility function over consumption, $u(c_t)$, with $u'(c_t) > 0$ but $u''(c_t) < 0$, just as we discussed in the life-cycle model of chapter 2. We denote the discount factor by $\beta = 1/(1+\rho)$, where $0 < \rho < 1$ describes the representative landowner's rate of time preference. The landowner maximizes utility from periodic consumption over an infinite time horizon by choosing the shares to be harvested in each age class at each point in time. Let p denote the timber price, r denote the real interest rate, and k denote the planting costs for each stand, all of which are assumed to be constant over time. The landowner's dynamic optimization problem can be expressed as one of choosing harvest shares according to:

$$
\max_{a_{ti}} \sum_{t=0}^{\infty} \beta^t u(c_t), \text{ s.t.} \tag{8.1}
$$

$$
c_0 = \sum_{i=1}^{n} a_{0i} x_{0i} (pq_i - k) + w_0 - w_1 \tag{8.2}
$$

$$
c_t = \sum_{i=1}^{n} a_{ti} x_{ti} (pq_i - k) + w_t(1 + r) - w_{t+1}, \quad \forall t \geq 1 \tag{8.3}
$$

together with $a_{ti} \geq 0$, $1 - a_{ti} \geq 0$, $x_{0i} \geq 0$, as well as the definition $Q = qx_t$ and the transition equation $x_{t+1} = A_t x_t$, where w_t and w_{t+1} are exogenous income of the landowner in period 1. Thus the decision made for the harvesting shares in each age class reflects the tradeoffs of the marginal impacts of allocating land between any two consecutive age classes over infinitely repeated rotations for each stand. We next examine these choices in more detail.

Substituting constraints (8.2) and (8.3) into the utility function and accounting for the age class dynamics defined by matrix A_t we obtain

$$u\left[\sum_{i=1}^{n} a_{0i}x_{0i}(pq_i - k) + w_0 - w_1\right] + \sum_{t=1}^{\infty} \beta^t u$$

$$\times \begin{bmatrix} \sum_{i=1}^{n} a_{t1}a_{t-1,i}x_{t-1,i}(pq_i - k) \\ \\ + \sum_{i=2}^{n-1} a_{ti}(1 - a_{t-1,i-1})x_{t-1,i-1}(pq_i - k) \\ \\ + \sum_{i=n-1}^{n} a_{tn}(1 - a_{t-1,i})x_{t-1,1}(pq_n - k) + w_t(1+r) - w_{t+1} \end{bmatrix}.$$

$$(8.4)$$

The optimality condition for consumption requires $u'(c_t) = [(1+r)/(1+\rho)]u'(c_{t+1})$, or $u'(c_t)/u'(c_{t+1}) = (1+r)/(1+\rho)$, which is the same intertemporal consumption choice condition we solved for in the utility-based model of chapter 2. The first-order condition for harvesting a share a_{ti} in each age class i can be developed as follows. By differentiation one obtains:

$$u'(c_t)x_{ti}(pq_i - k) + \beta^i u'(c_{t+i})x_{ti}(pq_i - k) + \beta^{2i}u'(c_{t+2i})x_{ti}(pq_i - k) + \cdots$$

$$- \beta u'(c_{t+1})x_{ti}(pq_{i+1} - k) - \beta^{(i+1)+1}u'(c_{t+(i+1)+1})x_{ti}(pq_i - k)$$

$$- \beta^{2(i+1)+1}u'(c_{t+2(i+1)+1})x_{ti}(pq_{i+1} - k) - \cdots$$

Dividing this derivative by $u'(c_t)$, denoting $u'(c_t)/u'(c_{t+1}) = (1+r)/(1+\rho)$ and collecting the terms yields

$$(pq_i - k)(1+r)\frac{(1+r)^i}{(1+r)^i - 1} - (pq_{i+1} - k)\frac{(1+r)^{i+1}}{(1+r)^{i+1} - 1}\begin{Bmatrix}>0\\<0\end{Bmatrix} \text{ as } \begin{Bmatrix}a_{ti} = 1\\a_{ti} = 0\end{Bmatrix}. \quad (8.5)$$

Note first that owing to the discrete time assumption, setting the LHS expression equal to zero is not feasible.[2] However, a pair of time periods can be found where the signs of the expressions shift from negative to positive. Hence

---

2. If amenities were included in (8.1), then an additional possibility of a corner solution exists with $0 < a_{ti} < 1$.

according to (8.5), it is profitable to cut the stand in age class i when the marginal return from harvesting is greater than the marginal return from not harvesting, $a_{ti} = 1$. If the other case holds, then it is optimal to abstain from cutting and set $a_{ti} = 0$.

Equation (8.5) is in fact related to the Faustmann rotation age condition of chapter 2. We can see this by dividing and accounting for interest rate factors to obtain

$$\frac{p(q_{i+1} - q_i)}{(pq_i - k)} \geq (<) \, r \left(\frac{1}{1 - \frac{1}{(1+r)^t}} \right). \tag{8.6}$$

According to (8.6), it is profitable to cut the stand in age class i when the marginal return from harvesting is smaller than the marginal return from not harvesting ($a_{ti} = 0$). This is, in fact, a discrete time analogue to the Faustmann rule derived in continuous time.[3]

The model outlined here can be used as a basis for a numerical analysis of several policy questions relevant to multiple age class forests. We return to an example later in this chapter. Before proceeding, readers should be aware of some critical assumptions for age class modeling. The assumption of a utility-maximizing landowner facing an intertemporal budget constraint turns out to play an important role. Suppose instead that the landowner is assumed to maximize only the net present value of harvest revenues. Then, for any initial condition, it would be optimal for the landowner to apply the Faustmann rotation age solution to the whole stand, and a multiple-age stand structure would not emerge. Thus, the models set forth in this chapter have relevance only when a landowner is assumed to behave according to a utility function that is concave either in consumption, amenities, or both, as would be the case with most non-industrial forest landowners discussed earlier in the book. Next we study a simplified version of this model that has been used to evaluate various steady states and assess whether normal forests are an optimal outcome. The nature of these steady states depends on whether or not there is a competing use of land. We will therefore discuss the two cases separately.

8.2   A Model with No Competing Land Use

Following Wan (1994) and Salo and Tahvonen (2002a), the model of section 8.1 can be simplified so that it is composed of only two age classes or vintages. De-

---

3. Uusivuori and Kuuluvainen (2005) wrote the first paper to develop the discrete version of the Hartman harvesting rule in an age class setting.

note land areas at the end of period t in age class one and two before harvesting by x_{1t} and x_{2t}. The authors assume that the old age class is harvested at the end of each period to obtain x_{2t} units of timber. The young age class is harvested from an area z_t at the end of time t and yields az_t units of timber, with $a < 1$. The total harvest in period t therefore comes from both young and old age class and is defined as $c_t = az_t + x_{2t}$. Given initial levels of timber stocks for each age class $x_{10}, x_{20} \geq 0$ at the end of period 0 before harvesting, the landowner's problem is to choose harvesting from the young age class, z_t, and harvesting from areas of age classes of age 1 and 2 for period $t+1$, $x_{1,t+1}, x_{2,t+1}$. Under these assumptions, the decision problem in (8.1)–(8.3) simplifies to

$$U = \underbrace{\max}_{(z_t, x_{1,t+1}, x_{2,t+1})} \sum_{t=0}^{\infty} \beta^t u(c_t), \text{ s.t.} \tag{8.7}$$

$$c_t = az_t + x_{2t}, \tag{8.8}$$

$$x_{1,t+1} = x_{2t} + z_t \tag{8.9}$$

$$x_{2,t+1} = x_{1t} - z_t \tag{8.10}$$

$$z_t \geq 0, \quad z_t - x_{1t} \leq 0. \tag{8.11}$$

Equation (8.8) defines the volume of timber harvested and therefore consumption in time t. The land area available for the young age class in the next period is equal to the harvested land areas for both age classes in this period, according to equation (8.9). Equation (8.10) describes the land area for the old age class in the next period, which is equal to what is left after harvesting the young age class in the present period. Equation (8.11) places non-negativity restrictions on the harvested young age class, and it restricts harvesting of the young age class to the land area for that class.

Age class models used to evaluate steady states have often been analyzed using dynamic Lagrange multiplier methods (we will see another example of this technique in chapter 12). Substitution of the consumption constraint leads to a dynamic Lagrangian function for the problem (8.7)–(8.11):

$$L = \sum_{t=0}^{\infty} \beta^t [u(az_t + x_{2t}) + \lambda_{1t}(x_{2t} + z_t - x_{1,t+1}) + \lambda_{2t}(x_{1t} - z_t - x_{2,t+1})$$

$$+ p_t z_t + q_t(z_t - x_{1t})]. \tag{8.12}$$

The first-order conditions are

(a) $\beta^{-t}\dfrac{\partial L}{\partial z_t} = au'(az_t + x_{2t}) + \lambda_{1t} - \lambda_{2t} + p_t + q_t = 0$

(b) $\beta^{-t}\dfrac{\partial L}{\partial x_{1,t+1}} = -\lambda_{1t} + \beta\lambda_{2,t+1} - \beta q_{t+1} = 0$

(8.13)

(c) $\beta^{-t}\dfrac{\partial L}{\partial x_{2,t+1}} = \beta u'(az_{t+1} + x_{2,t+1}) + \beta\lambda_{1,t+1} - \lambda_{2t} = 0$

(d) $p_t \geq 0, \quad z_t \geq 0, \quad p_t z_t = 0$

(e) $q_t \leq 0, \quad z_t - x_{1t} \leq 0, \quad q_t(z_t - x_{1t}) = 0,$

where λ_{1t}, λ_{1t}, p_t, and q_t are the Lagrange multipliers for the constraints (8.9)–(8.11), respectively, and must be solved along with consumption and harvesting. These first-order conditions characterize an optimal solution (for details, see Salo and Tahvonen 2002a). The multipliers λ_{1t} and λ_{2t} describe the marginal values in utility terms of forest land in the young and old forest age classes at the beginning of period $t+1$, respectively, while p_t and q_t are the Lagrange multipliers for the non-negativity and upper limit constraints (8.11) concerning harvesting of the young age class.

The interpretation of the first-order conditions is fairly straightforward. From equation (8.13a), at an interior solution with (8.11) not binding, i.e. $p_t = q_t = 0$, harvesting from the young forest age class yields marginal utility from consumption equal to $au'(az_t + x_{2t})$ and a marginal benefit from the land area of the young age class given by the multiplier λ_{1t}. The corresponding marginal cost is given by a decrease in the land area of the old age class captured in λ_{2t}. According to equation (8.13b), the interior solution must have $\lambda_{1t} = \beta\lambda_{2t}$ when the upper limit constraint is not binding; i.e., $q_{t+1} = 0$. In this case the marginal value of the land area of the young age class is equal to the marginal value of the land area for the old age class one period later, which is equal to the discounted marginal value of the old age class. Finally, according to equation (8.13c), the shadow value of land area in the old age class λ_{2t} is equal to the present value of the next period's marginal utility from harvesting the old age class defined by $\beta u'(az_{t+1} + x_{2,t+1})$, plus the present value marginal benefit of the land area of the young age class in the next period defined by $\beta\lambda_{1,t+1}$.

In this literature, normally the first-order conditions are expressed in more compact form as (see appendix 8.1 for details)

$$-au'(c_t) + \beta(1-a)u'(c_{t+1}) - \beta p_{t+1} - p_t - q_t = 0 \tag{8.14a}$$

$$q_t \leq 0, \quad x_{2,t+1} \geq 0, \quad x_{2,t+1}q_t = 0 \tag{8.14b}$$

$$p_t \geq 0, \quad 1 - x_{2,t} - x_{2,t+1} \geq 0, \quad p_t(1 - x_{2,t} - x_{2,t+1}) = 0, \tag{8.14c}$$

where the consumption equations in period t and $t+1$ are given by $c_t = a + (1-a)x_{2,t} - ax_{2,t+1}$ and $c_{t+1} = a + (1-a)x_{2,t+1} - ax_{2,t+2}$. In equations (8.14a)–(8.14c), the young age class is no longer present and we can simply write $x_t \equiv x_{2t}$ and define two forest management possibilities as a solution:

Regime $z_t = 0$: $1 - x_t - x_{t+1} = 0$; $x_{t+1} \geq 0$

Regime $z_t > 0$: $1 - x_t - x_{t+1} > 0$; $x_{t+1} \geq 0$

In the first regime, the young age class of period t is not harvested and becomes the old age class of period $t+1$. In the second regime, part of the period t age class is harvested, and the period $t+1$ old age class is smaller than the period t young age class. In both cases one has to account for the feasibility constraint that the land area of the old age class in period $t+1$ must be non-negative. This simply means that harvesting of the young forest age class cannot exceed its land area.

The steady state of this model has been characterized and involves two possible solutions. First, an optimal stationary state can exist, where x_t or $c_t = a + (1-a)x_{2t} - ax_{2,t+1}$ is constant over time; this steady state represents a normal forest structure in the long run. However, there are other possibilities that follow from the stability analysis of the steady state. There could be optimal steady states that involve cycles in harvesting over time around a Faustmann rotation age solution, implying that a normal forest structure with constant consumption would not in general be an optimal long-run solution. Which of these outcomes is obtained depends on whether the Faustmann rotation age is one or two periods for each age class, and whether both periods yield the same land value. We should note that we will comment on the results without a detailed analysis of stability of the possible steady states, although we will show the evaluation of one steady state in the next section. For more details, the reader is referred to the literature cited in this section and to a brief discussion of stability in chapter 12.

More specifically, Salo and Tahvonen (2002a) prove using the model of (8.14a)–(8.14c) that when there is no competing use of land there can be optimal steady states where consumption is not smoothed. If the Faustmann rotation age is two periods for each age class, then harvesting so that there is a normal forest (i.e., perfectly smoothing consumption) cannot be optimal because, as can be seen from the first-order conditions above, it would force the landowner to harvest earlier than the Faustmann age for the marginal land unit; doing so would mean that the landowner incurs potentially high costs given by foregone shadow values of future rents.

That said, this result clearly follows from the fact that forest land adjustment is measured continuously but time is measured in discrete units. Salo and Tahvonen (2000, 2003, 2004), for example, find that allowing for continuous time and discrete forest land leads to a solution where equilibrium cycles vanish in the limit. Further, Salo and Tahvonen (2002b) consider the model earlier with more than two age classes and show numerically that cycles can exist that dispel normal forest optimality on a global scale. The literature has also established that whether discounting is assumed to be present or absent has an important bearing on the existence of cycles or normal forest states. Clearly, the results of these models are highly sensitive to the assumptions underlying the landowner's forest management problem.

8.3   A Model with Competing Land Uses

Suppose that the landowner can allocate forest land costlessly between forestry and some alternative land use. Assume that utility from this alternative land use is given by $v(y_t)$, where $y_t = 1 - x_{1t} - x_{2t}$ is the land area allocated to the nonforest use, with $v'(y_t) > 0$ and $v''(y_t) < 0$. Now, all harvested land need not be used for forestry purposes, and we need only to postulate an upper bound for replanting, i.e., $x_{1,t+1} \leq x_{2t} + z_t + 1 - x_{1t} - x_{2t}$. This means that replanting cannot exceed the sum of harvested forest land and the new nonforested land area that is brought into forest production. Using equation (8.10), we have $x_{2,t+1} = x_{1t} - z_t$, which can be used to eliminate z_t in the earlier equation, arriving at $x_{1,t+1} \leq 1 - x_{2,t+1}$. Eliminating z_t from (8.10) and (8.11) yields $x_{1t} \geq x_{2,t+1}$ and $x_{2,t+1} \geq 0$, respectively.

A representative landowner is now assumed to choose a land allocation between young and old forest age classes and between forest and the competing uses of land to solve the following optimal harvesting and land allocation problem:

$$\max_{(x_{1,t+1}, x_{2,t+1})} \sum_{t=0}^{\infty} \beta^t [u(c_t) + v(1 - x_{1t} - x_{2t})], \text{ s.t.} \tag{8.15}$$

$$c_t = x_{2t} + a(x_{1,t} - x_{2,t+1}), \tag{8.16}$$

$$x_{2,t+1} \leq x_{1t}; \tag{8.17a}$$

$$x_{1,t+1} + x_{2,t+1} \leq 1 \tag{8.17b}$$

$$x_{1,t+1}, x_{2,t+1} \geq 0. \tag{8.17c}$$

The dynamic Lagrangian function for this problem (8.15)–(8.17) can be written as

$$L = \sum_{t=0}^{\infty} \beta^t [u(c_t) + v(1 - x_{1t} - x_{2t})] + p_t(x_{1t} - x_{2,t+1}) + \lambda_t(1 - x_{1,t+1} - x_{2,t+1}), \quad (8.18)$$

where p_t and λ_t are the Lagrange multipliers associated with the constraints in (8.17).

The first-order conditions for (8.18) are

(a) $\beta^{-t} \dfrac{\partial L}{\partial x_{1,t+1}} = a\beta u'(c_{t+1}) - \beta v'(1 - x_{1,t+1} - x_{2,t+1}) + \beta p_{t+1} - \lambda_t \leq 0$

(b) $\beta^{-t} \dfrac{\partial L}{\partial x_{2,t+1}} = -au'(c_t) + \beta u'(c_{t+1}) - \beta v'(1 - x_{1,t+1} - x_{2,t+1}) - p_t - \lambda_t \leq 0$

(c) $x_{i,t+1} \geq 0, \quad x_{i,t+1}(\partial L/\partial x_{i,t+1}) = 0, \quad i = 1,2,$ (8.19)

(d) $p_t \geq 0, \quad p_t(x_{1,t} - x_{2,t+1}) = 0,$

(e) $\lambda_t \geq 0, \quad \lambda_t(1 - x_{1,t+1} - x_{2,t+1}) = 0.$

Clearly, one can see that a higher marginal utility of consumption for u' or v', respectively, shifts land allocation toward forestry or toward the alternative use. Otherwise, the interpretation is similar to the first-order conditions in (8.13).

This model leads to different results than discussed in the absence of an alternative non-forest land use. Now, there is an interesting new consideration. For example, if the Faustmann value for the marginal bare land unit when managed under a normal forest exceeds the present value of the price of the marginal land unit, then new non-forest land should be brought into production to keep consumption constant over time. Salo and Tahvonen (2002a) and (2002b) find, in the discrete time continuous land model of (8.15)–(8.17), that the normal forest is a unique steady state (a saddle point) where convergence to the normal forest is a stable path (a proof of this optimality is shown in appendix 8.2). This implies that cycles will vanish as an optimal solution in the long run and only the oldest age class will be harvested in each period, but land allocation between forestry and the alternative use will be continuously adjusted to keep consumption and harvesting constant. This is not the case when there is no alternative use of land, for then it is costly for a landowner to smooth consumption given he can only adjust the rotation ages of age classes in the forest to do so. This implies in most cases that any given age class would have to be harvested before their Faustmann age, which results in costs, as we have seen. When additional land is available at a given rental value, there remains the option of the landowner shifting land as an additional choice to smooth consumption. The desirability of doing so, and hence the ultimate size of the normal forest, is determined by equality between the marginal value of land in the alternative use and the Faustmann value for bare land that is brought into forest production.

A critical assumption in these sets of results and in the model of (8.15)–(8.19) is that the transfer of land is costless. Under a more realistic assumption of costly land conversion, Salo and Tahvonen (2004) provide numerical examples showing that harvesting can again become cyclical around the normal forest even timber flow solution. Furthermore, in models such as the one here, the normal forest is typically defined under the assumption of homogeneous forest land. In other words, all age classes are assumed to have equivalent growth possibilities on any given land unit. The alternative is to consider a discrete-time model in which each forest age class has its own (distinct) growth function (Salo and Tahvonen 2002b). They find under discounting and no alternative use of forest land that the discrete-time forestry model has a cyclical stationary state not characterized by a normal forest even flow of timber. This is because the marginal cost of smoothing the age class structure toward the normal forest is always positive, but since the state space is continuous, the marginal benefit from this type of smoothing approaches zero as the age class structure approaches a normal forest. Hence, even with heterogeneous forestland units, there will not be convergence to a normal forest when time is modeled in discrete units.

Finally, an important problem is to understand how amenities modify the basic results concerning normal forest optimality; indeed, preserving old-growth forests may not be compatible with even-flow harvesting. Some preliminary work here has been done by Tahvonen (2004a), who studied a forestland allocation problem between timber production and maintenance of old-growth forests. He shows in the long run that age classes subject to timber production proceed according to the Faustmann rotation solution, but the forest age class structure might not be representative of a normal forest. The reader interested in more details of these models is referred to a recent survey by Tahvonen (2004b).

8.4 Carbon Policies

Age class models provide a powerful tool for analyzing forest policy beyond the single-stand assumption used throughout most of the literature. In this section we focus on climate policies and especially the design of instruments to deal with the production of carbon dioxide and green house gas emissions. We first touched on this in chapter 3 when we considered modifications of the single-stand amenity problem. Now we trace how accounting for multiple age classes affects those findings. Our presentation closely follows the model in Uusivuori and Laturi (2007), who compare the impacts of carbon rental payments and silvicultural subsidies to forest landowners holding both single and multiple stands.

Abstracting from silvicultural investments and their impacts on forest growth, we can modify the basic age class model in (8.1)–(8.3) to include carbon rental payments made to landowners for holding unharvested forest stocks. Assuming some conversion factor for translating wood biomass volume to tons of carbon, denote the price of carbon by p_c. The value of land when forest grows on each land unit devoted to forestry can be expressed using the bare land value and the net revenue from harvesting the existing timber. For carbon policies, it makes sense to analyze the case where a growing forest exists because growth increases sequestered carbon stored on the site. Thus, we must define land values for bare land (i.e., the steady state) and for land with a growing forest.

We start with the value of bare land (x^0) and let a_1^0 equal the share of initial bare land harvested at the end of the first period after forest growth on the land occurs. On remaining bare land, $(1 - a_1^0)x^0$, it is assumed that the forest will be grown for at least two periods (i.e., not harvested in one period). If it is harvested in the second period, then a share a_2^0 of this land is cut leaving $(1 - a_2^0)(1 - a_1^0)x^0$ of the initial land left to continue with a growing forest. After the next harvest, we therefore have the size of the remaining forest equal to: $(1 - a_3^0)(1 - a_2^0)(1 - a_1^0)x^0$, and this pattern continues over time. Other than these definitions, the same notation will be used that is present in (8.1)–(8.3), so that r is the real interest rate, k is regeneration cost, and the periodic volume of forest that is grown for j periods is $q_{j,h}^0$, with $h = 1 \ldots j$. Let the timber price at time t equal p_t. Using all of these definitions, the value of bare land in the presence of carbon policies can be expressed in a condensed form as:

$$LV(x^0) = -kx^0 + (1+r)^{-1}(p_t q_{1,1}^0 - k)\frac{(1+r)}{r}a_1^0 x^0$$

$$+ \sum_{j=2}^{\infty}\left[(1+r)^{-j}(p_t q_{j,j}^0 - k) + \sum_{h=1}^{j}(1+r)^{-h}rp_c\delta q_{j,h}^0\right]$$

$$\times\left[\frac{(1+r)^j}{(1+r)^j - 1}a_j^0\prod_{h=1}^{j-1}(1 - a_h^0)x^0\right]. \tag{8.20}$$

In (8.26), the first term indicates the initial forest investment and subsequent terms are revenue from harvesting and carbon rental payments, respectively. Next we can express land value from the growing forest stock using (8.20). We denote the per hectare land value by $LV(x^0)/x^0$ and obtain

$$
LV(x^i) = \left[p_t q_{0,0}^i + \frac{LV(x^0)}{x^0} \right] a_0^i x^i
$$

$$
+ \sum_{j=1}^{\infty} \left[(1+r)^{-j} \left[p_t q_{j,j}^i + \frac{LV(x^0)}{x^0} \right] + \sum_{h=0}^{j-1} (1+r)^{-h} r p_c \delta q_{j,h}^0 \right]
$$

$$
\times \left[a_j^i \prod_{h=0}^{j-1} (1 - a_h^i) x^i \right]. \tag{8.21}
$$

The landowner maximizes utility from consumption subject to net revenue from forestry, and subject to constraints defining forest dynamics and the intertemporal budget constraint:

$$
\max_{a_{ti}, c_t} \sum_{t=0}^{\infty} \beta^t u(c_t), \text{ s.t.} \tag{8.22a}
$$

$$
\sum_{t=0}^{\infty} (1+r)^{-t} c_t \leq \sum_{i=0}^{n} LV(x^i). \tag{8.22b}
$$

The optimal choices of consumption and harvesting in the presence of carbon rental payments are obtained from the first-order conditions as before and become

$$
\frac{\lambda}{u'(c_t)} = \left(\frac{1+r}{1+\rho} \right)^t. \tag{8.23}
$$

This should be a familiar condition by now. From (8.23), if the rate of time preference is lower (or higher) than the real interest rate, then optimal consumption is increasing (or decreasing) in time.

For the optimal harvesting choice, we first solve for the bare land rotation age and then the rotation age assuming standing timber is present. The bare land rotation age represents the long-run solution, while the rotation age for standing forests provides the short-run solution. The bare land case yields

$$
\left[(1+r)^{-j} (p_t q_{j,j}^0 - k) + \sum_{h=1}^{j-1} (1+r)^{-h} r p_c \delta q_{j,h}^0 \right] \frac{(1+r)^j}{(1+r)^j - 1}
$$

$$
> (<)
$$

$$\left[(1+r)^{-(j+1)}\left(p_t q^0_{j+1,j+1} - k\right) + \sum_{h=1}^{j}(1+r)^{-h}rp_c\delta q^0_{j+1,h}\right]\frac{(1+r)^{j+1}}{(1+r)^{j+1}-1}$$

$$\Rightarrow a^0_j = 1 \ (a^0_j = 0). \tag{8.24}$$

Equation (8.24) extends the previous version of the discrete Faustmann model in this chapter to cover carbon rental payments. The optimal rule for harvesting in the case of a growing forest can be expressed as follows:

$$\left[(1+r)^{-j}\left[p_t q^i_{j,j} + \frac{LV(x^0)}{x^0}\right] + \sum_{h=1}^{j-1}(1+r)^{-h}rp_c\delta q^0_{j,h}\right]$$

$$> (<)$$

$$\left[(1+r)^{-(j+1)}\left[p_t q^i_{j+1,j+1} + \frac{LV(x^0)}{x^0}\right] + \sum_{h=1}^{j}(1+r)^{-h}rp_c\delta q^0_{j+1,h}\right]$$

$$\Rightarrow a^0_j = 1 \ (a^0_j = 0). \tag{8.25}$$

Equation (8.25) has the same interpretation as before. In the initial forest, a stand should be cut (or left uncut) after j periods if the marginal benefits of doing so are larger (or smaller) than those when the stand is cut after $j+1$. This condition applies to each stand $i = 1, \ldots n$. Carbon rental payments affect both terms. It can be shown by comparative statics that a higher price on carbon increases rotation age, provided that

$$\left[\sum_{h=1}^{j-1}(1+r)^{-h}q^0_{j,h}\right]\frac{(1+r)^j}{(1+r)^j-1} < \left[\sum_{h=1}^{j}(1+r)^{-h}q^0_{j+1,h}\right]\frac{(1+r)^{j+1}}{(1+r)^{j+1}-1}.$$

Uusivuori and Laturi (2007) resort to numerical analysis to shed further light on carbon policies. They demonstrate that carbon rental payments increase timber volume in both the short run and the long run, as expected. When sequestered carbon payments are discounted, the age class structure of the forest matters for carbon policies. Now, the supply of carbon from private forests is higher in forests with older age class structures. If carbon benefits to the landowner are not discounted, then the age class structure does not matter. This is an interesting finding and complements lessons derived from single stand-based analysis.

8.5   Uneven-Aged Forest Management

Thus far in this chapter we have discussed models based on multiple even-aged age classes. A landowner was assumed to manage multiple stands, but each distinct stand had trees of equal age. Throughout the history of forest science, problems unique to uneven-aged forests have mystified the minds of foresters and economists alike. Uneven-aged forests are those with each stand containing trees of different ages, heights, and diameters. In uneven-aged forests, there are certainly distinct age classes, or diameter classes to be more precise, but all grow together on the same unit of land. Further, how each of these age class grows depends on the presence of other age classes in the stand. This distinction for forest growth does not describe even-aged–based modeling, because each land unit has only a single age class that grows in isolation of other age classes. The questions of how to manage uneven-aged stands have existed for much longer than questions surrounding even-aged management. In fact, even-aged forestry as we have discussed in much of the book emerged only with the evolution of intensive agriculture a couple of hundred years ago.

Given its complexity, the rigorous study of uneven-aged management is quite recent. The first clear documentation of an uneven-aged management system occurred in the French Jura Mountains in the early 1700s. Even today, true uneven-aged management regimes are the exception rather than the rule in temperate and boreal forests (Buongiorno 1996). This is somewhat due to early failures of uneven-aged forest management in early 1900s in the United States and Scandinavia. Interest is now shifting back to uneven-aged management for several reasons, not the least because of environmental considerations. Since uneven-aged management tends to retain forest cover permanently on a site even with a series of harvests over time, the resulting stand structures are conducive to recreation, wildlife management, and biodiversity conservation among other amenities.

The intellectual history of economic analysis applied to uneven-aged stands is relatively young and dates from the 1970s. Prior to this, uneven-aged management was based on matrix models of population mathematics first described by Leslie (1945) and Usher (1966). The matrix model can explain how trees in a given diameter class move into the next highest diameter class, assuming an underlying growth function that accounts for how different diameter classes in the same stand interact with each other. The economics models built upon this type of transitional dynamic suppose that a landowner must choose the number of harvested trees in each diameter class and the frequency of harvesting over time. Combining the matrix of growth and harvesting fully characterizes the state of the uneven-aged stand at any point in time. When harvests are made, all trees on the site need not be removed, and instead trees can be felled from many diameter classes while some diameter classes will be left untouched. This "selective harvesting" of trees in the stand is a key concept of optimal uneven-aged management. Selective harvesting has many applications in both developed countries and even to deforestation problems, as we have seen in chapter 6 and will further see in chapter 12.

Early seminal articles on uneven-aged stand economics were Adams and Ek (1974), Adams (1976), and Buongiorno and Michie (1980). An interesting and equally seminal contribution is due to Haight (1985), who contrasts these early articles by comparing static equilibrium concepts to truly dynamic ones. Haight and Getz (1987) go further to provide a comprehensive dynamic analysis of uneven-aged management. Other early applications include Mendoza and Setyarso (1986) and Pukkala and Kohlström (1988). Recently, Tahvonen (2007) has returned to these questions by treating as endogenous the choice between optimal even-aged and uneven-aged management given an arbitrary initial state of the forest. Buongiorno et al. (1995) have also provided the first application of an uneven-aged forestry model to the optimal maintenance of forest biodiversity.

Uneven-aged models are inherently more complex than even-aged models. In fact, complete analytical solutions in these situations are often illusive. In this section we draw on the work of Haight (1985) to discuss optimality conditions for dynamic and static uneven-aged harvesting problems over an infinite time horizon and the presence of M diameter classes. We then show how to write an equivalent and easily tractable two-period version of Haight's model.

8.5.1   Dynamic Problem

The key piece of any uneven-aged management model is a formal description of diameter classes, the number of trees in each diameter class, and the transition of trees from one diameter class to another as the forest grows or is harvested. In this set-up, the diameter class replaces the age class as a key state describing the forest. Let M denote the number of diameter classes in a forest of any given size. The growth of one diameter class into the next higher one is called "upgrowth," but "ingrowth" also refers to the appearance of new trees into any diameter class. Thus, the emergence of young trees in the smallest diameter after this class is harvested is defined by ingrowth. Both growth and mortality are accounted for in specifying upgrowth and ingrowth.

Let X_{ij} denote the number of trees per acre in diameter class j at the beginning of period i before harvesting, and S_{ij} denote the number of trees per acre in the same diameter class j at the beginning of period i but *after* harvesting. The variable S_{ij} is sometimes called the residual distribution of the stand. The difference $(X_{ij} - S_{ij})$ therefore gives the number of trees harvested from diameter class j in period i. Forest upgrowth is explained by the following transition equation; $X_{i+1j} = S_{ij} + f_{ij}(S_{i1}, \dots S_{iM})$, where $f_{ij}(S_{i1}, \dots S_{iM})$ is the forest growth function. The fact that the cross derivative of this function does not vanish, $\partial f_{ij}/\partial S_{ij}S_{ik} \neq 0$, is one of the features that distinguishes these models from those discussed earlier in this chapter. Each diameter class is not necessarily harvested every year. Rather, it is harvested every t years, so that t is endogenous.

Following Haight and letting the number of trees in each age class after cutting be the choice variable, and letting p_j denote the price of trees in diameter class j, the infinite time horizon problem of the landowner is to choose the residual number of trees and the time period of cutting for each diameter class,

$$\underset{S_{ij}}{Max}\ PNW = \sum_{i=0}^{\infty} \sum_{j=1}^{M} \frac{(X_{ij} - S_{ij})p_j}{(1+r)^{it}}\ \text{s.t.} \tag{8.26}$$

$$X_{i+1j} = S_{ij} + f_{ij}(S_{i1}, \dots S_{iM}) \tag{8.27a}$$

$$S_{ij} \geq 0 \tag{8.27b}$$

$$X_{ij} - S_{ij} \geq 0 \tag{8.27c}$$

$$X_{0j} \tag{8.27d}$$

where t in the discount factor now represents the number of years in the cutting cycle for each diameter class j, i.e., t measures the frequency of cutting. Harvest revenue is maximized over an infinite time horizon by choosing the number of trees saved from cutting in each diameter class. Equation (8.27a) determines upgrowth and ingrowth for every diameter class, equations (8.27b) and (8.27c) are the relevant non-negativity constraints, and (8.27d) defines the initial state of the forest. The model can be solved using dynamic programming methods (see Haight 1985 for the details and chapter 12 for a discussion of these methods). This solution reveals a basic message of uneven-aged models explained through marginal production rules indicating how the standing stock in each diameter class is managed. Haight (1985) shows the following marginal production rule defines the set of all M diameter classes in any period i:

$$\frac{p_j}{(1+r)^{it}} - u_{ij} + v_{ij} = \sum_{k=1}^{M} \left[\frac{p_k}{(1+r)^{(i+1)t}} + v_{i+'1k}^* \right] \frac{\partial f_{ik}}{\partial S_{ij}} + \frac{p_j}{(1+r)^{(i+1)t}} + v_{i+1j};$$

$$j = 1, \dots M \tag{8.28}$$

In equation (8.28) u_{ij} and v_{ij} are Kuhn-Tucker multipliers associated with constraints (8.27b) and (8.27c), respectively. The LHS of equation (8.28) is the marginal opportunity cost of saving an additional tree in diameter class j. At an interior solution $(0 < S_{ij} < X_{ij})$ this must equal the discounted stumpage price (first term on LHS). When $X_{ij} - S_{ij} = 0$, it includes, in addition to the discounted price, a shadow price of harvesting at least one tree in the diameter class (third term on LHS). When $S_{ij} = 0$, the discounted price is now reduced by the shadow value of leaving (not harvesting) an additional tree in the diameter class (second term on the LHS). The RHS of (8.28) is the marginal revenue of leaving an additional tree (not cutting) in diameter class j. It is defined by the sum of the discounted value of growth resulting from the additional tree over all diameter classes, and the discounted value of the tree itself; the Kuhn-Tucker multipliers have the same interpretation as above. Assuming an interior solution over all diameter classes reduces (8.28) to the following rule

$$\frac{p_j}{(1+r)^{it}} = \frac{p_j}{(1+r)^{(i+1)t}} + \sum_{k=1}^{M} \left[\frac{p_k}{(1+r)^{(i+1)t}} \right] \frac{\partial f_{ik}}{\partial S_{ij}} \tag{8.29}$$

At an interior solution (8.29) determines the equilibrium residual forest stock in each diameter class and the harvested number of trees in each class defined by equilibrium ingrowth and upgrowth as functions of stumpage prices and the real interest rate.

8.5.2   Static Problem

An analytically more tractable static version of (8.26) can be developed as follows. Recall the constraint set (8.27a)–(8.27d). Suppose we add one additional condition:

$$X_{1+1j} = X_{ij} \quad \text{for all } i = 1, \dots \infty \text{ and } j = 1, \dots M \tag{8.30}$$

Equation (8.30) implies that an equilibrium harvest regime is achieved after the first harvest from the existing stand. Put differently, throughout time the landowner is required to leave the same amount of trees in each diameter class as he did with the first harvest. Using this constraint, the target function (8.26) can be reexpressed as

$$\underset{S_{1j}}{Max} \ PNW = \sum_{j=1}^{M} (X_{0j} - S_{0j}) p_j + \sum_{j=1}^{M} \frac{(X_{1j} - S_{1j}) p_j}{(1+r)^t - 1}, \tag{8.31}$$

where $(1+r)^t - 1$ is a discrete discount factor that serves to discount harvest revenue over an infinite time horizon (see chapter 2). Now, using the growth dynamics embodied in equations (8.27a), equation (8.31) can be written

$$\underset{S_{1j}}{Max} \ PNW = \sum_{j=1}^{M} (X_{0j} - S_{0j}) p_j + \sum_{j=1}^{M} \frac{f_{0j}(S_{01}, \dots, S_{0M}) p_j}{(1+r)^t - 1}. \tag{8.32}$$

The maximization problem (8.32) is constrained by the following relationships

$$S_{0j}(1+r)^t[(1+r)^t - 1] \geq 0 \tag{8.33a}$$

$$f_{0j}(S_{01}, \ldots S_{0M})/[(1+r)^t - 1] \geq 0 \tag{8.33b}$$

$$X_{ij} - S_{ij} \geq 0 \tag{8.33c}$$

$$X_{0j} \tag{8.33d}$$

Thus, the economic problem of the landowner is now to find a steady-state distribution of $S_{01}, \ldots S_{0M}$ that maximizes the value of the first harvest plus net present values of future harvests over infinite time horizon. Again, the constraints of the problem impose non-negative residual stocking level and harvests. Denoting the Kuhn-Tucker multipliers associated with the constraints by a, b, and c, Haight (1985) expresses the marginal production rules as

$$p_j - a_j^* + c_j^* = \sum_{k=1}^{M} \frac{(p_k + b_k)}{(1+r)^t} \frac{\partial f_{0k}}{\partial S_{0j}} + \frac{p_j}{(1+r)^t} + \frac{c_j^*}{(1+r)^t}; \quad j = 1, \ldots M \tag{8.34}$$

The interpretation of this rule is similar to the dynamic rule above. For example, the marginal cost of saving a tree in diameter class j should equal the marginal revenue it can provide when saved. At an interior solution this rule reduces to

$$p_j = \sum_{k=1}^{M} \frac{p_k}{(1+r)^t} \frac{\partial f_{0k}}{\partial S_{0j}} + \frac{p_j}{(1+r)^t}; \quad j = 1, \ldots M \tag{8.35}$$

Comparing equations (8.34) and (8.35) reveals a basic difference with respect to equations (8.29) and (8.28). The static optimum depends on the initial number of trees in each diameter class, whereas the dynamic optimum is invariant to this initial condition. Indeed, Haight demonstrates that, despite their common goal of maximizing the present discounted value of harvest returns, the dynamic model defines inter-temporal solutions for both transition and equilibrium management regimes, but the static model defines a production rule for equilibrium (steady state) management only. Further, the dynamic solution typically leads to a higher present discounted value and equals the static solution only under certain special cases.

8.5.3 Timber Market Considerations

The message to take from Haight (1985) is that the static uneven-aged problem is fairly restrictive. While Haight and Getz (1987) extend the dynamic model to incorporate many additional features, the static version may have some applications in the qualitative analysis of forest policies and forest landowner behavior (e.g., Laaksonen-Craig and Ollikainen 2008). Another interesting point is that the uneven-aged forest management model has much in common with the two-period model we discussed in chapter 4. For example, consider equation (8.32) and assume that the landowner harvests all diameter classes at the same intensity. Then, given an original state of the forest we can heuristically write $Q = \sum_{j=1}^{M} X_{0j}$ and $S_{0j} = S > 0$, for all $j = 1, \ldots M$, so that forest growth can be expressed in reduced form by the function $f(S)$. Assuming that the stumpage price differs in the current and future period, we can then write the forest management problem in terms of saved (residual) trees in current and in the future period as

$$\underset{S}{Max}\ PNW = (Q - S)p_1 + \frac{p_2 f(S)}{(1 + r)^t - 1} \qquad (8.36)$$

The first-order condition for the optimal number of saved trees is

$$V_S = -p_1 + p_2 \frac{g'(S)}{(1 + r)^t - 1} = 0 \qquad (8.37)$$

Using $R = (1 + r)^t - 1$ again as the discount factor, we obtain the following harvesting rule:

$$p_1 = p_2 R^{-1} g'(S) \quad \text{or} \quad Rp_1 = p_2 g'(S) \qquad (8.38)$$

This rule is consistent with what we found in chapter 4. It is optimal to save the trees up to the point where the price of saving the last tree equals the growth in value of that tree. Note that the number of saved trees also uniquely defines first $(x_1 = Q - S)$ and the future, steady-state timber supply. Despite the similarity to chapter 4, this two-period model has an uneven-aged forest management technology but extends to an infinite time horizon. That is, first period harvesting moves the forest with its diameter classes to a steady state, and future harvesting (in second and subsequent periods) is then defined by sustained steady-state forest growth.

8.6   Summary

The management of multiple even-aged stands, and the resulting questions concerning the optimal distribution of forest age classes, have been exceptionally challenging theoretical problems for decades. Early suggestions of harvesting to achieve normal forest structures were often based more on noneconomic than economic arguments. The most typical example of these arguments was the even-flow harvest requirement as sole justification for achieving normal steady state forests.

The theoretical work by Mitra, Wan, Salo, and Tahvonen discussed in this chapter characterizes optimal age-class structure in discrete time-continuous land area or continuous time-discrete land area based Faustmann models. Mitra and Wan (1985, 1986) first demonstrated that for a zero discount rate, a normal steady state emerges as the optimal long-run forest structure. However, Salo and Tahvonen (2000, 2003, 2004) find a different result, that in the absence of competing land uses, the optimal steady-state forest may be a cyclical solution around the Faustmann rotation age. These cycles would vanish if time were continuous and land area were discrete. They also find the same result if there are competing non-forest land use forms. The work by Uusivuori and Kuuluvainen (2005) extends this work by introducing multiple age classes and amenity benefits in a Hartman model.

Models that incorporate age-class distributions of forest stands are well suited to empirical analyses seeking to trace out timber supply on regional or national levels. This problem was originally studied by Lyon and Sedjo (1983). Postulating an arbitrary, or status quo distribution of age classes, one can use such a model to determine how prices, interest rates, planting costs, and other exogenous variables impact forest management, timber supply and amenity benefits in both the short and long run, thereby combining features of the long run Faustmann/ Hartman frameworks and the short-run two-period framework. We expect the number of age class-specific studies to increase in the future for these reasons.

The age class model has not been applied much to analyze policy instrument issues. However, there are clearly applications to biodiversity and forest taxation questions. It would be quite interesting to compare whether the policy conclusions derived in single stand problems are still valid for the more realistic age class models. An interesting study on the impacts of forest taxation is due to Uusivuori and Kuuluvainen (2008). They find that the results derived for a single stand hold for an age class-specific model when applied in the long run to an already achieved stable steady state. However, forest taxes tend to influence convergence of stand structures to steady states, and thus this role of taxation cannot be dismissed. This area of study remains quite open to this day, and many interesting policy problems remain unstudied in age class models.

Appendix 8.1   Reduction of First-Order Conditions (8.13a)–(8.13e)

From Salo and Tahvonen (2002a), equation (8.13a) can be written: $\lambda_{2t} = au'(az_t + x_{2t}) + \lambda_{1t} + p_t + q_t$. Substituting the RHS of this for λ_{2t} in equation (8.13c) yields

$$\beta u'(az_{t+1} + x_{2,t+1}) + \beta\lambda_{1,t+1} - au'(az_t + x_{2t}) - \lambda_{1t} - p_t - q_t = 0. \tag{A8.1}$$

Rewriting (8.13a) as one period ahead $\lambda_{1,t+1} = -au'(c_{t+1}) + \lambda_{2,t+1} - p_{t+1} - q_{t+1}$, we can express (A8.1) as follows: $-au'(c_t) + \beta u'(c_{t+1}) - a\beta u'(c_{t+1}) + \beta\lambda_{2,t+1} - \beta p_{t+1} - \beta q_{t+1} - \lambda_t - p_t - q_t = 0$. According to (8.13b), it must be true that $-\lambda_{1t} + \beta\lambda_{2,t+1} - \beta q_{t+1} = 0$ so that the above equation can be described as

$$-au'(c_t) + \beta u'(c_{t+1}) - a\beta u'(c_{t+1}) - \beta p_{t+1} - p_t - q_t = 0. \tag{A8.2}$$

Defining total forest land as $x_{1t} + x_{2t} = 1 \ \forall t$, equation (8.9) of the text can be written as $z_t = 1 - x_{2t} - x_{2,t+1}$, and using equation (8.8) of the text we have $c_t = a + (1-a)x_{2,t} - ax_{2,t+1}$. The first-order conditions (8.13a)–(8.13e) can now be written, as presented in equations (8.14a)–(8.14c).

Appendix 8.2   Proof of Normality as an Optimal Stable Steady State

The proof, taken from Salo and Tahvonen, proceeds as follows: when $z_t = 0$, $y_t > 0$, there is no harvesting of the young age class and part of the land area is allocated to a non-forest use. The first-order conditions (8.19a) and (8.19b) are re-expressed as follows:

$$au'(x_{2,t+1}) - v'(y_{t+1}) + p_{t+1} = 0 \tag{A8.3}$$

$$-au'(x_{2t}) + \beta u'(x_{2,t+1}) - \beta v'(y_{t+1}) - p_t = 0. \tag{A8.4}$$

For period $t-1$, equation (A8.3) is written as $-p_t = au'(x_{2t}) - v'(y_t)$. In the regime of no harvesting of the young forest age class, i.e., $z_t = 0$, we have $x_{1,t+1} = x_{2,t+1}$ and $x_{1t} = x_{2,t+1}$. Thus equation (A8.3) becomes

$$g \equiv \beta u'(x_{2,t+1}) - \beta v'(1 - x_{2,t+2} - x_{2,t+1}) - v'(1 - x_{2,t+1} - x_{2t}) = 0. \tag{A8.5}$$

This is a second-order nonlinear difference equation. Denoting $x_2 = x$, it can be written as

$$u'(x_{t+1}) - v'(1 - x_{t+2} - x_{t+1}) - \frac{1}{\beta}v'(1 - x_{t+1} - x_t) = 0. \tag{A8.6}$$

To evaluate stability of this nonlinear steady state, we take a Taylor series approximation around the stationary state $\bar{x}$ to obtain $v''(x_{t+2} - \bar{x}) + u''(x_{t+1} - \bar{x}) + v''(x_{t+1} - \bar{x}) + (1/\beta)v''(x_{t+1} - \bar{x}) + (1/\beta)v''(x_t - \bar{x}) = 0$. Or,

$$(x_{t+2} - \bar{x}) + \left(\frac{u''}{v''} + 1 + \frac{1}{\beta}\right)(x_{t+1} - \bar{x}) + \frac{1}{\beta}(x_t - \bar{x}) = 0. \tag{A8.7}$$

To solve this equation we form the characteristic equation of (A8.7),

$$\theta(s) = s^2 + n_1 s + n_2 = 0, \tag{A8.8}$$

where $n_1 = [(u''/v'') + 1 + (1/\beta)]$ and $n_2 = 1/\beta$. The roots of the polynomial in (A8.8) are

$$s_1, s_2 = \frac{-n_1 \pm \sqrt{n_1^2 - 4n_2}}{2}.$$

One can show that both roots are real with values strictly below and above -1. Therefore the steady state is a saddle point. Q.E.D.

III Advanced Topics

9   Uncertainty in Life-Cycle Models

Economic behavior is influenced by various types of uncertainties. Uncertainty is as likely to occur in timber markets as it is in other sectors of the economy. The long-term nature of forest production means forest landowners may never know the value of future parameters when they make management decisions. Future timber price is probably the easiest type of uncertainty to envision, but the landowner can hardly be expected to know future real interest rates and even the pattern of forest growth in the face of natural events such as fire. A landowner may also be uncertain about several parameters when making decisions. In this chapter we analyze the impacts of uncertainty using two-period models modified from chapter 4. This requires asking how uncertainty should enter a landowner's objective function and how decisions depend on the landowner's attitude toward risk.

Uncertainty in timber markets can certainly change the ways in which forest policy instruments such as taxes affect incentives for a landowner to provide timber and amenity services. As we have already discussed, governments can influence these incentives through the design of policies. We therefore also examine a basic question of whether forest policies can and should be tailored in such a way that they correct for the possible biases that uncertainty causes in the decisions of private landowners. Answering this question requires understanding how uncertainty enters the objective function of the government, and whether the government should be regarded as risk averse or risk neutral when forming its policies.

This chapter makes extensive use of the economic theory of risk-bearing behavior, which dates back to 1960s, notably to the work of Kenneth Arrow (1965). While his expected utility theory ideas have been widely applied throughout economics, a slightly different path is taken in this chapter that is more relevant to life-cycle forestry modeling. We will discuss models of uncertainty using the analytically simple and economically intuitive approach of nonexpected utility theory, which is a construct for distinguishing preferences for deterministic consumption from preferences for risky consumption in intertemporal models (see Kreps and Porteus 1978, Selden 1978, Epstein and Zin 1989).

The first articles in forest economics dealing with timber price uncertainty in two-period models are by Johansson and Löfgren (1985), Koskela (1989a, b), and Ollikainen (1990). Interest rate uncertainty and multiple sources of uncertainty were studied by Ollikainen (1991, 1993). Extensions to joint production of timber harvesting and amenities under price uncertainty are due to Koskela and Ollikainen (1997b), and Koskela and Ollikainen (1998) for forest growth risk. Uusivuori (2002) examines the effects of price uncertainty under nonconstant risk attitudes, while Gong and Löfgren (2003a) study timber supply when harvest revenue can be allocated between riskless and risky assets under uncertain future timber prices.

All of the papers cited here applied the expected utility hypothesis (EUH) approach to obtain results, with the exception of those that discuss the use of non-expected utility measures, namely, recursive preferences, in forest management decisions.

In this chapter the existing literature is modified and extended to provide a synthesis of uncertainty in two-period models using a Kreps-Porteus-Selden (K-P-S) nonexpected utility approach (see Peltola and Knapp 2001 and Koskela and Ollikainen 1999b for forestry examples). It is not our intention to argue that non-expected utility approaches are always superior to expected utility approaches when dealing with uncertainty. Rather, the K-P-S approach is useful because it invariably leads to simple and intuitive expositions of results. There is also an additional advantage in that one can produce an easily interpretable expression for risk attitudes of landowners and the government.

The rest of the chapter is structured as follows: In section 9.1 we outline and describe the basic types of uncertainties in detail and discuss how to represent the risk-bearing behavior of landowners. Section 9.2 is devoted to examining risk effects and the effects of forest taxation on harvesting decisions. The same issues are then considered when amenities are introduced in section 9.3. The socially optimal choice of taxation policies under various uncertainties is examined in section 9.4.

9.1 Uncertainties and Risk Preferences

Recall that the harvesting rules in two-period timber and amenity production models in chapter 4 were found to be functions of current and future timber prices, the real interest rate, forest growth rate, and in the amenity production model, the marginal substitution rate between consumption and amenity benefits. Uncertainty in the timber market can enter these harvesting rules through the future price of timber, the real interest rate, and the forest growth rate. Uncertainty will change the marginal rate of substitution; how it does so depends on the way that landowner risk attitudes enter the harvesting rule.

9.1.1 Types of Uncertainties

Since Irving Fisher and Frank Knight, it has become common to distinguish between "risk" and pure "uncertainty." The former refers to an uncertain event for
which a probability distribution is known. The latter cannot be approached with
any measures because the decision maker is unsure of the probability distribution
itself. In what follows, we characterize the many economic risks in forestry, focusing on their sources and the conventional assumptions concerning their stochastic
nature.

Timber Price Uncertainty Any unobserved factor affecting timber market equilibrium can render timber price movements unknown to the landowner. The
most typical factor is stochastic future demand for timber, which in turn depends
on general business cycles. Unexpected shocks to future timber supply as well as
future government policies targeting forestry are other possible sources.

A stochastic timber price is defined by its distribution governing the probability
of possible future price realizations. Denote the probability density function of $\tilde{p}$
by $f(p)$; this gives the probability that $\tilde{p} = p$. If timber price is normally distributed, then its distribution can be characterized solely by its mean and variance. In
what follows we use a bar above a stochastic variable to indicate its mean and a
sigma with a subscript of the variable symbol to indicate its variance. Hence the
normal distribution of the future timber price is described by the mean variance
pair $\{\bar{p}, \sigma_p^2\}$.

Interest Rate Uncertainty As we saw previously in this book, real interest rates
critically affect the intertemporal allocation of harvesting activities by changing
the opportunity costs of not harvesting, as well as other things. However, the
determination and sources of possible variation in interest rates occur outside of
timber markets. Interest rates depend on the state of financial markets, countries'
external positions in exchange markets, and various policies. These factors are exogenous to a landowner's decision making, and thus it makes sense to think of interest rate uncertainty as aggregative in nature.

Denote $\tilde{r}$ as a random variable defining the future interest rate and assume its
probability density function is $f(r)$. A normal distribution for the random variable
then implies that the interest rate is characterized by the mean variance pair
$\{\bar{r}, \sigma_r^2\}$.

Forest Growth Uncertainty There are many forms of biological uncertainty in
forestry problems. Uncertainty about forest stock concerns the amount of actual
volume in the forest at any point in time, while forest growth uncertainty concerns the rate at which the stock increases over time. While both types of uncertainty could be due to a landowner's insufficient knowledge concerning the forest

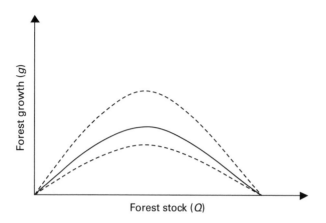

Figure 9.1
Multiplicative growth risk. Solid line, expected forest growth; dashed lines, stochastic variations.

ecosystem, they often arise because the forest stock decays over time in unforeseen ways; consider damage that results from insects, weather (ice, fire), animals, or anthropogenic sources such as acid rain and human-induced climate change.

One can describe forest stock and forest growth uncertainty in many alternative ways. Here we follow the agricultural economics literature and define two forms: multiplicative and additive risk (see Newbery and Stiglitz 1981, pp. 65–66). For a given forest stock and a given forest growth function, multiplicative and additive risks are written respectively as follows:

$$\tilde{\varepsilon}Q \text{ and } Q + \tilde{\varepsilon}; \quad \tilde{\theta}g(Q - x) \text{ and } \tilde{\theta} + g(Q - x) \tag{9.1}$$

where Q is forest stock (volume), x is volume harvested, and $g(.)$ is growth. We will assume a normal distribution for the random variables $\tilde{\varepsilon}$ and $\tilde{\theta}$ and, for simplicity, let the mean of $\tilde{\varepsilon}$ and $\tilde{\theta}$ equal one for multiplicative risk and zero for additive risk. The distributions of $\tilde{\varepsilon}$ and $\tilde{\theta}$ are then given by $N(1, \sigma_\varepsilon^2)$ and $N(0, \sigma_\varepsilon^2)$ for forest stock uncertainty, while the distributions for forest growth uncertainty are denoted by $N(1, \sigma_\theta^2)$ and $N(0, \sigma_\theta^2)$.

Multiplicative and additive growth risks are illustrated in figures 9.1 and 9.2. The solid lines denote expected forest growth and the dashed lines describe stochastic variations in growth resulting in a different position of the growth function. Under growth uncertainty, multiplicative risk preserves the expected maximum sustained yield point of the forest stock but shifts the growth function vertically upward or downward as a constant fraction of the stock. Additive growth risk is realized as a back-and-forth movement of the horizontal axis irrespective of the stock that results in a different MSY point.

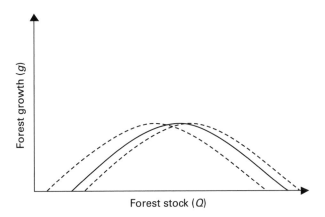

Figure 9.2
Additive growth risk. Solid line, expected forest growth; dashed lines, stochastic variation.

An uncertain forest stock is illustrated in figure 9.3. The expected values of the stock are drawn as the solid lines Q_1 and Q_2. The dashed lines describe stochastic variation in the stock. For additive risk, these variations are independent of the stock Q, but for multiplicative risk, the risk increases with Q.

9.1.2 Risk-Bearing Behavior
The next task is to describe the risk attitudes of forest owners. The most common way to model risk attitudes is using the expected utility hypothesis. The EUH assumes that one can define a probability associated with each possible realization of a stochastic event. An expected utility is obtained by multiplying the utility an agent receives from each realization by the (linear) probability of the realization and then adding these up over all possible realizations. The risk-bearing behavior of the agent is then easily defined by the properties (curvature) of the utility function. A concave utility function indicates risk-averse behavior whereas a linear utility function indicates risk-neutral behavior.

Although the EUH has been an influential approach for solving uncertainty problems, it has not been the only one suggested. There are two features of it that are regarded as undesirable. First, empirical studies of actual behavior provide evidence that the so-called independence axiom, necessary for EUH, is typically violated (see e.g., Gollier 2001, chapter 1). Second, and important to life-cycle models, under EUH it becomes impossible to distinguish between preferences for intertemporal substitution and preferences for risk. Under EUH, each one is simply the reciprocal of the other. As a result, several versions of nonexpected utility theories have been presented, with the most interesting one for forest economics being the ones by Kreps and Porteus (1978), Selden (1978), and Epstein and Zin

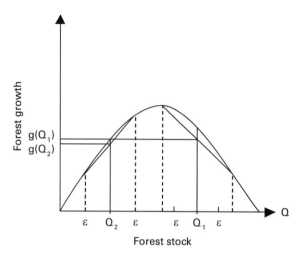

Figure 9.3
Additive and multiplicative stock risk.

(1989), all of whom distinguish preferences for deterministic consumption from preferences for risky consumption, thus overcoming the main interpretative problem of the EUH approach (for a detailed discussion of this, see appendix 9.1 and Gollier 2001, chapter 20).

9.2 Timber Production under Uncertainty

Drawing on Weil (1993) we will examine here a simple nonexpected utility specification that is exponential in risk.[1] When we add a normality assumption about random variables we obtain an expression for risk attitudes that is easy to interpret. We start by introducing uncertainty into the objective function of the two-period timber production model presented in chapter 4.

9.2.1 Timber Prices
Assume conventionally that any uncertainty occurs during the second period, and that the landowner makes decisions with perfect knowledge of all first-period parameters (this is commonly assumed even in more dynamic models). First-period consumption is deterministic and given as before by $c_1 = p_1^* x - s - T$, where x is current harvesting, $p_1^* = (1 - \tau) p_1$ is the after-tax current timber price, τ is the yield tax, s is saving, and T is a site productivity tax. Under a stochastic

---

1. Weil (1990) presents a parametric class of nonexpected utility preferences by distinguishing attitudes toward risk from preferences for intertemporal substitution.

future timber price, future consumption becomes stochastic from the landowner's period one perspective. We express (random) second-period consumption generally as $\tilde{c}_2 = \tilde{p}_2^* z + Rs - T$, where z is future harvesting, $\tilde{p}_2^* = (1 - \tau)\tilde{p}_2$ is the stochastic after-tax future timber price, and $R = 1 + r$ is the discount factor. To keep the analysis simple in other dimensions, assume, as in chapter 4, that consumption in the first period is a perfect substitute for consumption in the second period. Thus stochastic consumption can be expressed as a simple sum of consumption in both periods: $\tilde{c} = c_1 + R^{-1}\tilde{c}_2 = p_1^* x + R^{-1}\tilde{p}_2^* z - (1 + R^{-1})T$. Note that because of our assumption of perfect substitutability, savings vanish from the present value of consumption expression.

The landowner's economic problem is now to choose current harvesting to maximize his expected utility from consumption. This requires the landowner to allocate his forest wealth between a certain return (current harvest revenue) and an uncertain future return (future harvest revenue). To solve this problem, we must distinguish between the landowner's preferences over consumption and risk. Preferences over consumption are defined conventionally by a concave utility function $U = u(c)$. Following the K-P-S approach and using the parameterization of Weil (1993), the risk preferences of the landowner are then described by an exponential utility function of the form

$$W(\tilde{c}) = -\exp(-A\tilde{c}), \tag{9.2}$$

where $A = -W''(\tilde{c})/W'(\tilde{c})$ is the Arrow-Pratt measure of absolute risk aversion (this is described in more detail in Gollier 2001, chapter 2). The sign of A depends on landowner risk attitudes. Constant absolute risk aversion is consistent with A being constant as consumption increases, while decreasing absolute risk aversion is consistent with A being a decreasing function of consumption given a concave utility function. Given that the future timber price was assumed to be normally distributed, $\tilde{c}$ is also normally distributed (being a linear function of price) with mean $\bar{c}$ and variance σ_c^2 (for a good discussion of deriving means and variances in the presence of several random variables, see Mood et al. 1974, chapter 5).

Using the mean and variance, we can now define a certainty equivalent consumption level

$$\hat{c} = \bar{c} - \tfrac{1}{2}A\sigma_c^2. \tag{9.3}$$

Here $\hat{c}$ is a value for consumption that makes a risk-averse landowner indifferent between receiving it and random consumption. In other words, $\hat{c}$ indicates the level of certain consumption that is equivalent to uncertain consumption $\tilde{c}$ in terms of the utility obtained from uncertain consumption. For a risk-averse landowner, the certainty equivalent is less than the expected value of consumption because this agent prefers to reduce risk; thus he would take a lower certain consumption to avoid the risk associated with receiving a random consumption (how much less depends on the concavity of the utility function through the magnitude of the Arrow-Pratt measure A in 9.3). For a risk-neutral landowner, the certainty equivalent reduces to the expected value of random consumption $E(\tilde{c}) = \bar{c}$, because this landowner is indifferent between certain consumption and an expected value of a "fair gamble." Risk neutrality also implies that risk attitudes are the same for all levels of consumption.[2]

Under these assumptions, maximizing $Eu(\tilde{c})$ is equivalent to maximizing the utility received from certainty equivalent consumption, $u(\hat{c})$, where $\hat{c} = \bar{c} - \frac{1}{2}A\sigma_c^2 = p_1^*x - (1 + R^{-1})T + R^{-1}\bar{p}_2^*z - \frac{1}{2}AR^{-2}(1 - \tau)^2\sigma_{p_2}z^2$. Using this objective function, the economic problem of the landowner is now to choose current harvesting subject to the forest growth technology. The decision problem is formally described by choosing current harvesting x and thereby future harvesting z to maximize

$$V = u(c) = u\left[p_1^*x - (1 + R^{-1})T + R^{-1}\bar{p}_2^*z - \frac{1}{2}AR^{-2}(1 - \tau)^2\sigma_p^2z^2\right], \qquad (9.4)$$

where $u'(c) > 0$ and $u''(c) < 0$ and $z = (Q - x) + g(Q - x)$. The first- and second-order conditions are $V_x = u'(c)[p_1^* - R^{-1}\bar{p}_2^*(1 + g') + AR^{-2}(1 - \tau)^2\sigma_p^2z(1 + g')] = 0$, and $V_{xx} = u'(c)[R^{-1}g''[\bar{p}_2^* - AR^{-1}(1 - \tau)^2\sigma_p^2z] - AR^{-2}(1 - \tau)^2\sigma_p^2(1 + g')^2] < 0$, respectively. Arranging the first-order condition and dividing it by $(1 - \tau)$ produces a harvesting rule under future timber price uncertainty:

$$Rp_1 - (1 + g')[\bar{p}_2 - AR^{-1}(1 - \tau)\sigma_p^2z] = 0. \qquad (9.5)$$

Uncertainty appears as a risk term in the usual opportunity cost term. This risk term is a product of the degree of risk aversion and variance of future price; it serves to reduce the opportunity cost of harvesting in the current period by decreasing the future timber price, inducing the landowner to increase period one harvesting up to the point where the marginal revenue of the last harvested unit equals the opportunity cost of its certainty equivalent. It is worth noting that the higher absolute risk aversion is, the greater is this reduction in opportunity cost through the A term. Hence we have proposition 9.1.

Proposition 9.1 Under risk aversion, uncertainty in future timber price increases current harvesting and short-term timber supply.

The economic interpretation of proposition 9.1 is obvious. When facing future uncertainty, a risk-averse landowner hedges against price risk by harvesting more

---

2. See e.g., Newbery and Stiglitz (1981, pp. 85–92).

timber during period one at a certain price. The role of risk aversion can be seen by comparing the harvesting rule in (9.5) with that of a risk-neutral harvester who equalizes the marginal revenue of harvesting to the expected value of future forest growth, i.e., $Rp_1 - (1 + g')\bar{p}_2 = 0$.

The comparative statics of short-term harvesting can be solved by totally differentiating (9.5). It is easy to show that some of the chapter 4 results continue to hold. In particular, we find that $x_{p_1} > 0$, $x_{\bar{p}_2} < 0$, and $x_r > 0$. The effects of risk and risk aversion are the new aspects of the model and are thus of special interest:

$$x_{\sigma_p} = -\frac{AR^{-2}(1 - \tau)^2 z(1 + g')}{V_{xx}} > 0 \quad \text{and} \quad x_A = \frac{\sigma_p}{A} x_{\sigma_p} > 0. \tag{9.6a}$$

According to (9.6a), higher risk, defined by a higher variance of future prices σ_p, increases current-period harvesting. The same is true for a higher rate of absolute risk aversion, A. Both higher risk and risk aversion encourage the landowner to hedge against future price risk by harvesting more in the current period. For the site productivity tax and the yield tax we have

$$x_T = 0 \quad \text{and} \quad x_\tau = \frac{AR^{-2}z(1 + g')\sigma_p^2}{V_{xx}} < 0. \tag{9.6b}$$

The site productivity tax turns out to be neutral under price uncertainty, but this result is not robust. A more general risk analysis reveals that the effect of the site productivity tax depends on the properties of the landowner's risk-aversive behavior. For decreasing absolute risk aversion, a higher site productivity tax increases current harvesting: $x_T > 0$. Equally interesting is that the yield tax is no longer neutral under price uncertainty as it was under certainty in chapter 4 because this tax decreases the variance term in the harvesting rule (in other words, a higher yield tax reduces future price risk, reducing the landowner's incentive to hedge by harvesting more in the short term).

9.2.2 Real Interest Rates

When the interest rate is stochastic but timber price is certain, we would write the consumption equations as: $c_1 = p_1^* x - s - T$ and $c_2 = p_2^* z + \tilde{R}s - T$. A stochastic real interest rate has the effect of making the periodic weight on consumption flows stochastic, and thus the present value of consumption is also stochastic. This becomes clear by defining the sum of consumption over both periods as $\tilde{c} = c_1 + \tilde{R}^{-1}c_2 = p_1^* x + \tilde{R}^{-1}p_2^* z - (1 + \tilde{R}^{-1})T$.

Preferences of the landowner over uncertain interest rate outcomes and consumption can be defined in a way similar to that of the previous case. Thus if the stochastic real interest rate is normally distributed, then the distribution of $\tilde{c}$ is also normal with a mean $\bar{c}$ and variance σ_c^2. Following (9.3), the certainty equivalent value of random consumption is again given by $\hat{c} = \bar{c} - (1/2)A\sigma_c^2$. For the interest rate factor, $\tilde{R} = 1 + \tilde{r}$, a Taylor approximation is useful for finding its expected value $E(\tilde{R}^{-1}) = \bar{R}^{-1}(1 + \delta_r^2)$, where δ_r^2 is the coefficient of variation and $\bar{R}$ is the mean value of R. Under a random real interest rate, the mean and the variance of the sum of both periods' consumption are defined respectively as follows see appendix 9.2 for a definition of the variance term in (9.7b):

$$\bar{c} = p_1^* x - [1 + R^{-1}(1 + \delta_r^2)]T + R^{-1}(1 + \delta_r^2)p_2^* z \tag{9.7a}$$

$$\sigma_c^2 = R^{-2}\delta_r^2(1 - \delta_r^2)[T^2 + (p_2^* z)^2]. \tag{9.7b}$$

The economic problem of the landowner now is to choose current harvesting subject to (9.7a) and (9.7b) and the forest growth technology. The utility function is

$$V = u(\hat{c}) = u\left(\bar{c} - \tfrac{1}{2}A\sigma_c^2\right). \tag{9.8}$$

Differentiating (9.8) with respect to x produces the following first-order condition defining optimal current harvesting:

$$V_x = p_1^* - \bar{R}^{-1}(1 + \delta_r^2)p_2^*(1 + g') + A\bar{R}^{-2}\delta_r^2(1 - \delta_r^2)p_2^{*2}z(1 + g') = 0. \tag{9.9}$$

The second-order condition holds, $V_{xx} < 0$. The following harvesting rule emerges from rearranging (9.9):

$$\frac{\bar{R}}{(1 + \delta_r^2)}p_1 - (1 + g')p_2\left[1 - A\bar{R}^{-1}\delta_r^2 p_2(1 - \tau)\frac{(1 - \delta_r^2)}{(1 + \delta_r^2)}z\right] = 0. \tag{9.10a}$$

This harvesting rule clearly differs from those we have previously shown. Both the marginal return from harvesting in the current period and the opportunity cost of harvesting are affected by the coefficient of variation. Thus a stochastic interest rate reduces both the marginal return and the opportunity cost of not harvesting. The return to harvesting decreases relative to the value of certain forest growth, but the present value of future forest growth is also decreased. The net effect can be seen by rearranging the harvesting rule to obtain $\bar{R}p_1 - (1 + g')p_2 = (1 + g')p_2\delta_r^2[1 - A\bar{R}^{-1}\delta_r^2 p_2^*(1 - \sigma_r^2)z] > 0$. Given the concave forest growth function, the net effect of interest rate risk is to decrease short-term harvesting and timber supply relative to the case with a certain interest rate.

For a risk-neutral landowner, (9.10a) reduces to

$$\frac{\bar{R}}{(1+\delta_r^2)} p_1 - p_2(1+g') = 0. \tag{9.10b}$$

Now, the mean of the stochastic interest rate decreases the return of current harvesting relative to the certainty case but does not affect the opportunity cost term. The landowner unambiguously decreases short-term harvesting. It is interesting that the yield tax now vanishes from the harvesting rule; thus it is a neutral tax in the risk-neutral case. We condense these findings in proposition 9.2.

Proposition 9.2 Uncertainty in the real interest rate decreases current harvesting for both a risk-neutral and a risk-averse landowner relative to a certainty case.

Once again we should emphasize that this result is not entirely general. By focusing on the present value of consumption, we are implicitly ignoring the landowner's consumption-saving decision, which clearly is affected by an uncertain interest rate. Ollikainen (1991, 1993) demonstrates in this case that the reaction of short-term harvesting depends on the landowner's position in the capital market. If a risk-averse landowner is a net borrower, then he will increase harvesting. Thus proposition 9.2 is therefore consistent with the case of a net lender.

Gong and Löfgren (2003a) provide additional insight regarding these findings. They modify the two-period model by allowing the landowner to invest his harvest revenue in both a riskless and risky asset under future timber price uncertainty. They further assume that timber price and the rate of return on the risky asset are independently and normally distributed. Their main finding is that a risk-averse landowner harvests more (or less) than a risk-neutral landowner when timber price risk is high (or low) relative to the rate of return the landowner can receive with the risky asset, or when the degree of risk aversion is high (low). These results are complementary to the ones derived in this section.

The comparative statics of the two-period model under a stochastic real interest rate are more involved than under timber price uncertainty. As before, current harvesting depends positively on the current timber price and negatively on the mean of the future timber price, i.e., $x_{p_1} > 0$ and $x_{p_2} < 0$. Of special interest are the effects of the mean and the variance of the real interest rate, and of absolute risk aversion. Recall that the expected value of the interest rate is $E(\tilde{R}^{-1}) = \bar{R}^{-1}(1+\delta_r^2)$. Here we approximate the mean and variance simply by $\bar{R}$ and δ_r^2. The comparative statics results are written as

$$x_{\bar{R}} = -\frac{\bar{R}^{-2}p_2^*(1+g')[(1+\delta_r^2) - 2A\bar{R}^{-1}\delta_r^2(1-\delta_r^2)p_2^*z]}{V_{xx}} > 0 \tag{9.11a}$$

$$x_{\delta_r^2} = \frac{\bar{R}^{-1}p_2^*(1+g')[1 - A\bar{R}^{-1}p_2^*z(1-2\delta_r^2)]}{V_{xx}} > 0 \qquad (9.11b)$$

$$x_A = -\frac{\bar{R}^{-2}\delta_r^2(1-\delta_r^2)(p_2^*)^2(1+g')z}{V_{xx}} > 0. \qquad (9.11c)$$

The numerators in (9.11a) and (9.11b) contain two opposing terms. Using the first-order condition for harvesting, one can see that the numerator of (9.11a) is positive while the numerator of (9.11b) is negative. Intuitively, a higher expected value of the real interest rate and a higher variance of the interest rate both decrease the present-value opportunity cost of harvesting (written as a function of forgone second-period harvest revenues). Thus the landowner increases current harvesting as a response to increased profitability (through the mean) or to hedging against higher risk (through the coefficient of variation). Higher risk aversion works as we expect and increases current harvesting.

9.2.3 Forest Growth

When forest growth is stochastic, the periodic flows of consumption are expressed as $c_1 = p_1^*x - s - T$ and $\tilde{c}_2 = p_2^*\tilde{z} + Rs - T$. Second-period consumption is uncertain because of uncertain forest growth between the two periods. The intertemporal sum of consumption in each period is $\tilde{c} = c_1 + R^{-1}\tilde{c}_2 = p_1^*x + R^{-1}p_2^*\tilde{z} - (1 + R^{-1})T$. The exact nature of the stochastic variable $\tilde{z}$ depends on specific assumptions concerning the nature of growth uncertainty. Recall from equation (9.1) that we can define multiplicative and additive risky forest stocks as $\tilde{\theta}g(Q - x)$ and $\tilde{\theta} + g(Q - x)$, with distributions $N(1, \sigma_\theta^2)$ and $N(0, \sigma_\theta^2)$, respectively. While the certainty equivalent, $\hat{c} = \bar{c} - (1/2)A\sigma_c^2$, contains the same mean for both multiplicative and additive cases, the variance term is different, i.e.,

$$\bar{c} = p_1^*x + R^{-1}p_2^*[(Q - x) + g(Q - x)] - (1 + R^{-1})T \qquad (9.12a)$$

$$\sigma_c^2 = R^{-2}p_2^{*2}\sigma_\theta^2 g^2 \qquad (9.12b)$$

$$\sigma_c^2 = R^{-2}p_2^{*2}\sigma_\theta^2. \qquad (9.12c)$$

The landowner now chooses x to maximize the utility of consumption subject to (9.12a) and either (9.12b) or (9.12c). The first-order condition under multiplicative risk is $V_x = u'(c)(p_1^* - R^{-1}p_2^*[1 + g'(1 - AR^{-1}p_2^*g\sigma_\theta^2)]) = 0$ and for additive risk is $V_x = u'(c)[Rp_1^* - p_2^*(1+g')] = 0$ (the second-order conditions for both cases can be shown to hold). The last condition indicates that under additive stock risk, current harvesting is unchanged relative to the certainty case because harvesting has no effect on this risk. Therefore for additive growth risk, the comparative statics results reduce to the results obtained under certainty. In what follows we will focus only on the case of multiplicative risk.

For multiplicative risk we obtain the following harvesting rule:

$$Rp_1 - p_2g'[1 - AR^{-1}p_2(1 - \tau)g\sigma_\theta^2] = 0. \tag{9.13}$$

According to (9.13), the risk-adjusted growth rate is lower than the certainty rate. Thus multiplicative growth risk increases current harvesting. We summarize these findings in proposition 9.3.

Proposition 9.3 A multiplicative stochastic forest growth rate increases current harvesting and short-term timber supply, but an additive stochastic growth rate has no effect on harvesting.

The interpretation of proposition 9.3 follows lines similar to those of proposition 9.1. The landowner hedges against multiplicative growth risk by harvesting more at the certain current price. Additive growth uncertainty, in turn, cannot be reduced by changing harvesting and therefore it has no effect.

The comparative statics for multiplicative risky growth are straightforward to solve and confirm our previous results, $x = x(p_1, p_2, r, T, A)$. The effects of the variance of forest growth and the yield tax on current harvesting are given by

$$x_{\sigma_\theta^2} = -\frac{R^{-2}(p_2^*)^2g'gA}{V_{xx}} > 0; \quad x_\tau = \frac{R^{-2}p_2^2gg'A\sigma_\theta^2}{V_{xx}} < 0. \tag{9.14}$$

The new finding here is that a higher growth risk increases current harvesting. Finally, a higher yield tax decreases the effect of forest growth risk on the landowner (by reducing revenue net of the tax). Thus current harvesting decreases.

9.2.4 Forest Stock

Suppose now that the original forest stock, Q, is uncertain. Recall we previously defined multiplicative and additive stock risk as follows: $\tilde{\varepsilon}Q$ and $Q + \tilde{\varepsilon}$ with $\tilde{\varepsilon}$ distributed $N(1, \sigma_\varepsilon^2)$ and $N(0, \sigma_\varepsilon^2)$, respectively. Now, the certainty equivalent consumption, $\hat{c} = \bar{c} - (1/2)A\sigma_c^2$, contains a different mean and variance, depending on the type of stock uncertainty. Under multiplicative stock uncertainty we have

$$\bar{c} = p_1^*x + R^{-1}p_2^*[(Q - x) + g(Q - x) + \tfrac{1}{2}g''Q^2\sigma_\varepsilon^2] - (1 + R^{-1})T \tag{9.15a}$$

$$\sigma_c^2 = R^{-2}p_2^{*2}\sigma_\varepsilon^2Q^2g'^2. \tag{9.15b}$$

Additive uncertainty implies the following mean and variance interpretation of certainty equivalent consumption:

$$\bar{c} = p_1^* x + R^{-1} p_2^* [(Q - x) + g(Q - x) + \tfrac{1}{2} g'' Q^2 \sigma_\varepsilon^2] - (1 + R^{-1}) T \tag{9.16a}$$

$$\sigma_c^2 = R^{-2} p_2^2 (1 - \tau)^2 \sigma_\varepsilon^2 g'^2. \tag{9.16b}$$

The landowner chooses current harvesting to maximize $V = u(\hat{c}) = u[\bar{c} - (1/2) A \sigma_c^2]$. If the stock uncertainty is multiplicative, then the first-order condition with respect to x is $V_x = u'(\hat{c})[p_1^* - R^{-1} p_2^*(1 + g') + AR^{-2} p_2^{*2} Q^2 \sigma_\varepsilon^2 g' g''] = 0$, whereas for additive uncertainty we obtain $V_x = u'(\hat{c})[p_1^* - R^{-1} p_2^*(1 + g') + AR^{-2} p_2^2 (1 - \tau)^2 \sigma_\varepsilon^2 g' g''] = 0$. (Again, the second-order conditions hold.) The first-order conditions yield the following harvesting rules for multiplicative and additive risks, respectively:

$$Rp_1 - p_2(1 + g'^*) = 0; \quad Rp_1 - p_2(1 + g'^{**}) = 0, \tag{9.17}$$

where $g'^* = g'[1 - AR^{-1} p_2 (1 - \tau) Q^2 \sigma_\varepsilon^2 g' g'']$ and $g'^{**} = g'^* Q^2$ are risk-adjusted growth rates under an uncertain stock. The size of the risk terms in the forest growth rate function depends on the landowner's risk-aversion parameter A, after-tax future timber price, growth and the variance of the future forest stock. These now work in opposite directions because unlike growth risk, stock risk increases the opportunity cost of harvesting through a risk-adjusted growth factor. This is true for both types of risks in the harvesting rules of (9.17). Consequently, current harvesting decreases. Thus, we have proposition 9.4.

Proposition 9.4 Uncertain initial forest stock in the form of both multiplicative and additive risk decreases current harvesting and thereby decreases the short-term supply of timber.

Proposition 9.4 indicates that the landowner facing stock uncertainty decreases harvesting relative to the case of certainty. This precautionary behavior reduces stock risk and yields higher harvest volumes during the second period.

The comparative statics for price, real interest rate, and site productivity tax under both additive and multiplicative stock uncertainty are qualitatively the same as under multiplicative growth risk: $x = x(p_1, p_2, r, T)$. Of special interest here are the effects of risk, risk aversion, and the yield tax. For multiplicative stock risk we have

$$x_{\sigma_\varepsilon^2} = -\frac{AR^{-2} (p_2^*)^2 Q^2}{V_{xx}} < 0; \quad x_A = \frac{\sigma_\varepsilon^2}{A} x_{\sigma_\varepsilon^2} < 0; \quad x_\tau = -\frac{\sigma_\varepsilon^2}{(1 - \tau)} x_{\sigma_\varepsilon^2} > 0. \tag{9.18}$$

A higher variance in the forest stock and risk aversion decrease current harvesting, whereas a higher yield tax increases it. The effects of variance and risk aversion are obvious. The only way of reducing risk (either via a higher variance or the landowner's risk perceptions) is to capture greater revenues by allowing the stock to grow to a greater volume. However, it is now immediately obvious why the yield tax increases current harvesting. This results for the same reason as before, i.e., a higher yield tax reduces the risk-adjusted forest growth factor and therefore reduces the marginal benefit of future harvesting. Thus current harvesting becomes more profitable and short-term timber supply increases. Finally, note that the comparative statics for additive stock uncertainty are obtained from (9.18) by simply dividing all effects reported in (9.18) by $1/Q^2$.

9.3 Amenity Production under Uncertainty

We now return to the two-period amenity production model of chapter 4, assuming as before that the landowner maximizes a sum of the present value of consumption and (concave) utility obtained from amenity production over both periods. Also, we continue to assume that amenity services are positively related to the volume of timber. We will examine the same four forms of uncertainty; timber price, interest rate, and uncertainties in forest growth and stock. However, we will abstract from some of the comparative statics derived in section 9.2 for the sake of brevity.

9.3.1 Timber Prices

In the previous section we defined the present value of consumption under an uncertain timber price to be $\hat{c} = \bar{c} - (1/2)A\sigma_c^2$. Thus the landowner chooses x and z to maximize (9.19) subject to periodic forest stocks $k_1 = Q - x$ and

$$k_2 = k_1 + g(k_1) - z.$$

$$V = p_1^* x - (1 + R^{-1})T + R^{-1}\bar{p}_2^* z - \tfrac{1}{2}AR^{-2}(1 - \tau)^2\sigma_{p_2}z^2 + v(k_1) + \beta v(k_2). \qquad (9.19)$$

The first-order conditions for this problem are

$$V_x = p_1^* - v'(k_1) - (1 + g')\beta v'(k_2) = 0 \qquad (9.20a)$$

$$V_z = R^{-1}\bar{p}_2^* - AR^{-2}(1 - \tau)^2\sigma_p^2 z - \beta v'(k_2) = 0. \qquad (9.20b)$$

As in chapter 4, we can solve $V_z = 0 \Leftrightarrow \beta v'(k_2) = R^{-1}\bar{p}_2^* - AR^{-2}(1 - \tau)^2\sigma_p^2 z$ from (9.20a) and use it in (9.20b) to obtain the following harvesting rule:

$$Rp_1 - (1 + g')[p_2 - AR^{-1}(1 - \tau)\sigma_p^2 z] = (1 - \tau)^{-1}Rv'(k_1). \tag{9.21}$$

The LHS of (9.21) is familiar from our previous discussion, but the RHS is different and represents (modified) marginal amenity benefits under uncertainty. According to (9.21), the landowner harvests up to the point where the difference between the marginal return from harvesting and the risk-adjusted opportunity cost of not harvesting equal the modified marginal valuation of amenity benefits. Accounting for amenity benefits therefore decreases short-term harvesting relative to the timber production model under uncertainty (compare equation 9.21 with equation 9.5).

For many parameters, the comparative statics of this model are similar to the case of pure timber production under uncertain future timber price. Specifically, we have $x_{p_1} > 0$, and $z_{p_1} < 0$; $x_{\bar{p}_2} < 0$, $z_{\bar{p}_2} > 0$; and $x_r > 0$ $z_r < 0$. For the timber price variance we obtain

$$x_{\sigma_p^2} = -\Delta^{-1}\{AR^{-2}(1 - \tau)^2 z(1 + g')\beta v''(k)\} > 0,$$

$$\tag{9.22}$$

$$z_{\sigma_p^2} = -(1 + g')x_{\sigma_p^2} + \Psi < 0,$$

where $\Delta > 0$ is the determinant of the Hessian matrix of the equation system (9.20a) and (9.20b) and $\Psi = \Delta^{-1}AR^{-2}(1 - \tau)^2 z[v''(k_1) + g''\beta v'(k_2)] < 0$. This shows that the landowner hedges against increased amenity risk in the future by harvesting more today at the certain timber price.

As for forest taxes, the site productivity tax remains neutral, $x_T = 0$, while the yield tax again has reinforcing effects on current harvesting and serves to decrease it. The effect of the yield tax on future harvesting is complicated and consists of price and risk effects. Moreover, it requires us to consider that reduced harvesting in the first period allows increased future harvesting, owing to a higher forest stock.

9.3.2   Real Interest Rates

Under a stochastic interest rate, we define the present value of consumption as in the previous section as $\tilde{c} = c_1 + \tilde{R}^{-1}c_2 = p_1^*x + \tilde{R}^{-1}p_2^*z + (1 + \tilde{R}^{-1})T$. The certainty equivalent value of random consumption was defined in equations (9.7a) and (9.7b). The landowner therefore chooses x and z to maximize (9.23) subject to the previous stock equations:

$$V = p_1^*x - [1 + \bar{R}^{-1}(1 + \delta_r^2)]T + \bar{R}^{-1}(1 + \delta_r^2)p_2^*z$$

$$- \tfrac{1}{2}A\bar{R}^{-2}\delta_r^2(1 - \delta_r^2)[T^2 + (p_2^*z)^2] + v(k_1) + \beta v(k_2). \tag{9.23}$$

The first-order conditions are

$$V_x = p_1^* - v'(k_1) - (1 + g')\beta v'(k_2) = 0 \tag{9.24a}$$

$$V_z = \bar{R}^{-1}(1 + \delta_r^2)p_2^* - A\bar{R}^{-2}(1 - \tau)^2(1 - \delta_r^2)\delta_r^2 p_2^{*2}z - \beta v'(k_2) = 0. \tag{9.24b}$$

From (9.24b), $V_z = 0 \Leftrightarrow \beta v'(k_2) = \bar{R}^{-1}(1 + \sigma_r^2)p_2^* - A\bar{R}^{-2}(1 - \tau)^2(1 - \sigma_r^2)\sigma_r^2 p_2 z$.

Using this in (9.24a) we obtain the following harvesting rule:

$$\frac{\bar{R}}{(1 + \delta_r^2)}p_1 - (1 + g')p_2 \left[1 - A\bar{R}^{-1}\frac{(1 - \delta_r^2)}{(1 + \delta_r^2)}\delta_r^2 p_2^* z\right] = (1 - \tau)^{-1}\bar{R}v'(k_1). \tag{9.25}$$

The LHS of (9.25) is familiar from our previous discussion and the RHS represents modified marginal amenity benefits. This RHS term is positive, tending to decrease current harvesting in favor of amenity production. Amenities here reinforce the risk associated with an uncertain interest rate. Thus the landowner reduces short-term harvesting.

The comparative statics results here are straightforward. A higher expected interest rate increases the profitability of timber production relative to amenity production and current harvesting, increasing current harvesting. The effects of the risk terms are evident. A higher coefficient of correlation increases current harvesting and decreases future harvesting. Finally, as with the case of price uncertainty, we have $x_\tau < 0$ and $z_\tau \gtrless 0$ with similar interpretations.

9.3.3   Forest Growth
Uncertainty of either forest growth or forest stock changes the nature of our amenity production model. Now, the present value of harvest revenue is deterministic but amenity production becomes stochastic because the latter depends directly on future forest stock. Under forest growth risk, the future stock is stochastic but the current stock is deterministic. Using the special form of multiplicative forest growth risk, future forest stock is given by $\tilde{k}_2 = Q - x + \tilde{\theta}g(Q - x) - z$. Under the additive risk form, it is given by $\tilde{k}_2 = Q - x + \tilde{\theta}g(Q - x) + \theta Q - z$.

The decision problem of the landowner under multiplicative risk is to choose x and z to maximize

$$V = p_1^* x - (1 + R^{-1})T + R^{-1}p_2^* z + v(k_1) + \beta v(\hat{k}_2), \text{ s.t.} \tag{9.26a}$$

$$k_1 = Q - x \tag{9.26b}$$

$$\hat{k}_2 = \bar{k}_2 - \tfrac{1}{2}Ag^2\sigma_\theta^2, \tag{9.26c}$$

where $\hat{k}_2$ is the certainty equivalent for the random future forest stock in the expected utility sense, and $\bar{k}_2 = (Q - x) + g(Q - x) - z$. If forest growth risk is additive, then the maximization problem remains the same as above, but the constraint (9.26c) is replaced by $\hat{k}_2 = \bar{k}_2 - (1/2)A\sigma_\theta^2$. As we showed in the previous section, additive growth uncertainty does not matter to the decision problem and we do not discuss this case in what follows.

When multiplicative risk is associated with forest growth, the first-order conditions for maximization of nonexpected utility with respect to x and z are

$$V_x = p_1^* - v'(k_1) - \beta v'(\hat{k}_2)[1 + g'(1 - Ag\sigma_\theta^2)] = 0 \tag{9.27a}$$

$$V_z = R^{-1}p_2^* - \beta v'(\hat{k}_2) = 0. \tag{9.27b}$$

Eliminating $\beta v'(\hat{k}_2)$ this can be solved for the following harvesting rule:

$$Rp_1 - p_2(1 + g'^*) = (1 - \tau)^{-1}Rv'(k_1), \tag{9.28}$$

where $g'^* = g'(1 - Ag\sigma_\theta^2)$ is the risk-adjusted growth rate under uncertain growth.

Current harvesting reflects a tradeoff between harvest revenue, amenity valuation, and risk aversion of the landowner. Harvesting is carried out to the point where the difference between the marginal return from current harvesting (first LHS term) and the opportunity cost of current harvesting (second LHS term) equal marginal amenity services (RHS term). The size of risk and risk aversion affect this choice through the opportunity cost of the harvesting term. An increase in forest growth risk or absolute risk aversion of the landowner decreases the opportunity cost of not harvesting, making current harvesting more attractive.

Given that the present value of harvest revenue is deterministic, the comparative statics of prices and interest rate remain equivalent to those we derived in chapter 4. We can show that a rise in forest growth risk increases current harvesting and decreases future harvesting, respectively. This reflects precautionary behavior. A higher forest growth uncertainty has an effect like that of a lower expected return, which increases the relative attractiveness of current harvesting. The site productivity tax continues to be neutral, while the yield tax effect is indeterminant.

9.3.4 Forest Stock

If uncertainty is associated with the initial forest stock, then both k_1 and k_2 become stochastic since $\tilde{k}_1 = \tilde{\varepsilon}Q - x$ and $\tilde{k}_2 = \tilde{\varepsilon}Q - x + g(\tilde{\varepsilon}Q - x) - z$ with multiplicative risk and $\tilde{k}_1 = \tilde{\varepsilon} + Q - x$ and $\tilde{k}_2 = \tilde{\varepsilon} + Q - x + g(\tilde{\varepsilon} + Q - x) - z$ with additive risk. Defining the certainty-equivalent forest stock here becomes more complicated than before because the uncertainty parameter, ε, affects forest growth nonlinearly.

The expected value of $\tilde{k}_2$ can be obtained using a second-order Taylor approximation (see appendix 9.2 for approximating moments of a distribution when the functions depend nonlinearly on stochastic variables). For multiplicative stock uncertainty, the mean and the variance are

$$\hat{k}_1 = \bar{k}_1 - \tfrac{1}{2}AQ^2\sigma_\varepsilon^2 \tag{9.29a}$$

$$\hat{k}_2 = \bar{k}_2 - \tfrac{1}{2}Ag'^2Q^2\sigma_\varepsilon^2, \tag{9.29b}$$

where $\hat{k}_1$ and $\hat{k}_2$ are the certainty equivalents of the random stocks and $\bar{k}_1 = Q - x$ and $\bar{k}_2 = (Q - x) + g(Q - x) + (1/2)g''Q^2\sigma_\varepsilon^2 - z$ are the expected values, so that the certainty-equivalent future forest stock is $\hat{k}_2 = (Q - x) + g(Q - x) - z + (1/2)(g'' - Ag'^2)Q^2\sigma_\varepsilon^2$.

For additive risk, the certainty-equivalent forest stocks are

$$\hat{k}_1 = \bar{k}_1 - \tfrac{1}{2}A\sigma_\varepsilon^2 \tag{9.30a}$$

$$\hat{k}_2 = \bar{k}_2 - \tfrac{1}{2}Ag'^2\sigma_\varepsilon^2, \tag{9.30b}$$

where $\bar{k}_1 = Q - x$ and $\bar{k}_2 = (Q - x) + g(Q - x) + (1/2)g''\sigma_\varepsilon^2 - z$. Thus, the certainty-equivalent future forest stock is $\hat{k}_2 = (Q - x) + g(Q - x) - z + (1/2)(g'' - Ag'^2)\sigma_\varepsilon^2$.

As the certainty equivalent stocks are almost identical for additive and multiplicative risks, we will analyze only the latter case. Under multiplicative forest stock risk, the problem of the forest owner is to choose x and z to maximize (9.31) subject to (9.29a) and (9.29b):

$$V = p_1^*x - (1 + R^{-1})T + R^{-1}p_2^*z + v(\hat{k}_1) + R^{-1}v(\hat{k}_2). \tag{9.31}$$

The first-order conditions are

$$V_x = p_1^* - v'(\hat{k}_1) - \beta v'(\hat{k}_2)[1 + g' - Ag''g'Q^2\sigma_\varepsilon^2] = 0 \tag{9.32a}$$

$$V_z = R^{-1}p_2^* - v'(\hat{k}_2) = 0. \tag{9.32b}$$

These yield the following harvesting rule:

$$Rp_1^* - p_2^*(1 + g'^{**}) = Rv'(\hat{k}_1), \tag{9.33}$$

where $g'^{**} = g'(1 - Ag''Q^2\sigma_\varepsilon^2) > 0$ is the risk-adjusted growth rate under the uncertain stock. As with forest growth risk, the harvesting rule reflects a tradeoff between the difference in the marginal return and the opportunity cost of harvesting, the marginal amenity benefits, and landowner risk aversion. Unlike with growth risk, stock risk raises both the opportunity cost of harvesting and the marginal valuation of the current forest stock. Consequently, current harvesting is reduced, not increased.

Most of the comparative statics results are similar to the previous case. A higher stock risk decreases current harvesting but the effect on future harvesting is ambiguous. The certainty equivalence of random forest stocks will decrease, so that the marginal value of amenity services increases. The forest owner reacts by harvesting less both today and in the future in order to diversify the effect of risk on his returns. The lower current harvesting, however, again gives a higher future volume and allows increased future harvests. Thus, one cannot a priori determine the total effect of initial forest stock risk on future harvesting. Also, once again, timber supply behaves slightly differently under additive stock risk than under multiplicative risk.

9.4   Modifications

The question of the design of forest taxation takes on special importance in the presence of uncertainty. In chapter 5 we found that the presence of public goods (amenities) and the government's budget constraint can factor importantly in the optimal design of instruments. The question now becomes how uncertainty faced by a risk-averse landowner might be taken into account by the government's policy choices.

So far we have been discussing various risks that forest landowners face. We should point out that the government can also face risks. An important distinction is whether risk is idiosyncratic or aggregate, because these are important to tax policies. Idiosyncratic risks are those that are independent across forest owners, implying that some do well and others do poorly in the face of uncertainty. Under idiosyncratic risk, the government's tax revenue from the entire forest sector is deterministic because of the law of large numbers (i.e., there are many tax-paying landowners so that gainers and losers cancel each other out in terms of revenue effects of risk). This has an important implication for tax policies. When risks in the economy are purely private like this and uncorrelated across agents, taxation can provide a new social insurance role. The government can choose taxes to smooth the distortions we showed earlier that risk causes in landowner decisions, thereby shifting harvesting toward optimal risk-free solutions.

Aggregate, or business cycle risk, is risk that is common to all landowners, implying that all either gain or lose because of the risk. Aggregate risk is typically caused by volatility in the market price of timber over time. Under aggregate risk, the government's tax revenue, and thereby public spending, becomes stochastic.

Thus, while we do not need to worry about public spending under idiosyncratic risk, the design of taxes needs to be considered simultaneously with public spending under aggregate risk. Furthermore, the design of taxation will also depend on the associated substitutability between the harvest revenue and public spending, as well as on risk aversion of landowners.

Under aggregate risk, the government has no means to reduce the risk that private agents face. Private agents thus ultimately bear all of the risk, and taxation cannot provide an insurance role (for a detailed discussion concerning optimal design of taxation in the general case under both types of risks, see Myles 1995, chapter 7; the initial contribution of these theories is due to Chistiansen 1993). However, the government still can shift the risk between private and public spending through its policy choices. The case of idiosyncratic and aggregate risk in a forestry context is analyzed in Koskela and Ollikainen (1998a).

In this section we specifically focus on the government's choice of yield and site productivity taxes under uncertainty and a binding tax revenue requirement. We analyze the impact of risk on tax policies in the simplest (second-best) case where the governments face a budget constraint but there are no externalities inherent in landowner decision making (see chapter 5). Policy design under both idiosyncratic and aggregate risks is compared. For brevity we conduct our analysis only for the timber production model and we consider only price uncertainty.

9.4.1 Idiosyncratic Timber Price Risk

Consider a risk-neutral planner who maximizes social welfare subject to a given tax revenue target for a forest economy. Suppose that the planner has available a site productivity tax and a yield tax. In the absence of externalities, the social welfare function consists simply of the indirect utility function of the landowner. To further simplify, let utility be linear, so that the indirect net present harvest revenue function, V^*, can be expressed as a function of optimal harvesting choices (see equation 9.4). Because the social planner is always assumed to be risk neutral, he chooses policies to maximize this utility function subject only to expected tax revenue (the variance is disregarded). Thus the problem of the social planner is

$$\max_{T,\tau} SW = V^*(T,\tau\ldots.), \text{ s.t.} \tag{9.34a}$$

$$G = (1 + R^{-1})T + \tau[p_1 x + R^{-1}\bar{p}_2 z]. \tag{9.34b}$$

The Lagrangian is $\Omega = V^* + \lambda[G - \bar{G}]$, where $\bar{G}$ is the exogenous budget revenue target for the forest sector. The first-order condition for each tax instrument is

$$\Omega_T = V_T^* + \lambda(1 + R^{-1}) = 0, \tag{9.35a}$$

$$\Omega_\tau = V_\tau^* + \lambda[(p_1 x + R^{-1}\bar{p}_2 z) + \tau(p_1 x_\tau + R^{-1} p_2 z_\tau)] = 0. \tag{9.35b}$$

As in chapter 5, for the site productivity tax the first order condition is: $V_T^* = -(1 + R^{-1})$, which implies by (9.35a) that $\lambda = 1$. The site productivity tax causes no distortions in landowner harvesting decisions; it is the perfect means for collecting the required tax revenue.

The yield tax has a role that goes beyond revenue generation though, because uncertainty is inherent in our problem. Assume that the site productivity tax is set at its optimal level $T = T^*$. Upon differentiating the indirect utility function, the welfare effect of the yield tax is then given by $V_\tau^* = -(p_1 x + R^{-1}\bar{p}_2 z) + AR^{-2}(1 - \tau)z^2\sigma_p^2 < 0$. This shows that a risk effect is present in the welfare effect; it follows because the landowner is risk averse and thus changes in the yield tax move the landowner to a different point on his utility function. Recall, finally, that $z_\tau = -(1 + g')x_\tau$. Using these formulas in (9.35b) yields a first-order condition governing the optimal yield tax,

$$\Omega_{\tau|T=T^*} = AR^{-2}(1 - \tau)z^2\sigma_p^2 + \tau[p_1 + R^{-1}p_2(1 + g')]x_\tau = 0. \tag{9.36}$$

To investigate if the yield tax is optimally positive or equal to zero, we examine the properties of this first-order condition at the corner solutions ($\tau = 0, 1$):

$$\Omega_\tau|_{\substack{T=T^* \\ \tau=0}} = AR^{-2}z^2\sigma_p^2 > 0, \quad \Omega_\tau|_{\substack{T=T^* \\ \tau=1}} = [p_1 + R^{-1}p_2(1 + g')]x_\tau < 0. \tag{9.37}$$

This clearly shows that the optimal yield tax rate must be positive, $0 < \tau^* < 1$. Its role here is one of reducing risk and providing social insurance to landowners facing risk. From (9.37) we can solve for the implicit tax rate to find

$$\tau^* = -\frac{AR^{-2}z^2\sigma_p^2}{[p_1 - R^{-1}p_2(1 + g')]x_\tau - AR^{-2}z^2\sigma_p^2}. \tag{9.38}$$

The optimal tax rate depends on the real interest rate, the future price of timber and its variance, the level of landowner risk aversion, and forest growth. By partial differentiation of (9.38), a higher risk aversion and price risk imply a higher yield tax rate. This eliminates some uncertainty and reduces precautionary harvesting on the part of the landowner.

9.4.2 Aggregate Risk

The tax policy design implications of aggregate risk depend on substitutability between the present value of private income V and publicly provided consumption through the size of G, as well as on risk attitudes associated with variability in these. A special case obtains when V and G are perfect substitutes and the landowner's risk attitudes toward each are similar. Now only their sum, $V + G$, matters. For instance, under an uncertain timber price, we have $V + G = p_1 x + R^{-1} \hat{p}_2 z$, so that neither the site productivity tax nor the yield tax matter at all. Consequently, they can have no insurance role, and risk cannot be shifted between private and public consumption through use of policies. This is a stochastic forestry example of the Ricardian equivalence theorem from macroeconomics (see Barro 1989).

It is more constructive to assume that private and public consumption are imperfect substitutes. A simple way to do this is to use an additively separable utility function like in Koskela and Ollikainen (1998a). Suppose the preference associated with stochastic tax revenues G, distributed to forest owners as public consumption, is given by an exponential utility function $u(G) = -\exp(-A_g G)$, where $A_g = -u''(G)/u'(G)$ is the Arrow-Pratt measure of constant absolute risk aversion. Since future timber price is normally distributed, the public consumption part of expected utility is $EU = -\exp(A_g N)$, where $N = -\bar{G} + (1/2) A_g R^{-2} z^2 \tau^2 \sigma_p^2$. Thus the landowner's utility function is now $EU^0 = -\exp(AV) - \exp(A_g N)$, where V and N are defined as

$$V = p_1^* x - (1 + R^{-1})T + R^{-1} \bar{p}_2^* z - \tfrac{1}{2} A R^{-2}(1 - \tau)^2 \sigma_p^2 z^2, \tag{9.39a}$$

$$N = \tau p_1 x + (1 + R^{-1})T + R^{-1} \tau \bar{p}_2 z - \tfrac{1}{2} A R^{-2} \tau^2 \sigma_p^2 z^2. \tag{9.39b}$$

The problem of the social planner is now to choose site productivity and yield taxes to maximize $SW = EU^{0^*}$. Choosing the site productivity tax yields

$$SW_T = -\exp(-AV)A(1 + R^{-1}) + \exp(-A_g N)A_g(1 + R^{-1}) = 0. \tag{9.40a}$$

Thus the site tax has been chosen optimally when the marginal utility from private income is equal to the marginal utility the forest owner derives from public consumption. Equivalently, we can express (9.40a) as

$$T : \frac{e^{-AV}}{e^{-A_g N}} = \frac{A}{A_g}, \tag{9.40b}$$

so that optimality requires that the marginal rate of substitution between private and public consumption equal the ratio of the respective risk-aversion parameters.

Let us next ask if the yield tax can have any role under aggregate uncertainty. Assuming that $T = T^*$ and differentiating EU^{0^*} with respect to x, we first develop the following harvesting rule under aggregate uncertainty:

$$p_1 - (1 + g')\{\bar{p}_2 - R^{-1} \sigma_p^2 z[A(1 - \tau)^2 + A_g \tau^2]\} = 0. \tag{9.41}$$

From (9.41) we find by differentiation that the effect of yield tax is given by

$$x_\tau = (EU_{xx}^0)^{-1}R^{-1}z\sigma_p^2 2[A(1-\tau) + A_g\tau] \geq (<)\, 0 \text{ as } \tau \geq (<)\, A/(A + A_g). \qquad (9.42)$$

The choice of the optimal yield tax rate when $T = T^*$ follows from

$$SW_{\tau|T=T^*} = 0 \Leftrightarrow R^{-2}z^2\sigma_p^2[A_g\tau - A(1-\tau)]$$

$$+ x_\tau\{p_1 - (1+g')[\bar{p}_2 - R^{-1}\sigma_p^2 z[A(1-\tau)^2 + A_g\tau^2]]\} = 0. \qquad (9.43)$$

Noting that the last braced term is equal to zero owing to the first-order condition (9.41) yields

$$\tau_{|T=T^*} = \frac{A}{A + A_g} > 0. \qquad (9.44)$$

The optimal yield tax rate reflects only landowner risk attitudes toward private and public consumption. Thus it affects the division of risk between after-tax timber revenues and government consumption. However, it does not contain a timber price variance term or any of the related terms we found under idiosyncratic risk. In the special case where $A = A_g$, we have the yield rate equal to 50%. Also, from (9.44), when the yield tax is set at the optimal level, it becomes nondistortionary because $x_\tau = 0$.

Hence the optimal yield tax is nondistortionary and depends only on the risk attitudes of landowners toward the variability of after-tax timber revenue vis-à-vis government-financed consumption. If forest owners are more (less) averse to private income risk than to public consumption risk, then the optimal yield tax is higher (less) than 50%. This is a natural result. If landowners are worried about variability in their private income, then a tax system that lowers private income risk at the expense of public consumption risk is preferred. Hence the difference with the optimal tax under idiosyncratic risk is striking. Under idiosyncratic risk, we found that the yield tax was distortionary under price and forest growth uncertainty; moreover, a subsidy (negative tax rate) was optimal under stock uncertainty.

9.5   Summary

In this chapter we have introduced the basic forms of uncertainties that landowners face concerning short-term harvesting decisions, using a two-period model of landowner behavior. The analyses required proposing a description of a landowner's behavior under uncertainty. We did this by applying a variant of nonexpected utility called the Kreps-Porteus-Selden certainty-equivalent approach. The advantage to such an approach is to make the role of risk more visible and easy to interpret. While this reduces the generality of analysis, the results are clearer than under the alternative expected utility approach and are especially useful for life-cycle cycle models.

A general theme of our results is that landowners exhibit precautionary harvesting behavior under risk. Precautionary harvesting is the landowner's behavioral means for hedging against future unknown parameters. Depending on the nature of the risk, precaution implies either increased or decreased short-term harvesting. This is a general result, which does not depend on possible assumptions concerning the distribution of the stochastic variable, or the behavior of absolute risk aversion.

Our results concerning the impacts of forest taxes relied on the assumption of constant absolute risk aversion and normally distributed stochastic variables. These produced results that are not necessarily general. Our model reflects the mere substitution effects of taxes in a general expected utility theory framework. That said, the findings concerning the optimal design of forest taxes are robust. In the presence of economic risk, the optimal taxation of instruments in particular takes on a new role. There is the possibility that the government can use policy instruments as a way of providing insurance against risks faced by landowners. This could be important, because in many countries, such as the United States, forest land investment is difficult to insure through private institutions. This is not all that surprising given that risk moves decisions away from optimal levels, with the level of distortion depending on the degree of risk aversion. We also find an important role for yield taxes that calls into question the lack of these tax instruments for most U.S. States at least. That all said, this role is highly dependent on the nature of risk, i.e., whether it is idiosyncratic or aggregate.

Appendix 9.1   Disentangling Risk and Time

If agents live for two periods, the conventional model of random future consumption has the following functional form:

$$U(c_1, \tilde{c}_2) = u_1(c_1) + Eu_2(\tilde{c}_2), \tag{A9.1}$$

where the degree of aversion to a change in consumption over time and the degree of aversion to risk are identical. Kreps and Porteus (1978) and Selden (1978) presented an alternative model to disentangle the attitudes toward smoothing consumption over time and across possible states of nature. Their model is a direct extension of (A9.1) and is expressed as

$$U(c_1, \tilde{c}_2) = u_1(c_1) + u_2(v^{-1}[Ev(\tilde{c}_2)]), \tag{A9.2}$$

where u_1, u_2, and v are increasing functions. Using the certainty-equivalent function $m(\tilde{c}_2)$, we can then re-express U as

$$U(c_1, \tilde{c}_2) = u_1(c_1) + u_2[m(\tilde{c}_2)]. \tag{A9.3}$$

First, according to (A9.3) one computes the certainty equivalent m of future uncertain consumption $\tilde{c}_2$ by using the utility function v. The concavity of v measures the degree of risk aversion alone. Second, one evaluates intertemporal utility by summing the utility of current consumption and the utility of future certainty equivalent consumption using the functions u_1 and u_2.

The specific cases of these preferences are as follows: (1) if v and u_2 are identical, then we have the specification (A9.1). (2) If v is the identity function but u_1 and u_2 are concave, then the agent is willing to smooth expected consumption over time even though he is risk neutral. (3) An opposite possibility is to assume that the agent is risk averse but indifferent toward (certainty equivalent) consumption smoothing. This is the case if v is concave, but u_1 and u_2 are linear. We use this specification in this chapter.

Appendix 9.2   Taylor Approximations and Distributions of Functions of Random Variables

Taylor Approximations for (9.7a) and (9.7b)

We apply the following theorem in this chapter (see Dudewicz and Mishra pp. 263–264):

Theorem   If $\tilde{X}$ is a random variable for which the kth moment $\mu_k = E[(\tilde{X} - \bar{X})^k]$ exists, where $\bar{X} = E(\tilde{X})$ and $\mathrm{var}[\tilde{X}] = \sigma_X^2 = E(\tilde{X} - \bar{X})^2$, and $g(.)$ is a k-times differentiable function, then

$$Eg(\tilde{X}) = g(\bar{X}) + \frac{g''(\bar{X})}{2!}\sigma_X^2 + \frac{g'''(\bar{X})}{3!}\mu_3 + \cdots + \frac{g^{(k-1)}(\bar{X})}{(k-1)!}\mu_{k-1} + \frac{E[g^{(k)}(\xi)(\tilde{X} - \bar{X})^k]}{k!}.$$

Corollary 1   Taking $k = 2$, we then have the following useful approximation for the expected value of a random variable, $g(\tilde{X})$:

$$Eg(\tilde{X}) = g(\bar{X}) + \frac{g''(\bar{X})}{2!}\sigma_X^2. \tag{A9.4}$$

Corollary 2   Taking $k = 2$ in the theorem, we have the following approximation for the variance of the random variable, $g(\tilde{X})$:

$$\text{var}[g(\tilde{X})] = [g'(\bar{X})]^2 \sigma_X^2. \tag{A9.5}$$

Proof From corollary 1 we obtain the following approximation: $E[g(\tilde{X})]^2 = g(\bar{X})^2 + \{[[g'(\bar{X})]^2 + g(\bar{X})g''(\bar{X})]\}\sigma_X^2$. This means that the variance is approximated by

$$\text{var}[g(\tilde{X})] = E[g(\tilde{X})]^2 - \{Eg(\tilde{X})\}^2 = g(\bar{X})^2 + [[g'(\bar{X})]^2 + g(\bar{X})g''(\bar{X})]\sigma_X^2$$

$$- \left\{ g(\bar{X}) + g''(\bar{X})\frac{\sigma_X^2}{2} \right\}^2 = [g'(\bar{X})]^2\sigma_X^2 - [g''(\bar{X})]^2\frac{\sigma_X^4}{4} \approx [g'(\bar{X})^2]\sigma_X^2. \tag{A9.6}$$

Q.E.D.

Distributions of Random Variables

Here we briefly characterize how to calculate the mean and variance of a product of two random variables (taken from Mood et al. 1974, chapter 5). First, let $\tilde{P}$ and $\tilde{R}$ be two random variables for which the covariance, $\text{cov}[\tilde{P}\tilde{R}]$, exists. The mean and variance of the product $\tilde{P}\tilde{R}$ is as follows:

$$E[\tilde{P}\tilde{R}] = \bar{P}\bar{R} + \text{cov}[\tilde{P}, \tilde{R}] \tag{A9.7a}$$

and

$$\text{var}[\tilde{P}\tilde{R}] = \bar{R}^2 \,\text{var}[\tilde{P}] + \bar{P}^2 \,\text{var}[\tilde{R}] + 2\bar{P}\bar{R} \,\text{cov}[\tilde{P}, \tilde{R}] - (\text{cov}[\tilde{P}, \tilde{R}])^2$$

$$+ E[(\tilde{P} - \bar{P})^2(\tilde{R} - \bar{R})^2] + 2\bar{R}E[(\tilde{P} - \bar{P})^2(\tilde{R} - \bar{R})] + 2\bar{P}E[(\tilde{P} - \bar{P})(\tilde{R} - \bar{R})^2], \tag{A9.7b}$$

where $\bar{P} = E[\tilde{P}]$, $\bar{R} = E[\tilde{R}]$, $\text{var}[\tilde{P}] = E[(\tilde{P} - \bar{P})^2]$, $\text{var}[\tilde{R}] = E[(\tilde{R} - \bar{R})^2]$ and $\text{cov}[\tilde{P}, \tilde{R}] = E[(\tilde{P} - \bar{P})(\tilde{R} - \bar{R})]$.

Proof $\tilde{P}\tilde{R} = \bar{P}\bar{R} + (\tilde{P} - \bar{P})\bar{R} + (\tilde{R} - \bar{R})\tilde{P} + (\tilde{P} - \bar{P})(\tilde{R} - \bar{R})$. Calculating $E[\tilde{P}\tilde{R}]$ and $E[(\tilde{P}\tilde{R})^2]$ gives the results (A9.7a)–(A9.7b). Hence the mean of the product can be presented in terms of the means and covariance, but the variance of the product requires higher-order moments.

If $\tilde{P}$ and $\tilde{R}$ are independent,

$$E[\tilde{P}\tilde{R}] = \bar{P}\bar{R} \tag{A9.8a}$$

and

$$\text{var}[\tilde{P}\tilde{R}] = \bar{R}^2 \,\text{var}[\tilde{P}] + \bar{P}^2 \,\text{var}[\tilde{R}] + \text{var}[\tilde{P}] \,\text{var}[\tilde{R}]. \tag{A9.8b}$$

Proof   The independence of $\tilde{P}$ and $\tilde{R}$ implies

$$E[(\tilde{P} - \bar{P})^2(\tilde{R} - \bar{R})^2] = E[(\tilde{P} - \bar{P})^2]E[(\tilde{R} - \bar{R})^2] = \text{var}[\tilde{P}]\,\text{var}[\tilde{R}] \qquad (A9.9)$$

$$E[(\tilde{P} - \bar{P})^2(\tilde{R} - \bar{R})] = E[(\tilde{P} - \bar{P})^2]E[(\tilde{R} - \bar{R})] = 0 \qquad (A9.10)$$

and

$$E[(\tilde{P} - \bar{P})(\tilde{R} - \bar{R})^2] = E[(\tilde{P} - \bar{P})]E[(\tilde{R} - \bar{R})^2] = 0. \qquad (A9.11)$$

Q.E.D.

10 Risk of Catastrophic Events

An important class of stochastic problems not covered yet concerns catastrophic natural events, such as fires, wind, ice, and pests. These events have always been part of the forest landscape, playing an important ecological role in maintaining forest structure and composition. Consider that fires are often crucial to the regeneration of new forests in the Pacific Northwest and Canadian boreal forests, as well as the forests of Scandinavia. Wind and ice storms have similar roles in boreal and temperate forests. By destroying trees, natural catastrophes also create open space that promotes biological diversity and uneven-aged stands.

Despite their ecological functions, catastrophic natural events obviously represent an economic loss for forest landowners. Not only are valuable forest stands destroyed, but costly effort may be required to salvage the loss from these events. In some cases, a landowner might be able to engage in activities early in a rotation to minimize damage from these events, but these actions are also costly. In all, natural hazards decrease rents and the value of a stand. Catastrophic events can also cause considerable impact on timber markets. Prestemon and Holmes (2000) find that weather events have led to both short- and long-term increases in stumpage prices, effectively transferring wealth to landowners whose forests are spared the events.

In this chapter we explore at length how the risk of catastrophic loss affects forest landowner decisions. A key question to ask is how landowners would adjust their decisions if they faced a positive probability of a natural disaster taking place in any future rotation. To answer this, we first need a workable description of how catastrophic natural events arrive during a rotation, and how they affect a stand of trees. At the theoretical level, this challenge was met in a seminal paper by Reed (1984). Focusing on forest fires, he extended the Faustmann framework to cover catastrophic events arriving in a stand independently over time. This allowed him to describe the arrival of these events using a Poisson process. Assuming that the entire stand is destroyed when an event occurs, Reed derived an expression for the expected bare land value of a stand.

The analysis of catastrophic risk has been an active area of interest in forest economics for the past 30 years. Even Reed has predecessors; among the earliest studies are those by Martell (1980) and Routledge (1980), who considered empirically the effect of forest fires on optimal rotation ages in a discrete-time model. Along similar lines, wind damage and forest management decisions were analyzed numerically by Lohmander and Helles (1987), while pest damage was studied by Anderson et al. (1987) and Williams and Nautiyal (1992), among others.

We start in section 10.1 by describing the most common types of stochastic catastrophic events landowners may face during a rotation, and by characterizing two different types of Poisson processes used to describe these events. In sections 10.2 and 10.3, we analyze rotation age decisions in expected Faustmann and Hartman frameworks. Finally, in section 10.4 we discuss modifications to allow for partial destruction of the stand, and to endogenize the protection effort a landowner can expend to reduce either the probability of hazards, or the damage to the stand should the hazard arrive before harvesting.

10.1   Stochastic Processes

The main catastrophic threats to forests in a given region depend largely on the interaction of climate, landform, and geology. The list of possible threats is large and includes pests and disease, windthrow, snow or ice and frost, fire, drought, land slippage, flooding, and even volcanic eruptions. The probability that any of these events will arrive and inflict damage on a forest stand depends on a set of key biological and ecological factors. Some factors, such as fuel loadings and tree form, can be controlled by a landowner. Other factors, such as weather, are not affected by landowner decisions.

Table 10.1, adapted from Gardiner and Quine (2000), lists some potential natural disasters that can cause catastrophic losses during any rotation. As the table shows, while the key processes leading to damage vary by event, variables such as location, tree dimension, and stand composition are common to determining the effects of most hazards.

Provided there is a history of forest management in a region, these risks can be characterized and the probabilities associated with them can in theory be determined. There are many possible ways to include these observed probabilities. Reed (1984) postulated that catastrophic loss events can be described using a Poisson stochastic process. A basic Poisson process assumes that a catastrophic event occurs independently and randomly over time. The average arrival rate of the natural hazard in any time period is given by $1/\lambda$. The parameter λ is known as a standard Poisson distribution parameter (Ross 1970).

Table 10.1
Abiotic Hazards to Forests and the Key Factors and Processes Involved

Type of hazard	Key factors	Key processes
Wind	Location, tree species, soil, spacing, tree dimensions	Uprooting, stem breakage, branch breakage, spread of damage within forest, speedup of wind by topography
Snow or ice	Location, tree species, tree dimension	Snow adhesion to crown, crown breakage, insect infestation following breakage
Frost	Location, tree species, ground vegetation	Seedling death, death of parts of live crown, splitting of trunk
Fire	Location, tree species, understory	Fuel buildup, fire ignition, fire spread, fire damage
Drought	Location, tree species, tree size, soil, ground vegetation	Unseasonable weather, stomatal shutdown, dessication of tree, death of tree
Land slippage	Location, soil, ground vegetation	Loss of soil cohesion, slippage caused by earthquakes
Flooding	Location	Unseasonable rainfall, river or lake overflow
Volcanoes	Location	Trees engulfed by volcanic ash or lava

Source: Gardiner and Quine (2000, p. 263).

Describing the arrival rate with a Poisson distribution parameter is convenient in the sense that it only requires data on the frequency of hazards. To illustrate the role of this parameter, consider the case of fires. In Finland the frequency of forest fires is typically one every 120 years, and in the United States and Canada there is probably one fire event on any given acre about once every 60 to 100 years. For a fire every 120 years and 60 years, the Poisson parameter would equal $\lambda = 120$ and $\lambda = 60$, respectively. Thus the probability of a forest fire arriving in any given year under the Markov property for these cases would equal $1/120$ and $1/60$, respectively.

Assumptions concerning λ define the type of Poisson process. If the probability of arrival for the catastrophic event is constant over time, then the process is called "homogeneous" and λ is constant over time. However, if the probability of arrival depends on calendar time, for instance, because it depends on the age or size of stems in a stand, then the process is called "nonhomogeneous" and λ varies over time. In the following definitions, we provide a rigorous description of these processes (see e.g., Ross 1970, chapter 2).

Definition 10.1   Homogeneous Poisson process   A stochastic process $[X(t), t \geq 0]$ is a homogeneous Poisson process with an arrival rate $\lambda > 0$ if the number of events in any interval of length t has a Poisson distribution with mean λt, and for all $s, t \geq 0$, $\Pr\{X(t+s) - X(s) = n\} = e^{-\lambda t}[(\lambda t)^n / n!]$, $n \geq 0$. Therefore $[X(t), t \geq 0]$ has stationary independent increments.

A homogeneous Poisson process represents a pure jump function in that the jump in the stochastic process at the time of the event is constant and does not depend on the mean value of the process. In the case of a nonhomogeneous Poisson process, we write the Poisson parameter as $\lambda = \lambda(t)$. If the probability of the arrival rate increases with time or stand age, then we assume $\lambda'(t) > 0$. This also implies that the probability of a loss occurring depends positively on the age of the forest. This is described in the next definition.

Definition 10.2 Nonhomogeneous Poisson process A stochastic process $[X(t), t \geq 0]$ is a nonhomogeneous Poisson process with an arrival rate $\lambda(t) > 0$ if the number of events in any interval of length t has a Poisson distribution with the mean value of the process equal to $m(t) = \int_0^t \lambda(s)\, ds$, and $\Pr\{X(t) = n\} = e^{-mt} m(t)^n / n!$, $n \geq 0$. Therefore $X(t)$, $t \geq 0$ does not have stationary independent increments.

It will become clear later how these definitions apply to forest management. Before proceeding, we should point out that Poisson processes belong to a larger set of useful and simple Markov processes. With Markov processes, the probability of an event in the next period depends only on the value of the system parameters in the current period (see e.g., Karlin and Taylor 1975, chapter 2). Because of their simple structure, Markov processes have been widely used in natural resource economics to study decision making under uncertainty. We will encounter them again in chapter 12.

10.2   Faustmann Interpretations

Following Reed's original contribution, we start by assuming that the arrival of a catastrophic loss event during a rotation period is independent of stand age and thus can be described by a homogeneous Poisson process. We then expand the model for the case where the arrival rate depends on stand age. Throughout this and section 10.4 we continue to assume that the natural hazard completely destroys the stand. There is no possibility of salvaging destroyed trees.

10.2.1   Arrival Rate Independent of Stand Age

Suppose a landowner faces risk caused by an uncertain catastrophic event during the current or any future rotation cycle. Consider an initial stand of age t. Let $V(t) = pF(t) - c_1$ denote the current (undiscounted) net value of this stand, where F is the time-dependent growth function, p is the timber price, and c_1 is the planting cost incurred to begin the next rotation when the current stand is harvested. A critical component is the timing and impact of a catastrophic event relative to the rotation age. If the event occurs before the stand is harvested, then the entire stand is destroyed and the landowner obtains zero rent. However, if the event does not occur and the rotation age T is reached, then the landowner harvests the trees and obtains the current value of the net economic rent equal to $V(T)$.

Let X define a random variable denoting the time between successive catastrophic events. By the earlier assumptions, X is also the stand age when and if an event occurs. Let the rate at which these events arrive be described by a Poisson parameter λ. Using the properties of a homogeneous Poisson distribution from definition 10.1, the relevant probabilities are

$$\Pr(X < T) = 1 - e^{-\lambda T} \tag{10.1a}$$

$$\Pr(X = T) = e^{-\lambda T}. \tag{10.1b}$$

The probability that a catastrophic event will occur before the rotation age is reached is given by (10.1a), while the probability that no event will occur and that the stand will reach the rotation age and be harvested is given by (10.1b). In the presence of catastrophic loss risk, the net current value of the stand over each rotation is also a random variable because rents received by the landowner depend on the realization of the random variable X during the rotation. We therefore define the stochastic economic value of the stand by a random variable, Y_n, which measures the landowner's net (undiscounted) return over one rotation cycle. This random net return depends on the relationship between the timing of the event and the rotation age. Indexing the time between stand destruction now as X_n for any rotation n, the random net *current* return for the nth rotation is

$$Y_n = \begin{cases} -c_2 & \text{if } X_n < T \\ pF(T) - c_1 & \text{if } X_n = T. \end{cases} \tag{10.2}$$

The first line describes the case where the catastrophic event occurs before the rotation age is reached. Here the landowner obtains zero forest rent, owing to the total destruction of the stand, and must clear the destroyed site and plant to begin a new rotation at a cost of c_2. The second line describes the case where the event does not occur before the rotation age is reached. Now, the landowner harvests and receives the stand's harvest value, but then incurs a cost c_1 to plant and begin the next rotation.[1] Even though it does not matter for the theory, in many cases, such as fire, we can think of c_2 as being higher than c_1. This process

---

1. Notice that owing to the initial stand assumption, the cost incurred to start the very first rotation does not appear in (10.2). It is regarded as sunk, or, alternatively, the stand is assumed to be naturally generated. We continue to assume this throughout the chapter. This is not an issue as long as it is a lump-sum amount and paid at time zero when there is no uncertainty (Conrad and Clark 1987, chapter 5). In fact, an appropriate subtraction could be made from (10.3) in order to include this cost, but it would drop out of the subsequent first-order condition for T.

of either harvesting or dealing with the catastrophic event continues over all future rotations ad infinitum.[2]

Under constant economic and forest growth parameters, the expected net present value of harvesting at the rotation age or clearing damaged trees and replanting over an infinite cycle of rotations is expressed as

$$J = E[e^{-rX_1}Y_1 + e^{-r(X_1+X_2)}Y_2 + \cdots] = E\left[\sum_{n=1}^{\infty} e^{-r(X_1+X_2+\cdots+X_n)}Y_n\right], \qquad (10.3)$$

where r is the real interest rate. The random variables $X_1, \ldots, X_n$ now denote times between successive destructions of the stand either by catastrophic loss or harvesting. Denote these events by i ($i = 1, 2, \ldots$). Then each X_i ($i = 1, 2, \ldots$) corresponds to the length of the ith rotation, given the assumption that a new stand is planted after reaching the rotation age or after a catastrophic event occurs. X_i is appropriately enough sometimes called a "waiting" time between successive destructions of a stand. The variables $Y_1, \ldots Y_n$ denote the rents defined in (10.2) from the successive rotations.

We next develop equation (10.3) further into an easier form to work with, and one that is essentially a Faustmann model extended to natural hazards. Note first that the waiting times X_i between stand destructions of any type are independent. Therefore the expected rent in the last rotation $E(e^{-rX_n}Y_n)$ is the same as in all other rotations. We can therefore factor $E(e^{-rX_n}Y_n)$ out of (10.3) as follows:

$$J = E\left[\sum_{n=1}^{\infty} e^{-r(X_1+X_2+\cdots+X_n)}Y_n\right] = E\left[\sum_{n=1}^{\infty}(e^{-rX_1}e^{-rX_2}\ldots e^{-rX_n})\right]E(e^{-rX_n}Y_n).$$

Next, notice that the expected waiting time is the same for each event, given they are independent and the probability distribution is stationary. This implies that $E(e^{-rX_i}) = E(e^{-rX})$ for all i. Thus we can modify the above formula to cover all future rotations by writing

$$J = E\left[\sum_{n=1}^{\infty}(e^{-rX_1}e^{-rX_2}\ldots e^{-rX_n})\right]E(e^{-rX_n}Y_n) = \sum_{n=1}^{\infty}\left[\prod_{i=1}^{n-1} E(e^{-rX})E(e^{-rX}Y_n)\right].$$

Finally, noting that the summation on the RHS of this equation is an infinite geometric series, we can finally write an expression for expected present value,

---

2. This model assumes that a new rotation is started after the event occurs. There is no possibility that the damaged trees can be salvaged, and the undamaged part of the stand is left to grow. For many cases this is too limiting an assumption. A more dynamic model, such as those reviewed in chapter 12, would be needed to allow for this possibility.

$$J = \sum_{n=1}^{\infty} \left[\prod_{i=1}^{n-1} E(e^{-rX}) E(e^{-rX} Y_n) \right] = \frac{E(e^{-rX} Y)}{1 - E(e^{-rX})}. \tag{10.4}$$

In equation (10.4), the summation adds up expected forest rent over an infinite time horizon, and the boundaries of the product term refer to different rotation ages given the assumption that a new stand is started after the arrival of every event or harvest. Equation (10.4) can be interpreted as the expected value of bare land, with stochastic events arriving randomly in all future rotations ad infinitum.

Although it appears simple enough, it is not trivial to precisely compute the expected-value term $E(e^{-rX})$ present in the numerator and denominator of (10.4). If the stand is harvested when it reaches the rotation age T in (10.2), then the distribution function for the random variable X_n is given by $\Gamma_X(t) = 1 - e^{-\lambda t}$, $\forall t < T$, and $\Gamma_X(t) = 1$, $\forall t \geq T$, where $\Gamma_X(t) = \Pr(X_n \leq t)$. The probability density function is in turn defined by the derivative of this distribution function: $d\Gamma_X(t)/dt = \lambda e^{-\lambda t}$. Using these relations and integrating by parts, one can develop the expectation for the denominator term as follows:

$$E(e^{-rX}) = \int_0^X e^{-rt} d\Gamma_X(t) = \int_0^T e^{-rt} \lambda e^{-\lambda t}\, dt + e^{-(r+\lambda)T} = \frac{\lambda + re^{-(r+\lambda)T}}{r + \lambda}. \tag{10.5}$$

Equation (10.5) then implies that the denominator of (10.4) becomes

$$1 - E(e^{-rX}) = \frac{r(1 - e^{-(r+\lambda)T})}{r + \lambda}. \tag{10.5'}$$

The expected value of $E(e^{-rX} Y)$ can be expressed as the sum of the possibility of destroying the stand by harvesting or by a natural hazard subject to their respective probabilities. Using (10.5) we have

$$E(e^{-rX} Y) = e^{-rT}[pF(T) - c_1]e^{-\lambda T} + \int_0^T (-c_2)e^{-rt} \lambda e^{-\lambda t}\, dt$$

$$= [pF(T) - c_1]e^{-(r+\lambda)T} - \frac{\lambda c_2(1 - e^{-(r+\lambda)T})}{r + \lambda}. \tag{10.6}$$

Substituting the right-hand sides of equations (10.5') and (10.6) into equation (10.4) yields a Faustmann-like formula for the expected bare land value of a stand in the presence of a risk of future catastrophic loss:

$$J = \frac{(r + \lambda)[pF(T) - c_1]e^{-(r+\lambda)T}}{r(1 - e^{-(r+\lambda)T})} - \frac{\lambda}{r} c_2. \tag{10.7}$$

Equation (10.7) is an extension of the deterministic Faustmann land value derived in chapter 2. The catastrophic risk here is realized through two channels. First, the effective discount rate is adjusted according to the Poisson risk parameter, $r + \lambda$, implying that the discount rate used by the landowner to make decisions is higher than before. This means that a risk-neutral landowner facing catastrophic loss perceives a higher opportunity cost of not harvesting than under certainty. Second, an additional term, $\lambda c_2/r$, is present and reflects the expected present-value cost of clearing the site and replanting after the catastrophic event, ad infinitum.

From (10.7) we can obtain a first-order condition by differentiating with respect to rotation age, which after simplifying (i.e., by adding together the two opportunity cost terms) is

$$J_T = pF'(T) - \frac{(r + \lambda)[pF(T) - c_1]}{(1 - e^{-(r+\lambda)T})} = 0 \tag{10.8a}$$

Taking the derivative of the unsimplified first-order condition, we obtain the following second-order condition,

$$J_{TT} = pF''(T) - (r + \lambda)pF'(T) < 0. \tag{10.8b}$$

According to (10.8), the presence of catastrophic loss adds a new term of size λ to the opportunity cost of waiting to harvest (the second term in 10.8). Because the opportunity cost of continuing a rotation is now higher, the optimal rotation age must be shorter. Setting $\lambda = 0$ in (10.8) yields the harvesting condition under certainty when the initial constant regeneration cost is ignored: $pF'(T) - r[pF(T) - c_1](1 - e^{-rT})^{-1} = 0$. Thus, owing to the expected risk, the risk-neutral landowner harvests sooner to reduce the exposure of his returns to the natural hazard. It is interesting that the cost of clearing and replanting the site after a catastrophic event does not affect rotation age at the margin because it enters the target function (10.7) in a lump-sum fashion (although it does decrease rents from forest management). Hence we have established proposition 10.1.

Proposition 10.1   In the Faustmann model, under risk of catastrophic loss and a risk-neutral harvester, the presence of risk increases the effective interest rate and shortens the optimal rotation age relative to the rotation age solved under the assumption of certainty.

According to proposition 10.1, the landowner harvests sooner to avoid potential future damage to his stand. For a fixed and exogenous hazard, this makes sense. The model here assumes there are no means for mitigating the hazard through landowner actions. The only means to reduce exposure to the risk is to shorten the rotation age according to the expected frequency of future hazards.

The comparative statics of (10.8a) confirm this. Recall that uncertainty shows up through the two new exogenous parameters, λ and c_2. Their effects on the rotation age are given by the following results:

$$\frac{\partial T}{\partial \lambda} = -\frac{J_{T\lambda}}{J_{TT}} < 0; \quad \frac{\partial T}{\partial c_2} = -\frac{J_{Tc_2}}{J_{TT}} = 0, \tag{10.9}$$

where

$$J_{T\lambda} = \frac{pF'(T)}{r + \lambda} [[1 + (r + \lambda)T]e^{-(r+\lambda)T} - 1] \quad \text{and} \quad J_{Tc_2} = 0.$$

The sign of $J_{T\lambda}$ is ambiguous. However, using $e^{-(r+\lambda)T} \approx [1 + (r + \lambda)T + (1/2)(r + \lambda)^2T^2]^{-1}$ as the second-order approximation, we obtain

$$J_{T\lambda} = -\frac{pF'(T)}{r + \lambda} \left[\frac{(1/2)(r + \lambda)^2T}{1 + (r + \lambda)T + (1/2)(r + \lambda)^2T^2} \right] < 0,$$

which establishes the first result in equation (10.9). Thus a higher probability of the catastrophic event shortens the rotation age, as suggested by proposition 10.1. In contrast, the site preparation cost has no impact on the rotation age.

The effects of other conventional economic parameters remain the same as in the deterministic case. As can be inferred from the first-order conditions, the interest rate has an effect (negative) similar to that of the hazard risk. A higher timber price shortens the rotation age, but an increased regeneration cost lengthens it for reasons we discussed in chapter 2.

10.2.2 Arrival Rate Dependent on Stand Age

The results here have been obtained under an assumption that the probability of a catastrophic event occurring in a stand is independent of time and thereby of stand age. This may not be the case for some events (such as fire) whose arrival might depend on the age of the forest stock. In other cases, only the damage to a stand once an event arrives depends on the age of the forest stock. Windthrow is an example. For our purposes, we can think of the arrival of the event and damage as equivalent factors that both depend on stand age. In some cases, though, the probability of an event arriving, and the severity of damage, could be controlled by different factors, and stand age may not play a role in both.

As alluded to by this discussion, there is an important point we must mention before continuing in this section. The appropriateness of a nonhomogenous

Poisson process is hazard-specific. Before one decides how to model the arrival of a natural hazard, it is important to first determine whether the probability of a hazard arriving and the damage once it arrives truly depend on stand age and forest stock conditions. The nonhomogeneous process makes sense only in the case where the hazard's arrival probability changes over time as the forest stock changes, that is, the probability of arrival itself is dependent on stand conditions that change over time. If, however, the hazard arrival probability depends on other factors such as weather (consider fire and lightning) that are not dependent on stand age, but once the hazard arrives the damage is dependent on the condition of the forest, then a better way to deal with this problem is to consider a homogenous Poisson process but have a salvage function that depends on stand conditions which can be controlled by landowner decisions. We will discuss endogenous salvage later in this chapter along with a return to assuming a homogenous Poisson process to describe the natural hazard events.

Here we focus on an example where the probability of the event arriving is dependent on stand age. Suppose now that for a stand of age t, the probability of a catastrophic event is $\lambda(t)h$ during an infinitesimal time interval of length h. If $\lambda'(t) > 0$, then the probability that the event will arrive increases over time as the forest ages. We will assume this is the case throughout this section.

A nonhomogeneous Poisson process (definition 10.2) is appropriate for this type of situation. The time between successive destruction, that is, to either the catastrophic event or to harvesting now has a different distribution function than before. This is because the probability that the event will arrive before the rotation age, i.e., $\Pr(X < T)$, depends on the time between events, meaning that we have a new distribution function, $\Gamma(X) = 1 - e^{-m(X)}$, where $m(.)$ is a function of the time since the last loss or harvesting event X. More specifically, $m(X) = \int_0^X \lambda(s)\,ds$ and is increasing in X, i.e., $dm/dX = \lambda(X)$. The corresponding probability density function for X is now modified to equal $d\Gamma(X)/dX = \lambda(X)e^{-m(X)}$.

As before, the distribution of X implies that for a given rotation of age T, the probability that the forest will be destroyed by a catastrophic event before the rotation age is reached is $\Pr(X < T) = 1 - e^{-m(T)}$. The probability that the rotation age will be reached and the forest harvested before the event occurs is $\Pr(X = T) = e^{-m(T)}$.

Recall that equation (10.4) defining the expected value of the land was $J = [E(e^{-rX}Y)]/[1 - E(e^{-rX})]$. The terms $E(e^{-rX})$ and $E(e^{-rX}Y)$ are now different. First define $E(e^{-rX})$ as follows: $E(e^{-rX}) = \int_0^T e^{-[rX+m(X)]}\,dx$. Using the assumption that there is no risk at time zero, $\lambda(0) = 0$, we obtain by integration

$$\int_0^T e^{-[rX+m(X)]}\,dx = \frac{re^{-[rT+m(T)]}}{r + \lambda(T)}.$$

This gives us

$$1 - E(e^{-rX}) = \frac{r(1 - e^{-[rT+m(T)]}) + \lambda(T)}{[r + \lambda(T)]}. \tag{10.10}$$

The expected net value in the numerator of the bare land value term is now defined by

$$E(e^{-rX}Y) = [pF(T) - c_1]e^{-[rT+m(T)]} - \int_0^T c_2\lambda(X)e^{-[rx+m(x)]}\,dx.$$

Using the same procedures as earlier we obtain a new expression for bare land value under a non-homogenous catastrophic risk identical to that reported in Reed (1984),

$$J = \frac{[pF(T) - c_1]e^{-[rT+m(T)]} - c_2(1 - e^{-[rT+m(T)]})}{r\int_0^T e^{-[m(x)+rx]}\,dx} + c_2. \tag{10.11}$$

Equation (10.11) is an extension of the previous homogenous Poisson-based bare land value formula (10.7), but again we see that the landowner applies a higher effective discount rate when making decisions. However, equation (10.11) is clearly more complicated than (10.7). Not only does the discount factor differ, but now c_2 (the cost of site preparation and regeneration once the natural hazard has occurred) is multiplied by the discount factor.

The first- and second-order conditions for optimal rotation age are now,

$$J_T = pF'(T) - [r + \lambda(T)][pF(T) - c_1 + c_2] - r(J - c_2) = 0 \tag{10.12a}$$

$$J_{TT} = [pF''(T) - pF'(T)(r + \lambda)] < 0 \tag{10.12b}$$

The rotation age determination from condition (10.12a) exhibits a trade-off similar to that found in (10.8a). The marginal return from continuing the rotation under risk is equal to the opportunity cost of delaying harvest, and this is increased by the risk of natural hazards. It is interesting that, unlike the case where catastrophic risk is independent of stand age, now the cost of clearing and replanting c_2 matters to the rotation age decision. We offer the following comparison across cases:

Proposition 10.2 In the Faustmann model under risk neutrality and an age-dependent catastrophic loss risk that increases in time, the optimal rotation is shorter than the rotation age when the probability of catastrophic loss is independent of stand age.

This is a natural finding of course. As the stand becomes older, the risk of natural hazards arriving increases. To cope with such a risk that also happens to increase with stand age, the landowner harvests at a shorter rotation age than in the constant risk case.

The comparative statics result for the effect of the new relevant term c_2 on rotation age is

$$\frac{\partial T}{\partial c_2} = -\frac{J_{Tc_2}}{J_{TT}} < 0, \tag{10.13}$$

as

$$J_{Tc_2} = -(r + \lambda(T)) + \frac{1 - e^{-[rT + m(T)]}}{\int_0^T e^{-[rx + m(x)]}\, dx}.$$

It can also be shown that $J_{c_2} < 0$. Thus, a higher cost of clearing and replanting the site after the event occurs decreases the optimal rotation age, which is opposite the effect we showed using the deterministic Faustmann model. The rest of the comparative statics findings are similar to those we derived under the case of independent risk.

10.3   Amenity Services

The case where forests produce amenities is, as we know from earlier chapters, perhaps more important for a large class of landowners. Reed (1993) was the first to study a catastrophic risk problem with amenities, but he combined this with stochastic stand value. Englin et al. (2000) incorporate amenity benefits into Reed's (1984) model presented here. Both Reed (1993) and Englin et al. (2000) use an assumption that catastrophic events reduce the amenity services of a stand. In this section we follow Englin et al. (2000). We also derive several results not found in either Reed (1984) or Englin et al. (2000) for a risk that is dependent on stand age.

10.3.1   Arrival Rate Independent of Stand Age

As before, the probability that a catastrophic event will arrive before the rotation age is reached is still given by $\Pr(X < T) = 1 - e^{-\lambda T}$, and the probability that the rotation age will be reached before the event occurs is given by $\Pr(X = T) = e^{-\lambda T}$ (see equations 10.1a and 10.1b). Amenity benefits are introduced by modifying the random periodic net return Y_n in any rotation associated with time X_n between destructions as follows:

$$
Y_n = \begin{cases} -c_2 + e^{rX_n} \int_0^{X_n} B(s)e^{-rs}\, ds & \text{if } X_n < T \\ pF(T) - c_1 + e^{rT} \int_0^T B(s)e^{-rs}\, ds & \text{if } X_n = T, \end{cases} \tag{10.14}
$$

where c_2 and c_1 continue to denote clearing and replanting costs after an event occurs before the rotation age, and replanting costs if the rotation age is reached without an event arriving, respectively. Amenity benefits at any time s, $0 < s \le \min\{X_n, T\}$, are denoted by $B(s)$. This valuation function exhibits properties similar to the one introduced in the Hartman model of chapter 3. Assume that amenity values are higher for older stands, $B'(s) > 0$, $B''(s) < 0$. Note that (10.14) determines the current value of amenity benefits accrued during the rotation up to the time the forest is destroyed either by harvesting or a natural hazard. As in section 10.3, each rotation period ends when the stand is either destroyed completely $(X_n < T)$ or when it is harvested $(X_n = T)$.

In the previous section we demonstrated that expected net present value can be re-expressed as

$$
J^H = \sum_{n=1}^{\infty} \prod_{i=1}^{n-1} E(e^{-rX_i}) E(e^{-rX_n} Y_n) = \frac{E(e^{-rX} Y)}{1 - E(e^{-rX})}, \tag{10.15}
$$

where

$$
E(e^{-rX}) = \frac{\lambda + re^{-(r+\lambda)T}}{r + \lambda}, \quad \text{so that } 1 - E(e^{-rX}) = \frac{r(1 - e^{-(r+\lambda)T})}{r + \lambda}.
$$

Using the properties of the Poisson process, the numerator of (10.15) is more complicated than in (10.6):

$$
E(e^{-rX} Y) = e^{-rT} \left[[pF(T) - c_1]e^{-\lambda T} + e^{-\lambda T} e^{rT} \int_0^T B(s)e^{-rs}\, ds \right]
$$

$$
+ \int_0^T e^{-rX} \left[-c_2 + e^{rX} \int_0^X B(s)e^{-rs}\, ds \right] \lambda e^{-\lambda X}\, dX
$$

$$
= \left[pF(T) - c_1 + e^{rT} \int_0^T B(s)e^{-rs}\, ds \right] e^{-(r+\lambda)T}
$$

$$
- \int_0^T \lambda e^{-(r+\lambda)X} \left[c_2 - e^{rX} \int_0^X B(s)e^{-rs}\, ds \right] dX.
$$

By rearranging and integrating we then end up with

$$\left[pF(T) - c_1 + e^{rT} \int_0^T B(s)e^{-rs}\,ds \right]e^{-(r+\lambda)T} - \frac{\lambda(1 - e^{-(r+\lambda)T})}{r+\lambda}c_2$$

$$+ \int_0^T \lambda e^{-(r+\lambda)X}\left[e^{rX} \int_0^X B(s)e^{-rs}\,ds \right]dX. \tag{10.16}$$

Using (10.16) and the expression of $E(e^{-rX})$ in (10.15) produces the following expected net present value:

$$J^H = \frac{(r+\lambda)([pF(T) - c_1 + e^{rT} \int_0^T B(s)e^{-rs}\,ds]e^{-(r+\lambda)T} + \int_0^T \lambda e^{-(r+\lambda)X}[e^{rX} \int_0^T B(s)e^{-rs}ds]dX)}{r(1 - e^{-(r+\lambda)T})}$$

$$- \frac{\lambda}{r}c_2 \tag{10.17}$$

Equation (10.17) differs from the deterministic Hartman model of chapter 3 in three respects. First, the interest rate r is replaced by a higher effective rate, $r + \lambda$. Second, the expected value of amenity benefits up to the harvesting age T is $(r + \lambda)e^{-(r+\lambda)T}[e^{rT} \int_0^T B(s)e^{-rs}\,ds][r(1 - e^{-(r+\lambda)T})]^{-1}$. The third difference is the presence of the expected value of amenity services up to the point at which the catastrophic event occurs, if it does at all (last term in numerator).

The first-order condition for the choice of the optimal rotation age is similar to Englin et al. (2000) and can be expressed as

$$pf'(T) + B(T) = r\left(J^H + \frac{\lambda}{r}c_2 \right) + (r + \lambda)(pf(T) - c_1). \tag{10.18}$$

Relative to the case of certainty, the first-order condition now contains a risk premium that again increases land rent. Moreover, since even in the presence of a natural hazard the landowner enjoys some amenity benefits, the "land rent component" (the first RHS term) contains the expected present value of amenity benefits over all future rotations. Consequently, and not surprisingly, the risk of a catastrophic event shortens the rotation age relative to the deterministic solution. As in the Faustmann case, the cost of clearing and replanting the site after the event (c_2) does not affect the choice of rotation age at the margin because the probability of catastrophic loss has been assumed to be independent of the age of the stand.

Proposition 10.3 Under catastrophic loss risk when amenities are valued by a risk-neutral landowner, the effective interest rate and land rent are increased, thus shortening the optimal rotation age relative to a rotation age defined under certainty, irrespective of the landowner's amenity valuation.

By destroying the stand, the natural hazard also reduces its amenity benefits. Hence the effect of this risk on timber and amenity production components is symmetric.

It is of interest to compare the expected Hartman rotation age with the expected Faustmann rotation age. Drawing on our findings in chapter 3, we can guess that despite the risk of natural hazards, the expected Hartman rotation age is longer than the expected Faustmann rotation age if the landowner values older stands for their amenities. This is indeed the case and can be shown as follows: Express the first-order condition (10.18) in terms of timber production and amenity production parts as $J_T^H = J_T + B_T = 0$, where J_T and B_T are the derivatives of the timber and amenity production parts of the first-order condition. From (10.18), we have the following derivatives:

$$J_T = pf'(T) - (r+\lambda)(pf(T) - c_1) - r[pf(T) - c_1]e^{-(r+\lambda)T}\left[\frac{(r+\lambda)}{r(1 - e^{-(r+\lambda)T})}\right]$$

and

$$B_T = \left[B(T) - \left(e^{rT}\int_0^T B(s)e^{-rs}ds\,e^{-(r+\lambda)T}\right.\right.$$

$$\left.\left. - \int_0^T \lambda e^{-(r+\lambda)X}\left[e^{rX}\int_0^X B(s)e^{-rs}ds\right]dx\right)\left[\frac{(r+\lambda)}{(1 - e^{-(r+\lambda)T})}\right]\right]$$

where J_T has been simplified. Then, we write $J_T^H = 0 \Leftrightarrow J_T = -B_T$, which yields after some further rearranging,

$$pf'(T) - [pf(T) - c_1]\left[\frac{(r+\lambda)}{(1 - e^{-(r+\lambda)T})}\right]$$

$$= -\left[B(T) - \left(e^{rT}\int_0^T B(s)e^{-rs}ds\,e^{-(r+\lambda)T}\right.\right.$$

$$\left.\left. + \left[\int_0^T \lambda e^{-(r+\lambda)X}e^{rX}\int_0^X B(s)e^{-rs}ds\right]dx\right)\left[\frac{(r+\lambda)}{(1 - e^{-(r+\lambda)T})}\right]\right].$$

Now, the LHS is simply the Faustmann condition we had in (10.8a) and is zero at that previous optimal rotation age solution. The RHS denotes the difference of the marginal amenity benefits at the harvesting time and the expected opportunity costs of amenity benefits over infinite cycles of rotations. If old stands are valued for amenities, then we can think of the value of the amenity benefit at harvest time as exceeding the interest cost; then the term inside the RHS brace is positive, making the RHS negative. Hence, the resulting rotation solution from the amenity problem is longer than the Faustmann solution derived under catastrophic risk. The opposite would hold if amenity benefits are such that young stands are valued.

The comparative statics results exhibit features similar to those found in the Faustmann model. Concerning the new parameters we have

$$\frac{\partial T^H}{\partial \lambda} = -\frac{J_{T\lambda}^H}{J_{TT}^H} \gtrless 0 \quad \text{and} \quad \frac{\partial T^H}{\partial c_2} = -\frac{J_{Tc_2}^H}{J_{TT}^H} = 0. \tag{10.19}$$

The partial derivative $J_{T\lambda}^H$ is a priori ambiguous, so whether or not a higher risk of stand destruction increases rotation age in the amenity model depends on specific parameters. Box 6.1 presents a discussion of numerical results presented in the literature concerning this point. The fact that the site preparation and regeneration cost, c_2, does not affect rotation age results from its lump-sum nature.

The impacts of timber price and regeneration costs (c_1) on rotation age are conventional. Whereas timber prices increase the profitability of timber production relative to amenity benefits, a higher cost reduces this relative profitability. When the landowner values amenity benefits from old stands, a higher timber price shortens the rotation age, while higher regeneration costs always increase the optimal rotation age. A higher interest rate affects both timber and amenity production in the same way as a higher risk of natural hazards in that it shortens the optimal rotation age.

10.3.2   Arrival Rate Dependent on Stand Age

Next we introduce a nonhomogeneous Poisson process into the Hartman model. From definition 10.2 the corresponding probability density function for X is equal to $\lambda(X)e^{-m(X)}$, where $m(X) = \int_0^X \lambda(s)\,ds$. Hence the probability that the stand will be destroyed by a catastrophic event before the rotation age is reached the first time is $\Pr(X < T) = 1 - e^{-m(T)}$, while the probability that the stand will be harvested before the catastrophic event occurs is $\Pr(X = T) = e^{-m(T)}$.

Recall that land rent formula was generally defined by $J = [E(e^{-rX}Y)]/[1 - E(e^{-rX})]$. We need to employ the earlier definitions of probabilities to the expectation terms $E(e^{-rX})$ and $E(e^{-rX}Y)$. For a random net return, the second expectation is slightly different than with a homogenous process. We obtain the following land value formula:

$$J^H =$$

$$\frac{\left[pF(T) - c_1 + e^{rT}\int_0^T B(x)e^{-rx}\,dx\right]e^{-[rT+m(T)]} - \left[c_2 - e^{rx}\int_0^x B(x)e^{-rs}\,ds\right](1 - e^{-[rT+m(T)]})}{r\int_0^T e^{-[m(x)+rx]}\,dx}$$

$$+\left(c_2 - e^{rx}\int_0^x B(s)e^{-rs}\,ds\right). \tag{10.20}$$

The first-order condition for the optimal rotation age can be expressed after some manipulation as:

$$J_T^H = [pF'(T) + B(T)] - [r + \lambda(T)]$$

$$\times \left[pF(T) - c_1 + c_2 - \left(\frac{\lambda(T)}{r + \lambda(T)}e^{rT}\int_0^T B(x)e^{-rx}\,dx + e^{rx}\int_0^x B(x)e^{-rs}\,ds\right)\right]$$

$$-r\left(J - c_2 + e^{rx}\int_0^x B(s)e^{-rs}\,ds\right) = 0 \tag{10.21}$$

This condition differs from that in the homogenous case primarily because site preparation and planting cost now matter at the margin. Also, amenity benefits enter the first-order condition in a more complicated manner than earlier. Other than this difference, the interpretation is similar to the one before: the marginal return from delaying harvest must equal the opportunity cost of that delay. The opportunity cost is again higher due to the higher effective discount rate and higher expected future amenity benefits from continuing a rotation.

Condition (10.21) is general in that it contains all previous versions of the model as special cases. It reduces to the non-homogeneous Faustmann condition (10.18) when amenity benefits are zero, and to the homogeneous Poisson condition (10.8) when in addition risk is independent of the age of the forest stand. The Hartman condition (10.25) is obtained when risk is assumed to be independent of the forest stand age. By comparing this first-order condition with (10.25) we have proposition 10.4.

Proposition 10.4 In the Hartman model with an age-dependent catastrophic loss risk that increases in time, the optimal rotation age is shorter than the rotation age when the probability of catastrophic loss is independent of forest stand age.

As for the comparative statics, one can verify that $\partial T^H/\partial c_2 = -J_{Tc_2}^H/J_{TT}^H < 0$ with

$$J_{Tc_2}^H = -\frac{[r + \lambda(T)]\lambda(T)}{r(1 - e^{-[rT+m(T)]}) + \lambda(T)} < 0.$$

Box 10.1
Risk of Fire Loss

There have been several simulation-based applications of Reed's basic model for the fire risk problem. In this box we summarize some of these results so that the reader can gauge the magnitude of changes in rotation ages that follow from different fire risks and the two different Poisson assumptions used in this chapter. This box relies on results from simulations presented in Englin et al. (2000) and Amacher et al. (2005). Amacher et al. solve the optimal rotation age under both homogeneous and nonhomogeneous Poisson assumptions using the Faustmann-based approach without amenities (they do include amenities, but only when fire prevention activities are included). Englin et al. examine a constant fire risk homogeneous Poisson process in a Hartman-based model. Both studies consider different fire arrival rates.

Amacher et al. use the growth function for southern U.S. pine that was presented in box 2.1. Englin et al. estimate a jack pine growth function from data in Manitoba, Canada, of the form: $Q(t) = 6.1192 - 66.6471(1/t)$, where $Q(t)$ is volume at time t. In terms of rotation ages, jack pine is typically grown on rotations of between 40 and 60 years, while southern pine has a range in more temperate climates and is grown on rotations of between 25 and 40 years. The amenity function assumed in Englin et al. comes from an estimate of visitor user data in Manitoba and is a piecewise continuous function that depends on time. This per-user amenity function is of the form $F(t) = -0.04101 + 0.0024T$ for $T < 64$, and $F(t) = 0.1133$ for $T \geq 65$. We will consider their results under an assumed 1,000 users.

For fire risk parameters, Englin et al. assume a constant fire risk through time and examine various percent risks (probabilities) related to the Poisson parameter, including probabilities of fire arrival in any one year of 2% and 4%. A fire risk of 1.6% would be equivalent to a Poisson parameter used in Amacher et al. of one fire every 60 years, i.e., $\lambda = 1/60$. For the constant process, Amacher et al. set the Poisson parameter to three different levels, $\lambda = 1/50$, or a 2% fire risk, which is quite comparable with Englin et al.'s assumption, and also $\lambda = 2/50 = 4\%$, and $\lambda = 3/50 = 6\%$. For their nonconstant cases, the Poisson parameter is assumed to increase or decrease smoothly over time (they present a cumulative risk parameter in this case). Both articles assume an initial planting cost and different cost of reforestation after a fire, should one arrive. In addition, unlike Englin et al., Amacher et al. also allow planting density as a choice along with rotation age. In the baseline numbers here, the planting density is set at 408 trees per acre.

Table 10.2 presents a summary of the results from these simulations for cases of constant and increasing risk over time (incidentally, there is a possibility of decreasing risk over time with southern pine because the forest naturally prunes itself). If this is significant enough, the fuel loadings may decrease over time, implying that for the Poisson parameter, $\lambda'(t) < 0$. The results are quite illustrative of the theoretical results developed in the text. An increasing risk of fire clearly reduces rotation ages, by as much as 20% for jack pine, owing to a doubling of the risk of fire during a rotation. The changes are smaller for southern pine, because it is grown on smaller rotations in the first place. However, a tripling of risk still leads to a nearly 10%

Box 10.1
(continued)

Table 10.2
Rotation Ages under Independent and Age-Dependent Fire Risk

Poisson assumption	Fire probability (%)	Faustmann-based (Amacher et al. 2005)	Faustmann-based (Englin et al. 2000)	Hartman-based (Englin et al. 2000)
Constant λ	2	21.2	31.0	51
	4	19.8	28.0	41
	6	18.7		
Increasing λ	3.2	18.0		
	8.4	15.3		

decrease in the rotation period. It is interesting, that the drop in rotation age as risk increases is much greater where the probability of destruction increases over time. The effect of risk increases on rotation age for both species appears similar, and is close to 10% in the Faustmann-based model. However, when the Hartman model is examined, the decrease in rotation age is close to 20%. This follows from the assumption that fire destroys amenity services. Whether or not it does is still an open question though and probably is a site-specific phenomenon.

This is qualitatively similar to what we found with the Faustmann model under a nonhomogeneous Poisson process. A higher cost of clearing and replanting the site decreases the optimal rotation age.

To make the quantitative role of our analyses of two different Poisson processes more visible, in box 10.1 we compare models using existing work relying on simulations of fire risk.

10.4   Modifications

Reed (1984) certainly ranks among the most widely cited work in forest economics within the past two decades. A large collection of articles now exist dealing with unknown catastrophic future shocks that are assumed to follow Poisson processes. In this section we present a fairly thorough review of this literature. A notable feature in the work done since Reed has been to endogenize the effort landowners take to mitigate the risk of destruction. We first present a case where a stand is partly destroyed and then we generalize Reed's model to handle an additional choice of preventive effort.

10.4.1   Partial Destruction of a Stand

Catastrophic events need not lead to total losses because some timber may be salvageable after the event subsides. It is therefore important to ask whether the possibility of partial destruction affects our results. Reed (1984) considered partial destruction in a specific way. To show this, let a random proportion $\tilde{k}(t)$ of the stand value at age t be salvageable if a catastrophic event arrives at time t. Denote the mean of this random variable by $\bar{k}(t)$, and assume for simplicity that the cost of replanting the site after the event, c_2, is independent of salvage.

Under this modification, the random net current return Y_n associated with a time X_n between successive destructions of the stand now becomes

$$Y_n = \begin{cases} \tilde{k}(X_n)pF(X_n) - c_2 & \text{if } X_n < T \\ pF(T) - c_1 & \text{if } X_n = T. \end{cases} \tag{10.22}$$

The only new aspect in (10.22) is in the first line, because the landowner receives some positive revenue from salvaging undamaged timber if the catastrophic event arrives before the rotation age is reached.

Recall that land rent or bare land value was defined by the expression: $J = E(e^{-rX}Y)/[1 - E(e^{-rX})]$. The term $[1 - E(e^{-rX})]$ is now the same as in (10.4). However, partial destruction means we have a different term in $E(e^{-rX}Y)$:

$$E(e^{-rX}Y) = [pF(T) - c_1]e^{-(r+\lambda)T} + \int_0^T e^{-rX}[\bar{k}(X)pF(X) - c_2]\lambda e^{-\lambda X}\,dX. \tag{10.23}$$

Using equation (10.23) and the definition of an expectation yields the following modification to expected land value under partial destruction:

$$J = \frac{(r+\lambda)[[pF(T) - c_1]e^{-(r+\lambda)T} + \phi(T,\dots)]}{r(1 - e^{-(r+\lambda)T})} - \frac{\lambda}{r}c_2, \tag{10.24}$$

where $\phi(T,\dots) = \int_0^T \lambda\bar{k}(X)[pF(X) - c_1]e^{-(r+\lambda)X}\,dX > 0$ is an additional stand value term that is due to part of the stand not being fully destroyed. This term is an increasing function of the choice of rotation age, $\phi_T(T,\dots) = \lambda\bar{k}(T)[F(T) - c_1]e^{-(r+\lambda)T} > 0$. The first-order condition for the optimal rotation age using (10.24) is

$$J_T = pF'(T) + \phi_T(T,\dots)e^{(r+\lambda)T} - \frac{(r+\lambda)[pF(T) - c_1 + \phi(T,\dots)]}{(1 - e^{-(r+\lambda)T})} = 0. \tag{10.25}$$

Catastrophic risk enters condition (10.25) through the effective discount rate term $r + \lambda$ (as before), the marginal return ϕ_T, and the opportunity cost term ϕ. Comparing (10.25) with (10.8a) reveals that both the marginal return and the opportunity cost terms become higher when total destruction of a stand is ruled out. The size of these new terms depend on the size of the Poisson risk parameter λ and the mean proportion of salvageable forest value, $\bar{k}$.

The optimal rotation age is clearly affected by salvage possibilities. To see how, we need to compare (10.25) and (10.8a). First we express (10.25) as

$$pF'(T) - \frac{(r+\lambda)[pF(T) - c_1]}{1 - e^{-(r+\lambda)T}} = \frac{(r+\lambda)\phi(T, \ldots)}{1 - e^{-(r+\lambda)T}} - \phi_T(T, \ldots)e^{(r+\lambda)T}. \tag{10.26}$$

The LHS of (10.26) is what we found as the first-order condition under total stand destruction in (10.8a). It is not equal to zero now, however. To judge its sign, we re-express the RHS as

$$\frac{(r+\lambda)\phi(T, \ldots)}{1 - e^{-(r+\lambda)T}} - \phi_T(T, \ldots)e^{(r+\lambda)T}$$

$$= \lambda\left\{(r+\lambda)\int_0^T \bar{k}(X)[pF(X) - c_1]e^{-(r+\lambda)X}\, dX - \bar{k}(T)[pF(T) - c_1]\right\} < 0. \tag{10.27}$$

Because the second term inside the braces is greater than the first term in the braces, equation (10.27) is negative, indicating that the marginal return term that is due to salvage, $\phi_T e^{(r+\lambda)T}$, is greater than the opportunity cost term. Therefore, compared with the total destruction condition (10.8a), the marginal return to harvesting increases and the rotation age is longer when salvage possibilities exist. This makes sense because the risk associated with continuing a rotation is now lower given that the loss in rents will not be total if the natural hazard arrives before the rotation age chosen.

This is also confirmed by a comparative statics result showing that an increase in the mean value of salvageable timber $\bar{k}$ lengthens the optimal rotation age:

$$\frac{\partial T}{\partial \bar{k}} = -\frac{J_{T\bar{k}}}{J_{TT}} > 0. \tag{10.28}$$

The derivative $J_{T\bar{k}}$ is defined as

$$J_{T\bar{k}} = \lambda\left[(1 - e^{-(r+\lambda)T})pF(T) - c_1 - (r+\lambda)\int_0^T [pF(X) - c_1]e^{-(r+\lambda)X}\, dX\right].$$

Integrating by parts yields

$$\int_0^T [pF(X) - c_1]e^{-(r+\lambda)X}\, dX = -\frac{[pF(T) - c_1]e^{-(r+\lambda)T}}{r+\lambda}.$$

This gives $J_{T\tilde{k}} = \lambda[pF(T) - c_1] > 0$, so that an increase in salvageable timber lengthens the optimal rotation age. We leave it to the reader to examine how other parameters affect the optimal rotation age under partial destruction.

10.4.2   Costly Protection

The approach in this chapter has been applied widely to any problem where natural events arrive independently over time, causing discrete changes (jumps) in forest rents when they occur. Fire risk has been a recurrent theme in Poisson-based models. An interesting result from this work has been to show the conditions under which landowners will undertake costly forest management to protect against losses should a fire arrive before the rotation period ends. This possibility changes the nature of the results concerning how rotation age depends on risk. Reed (1987) and Reed and Apaloo (1991) are examples of early work. Reed (1987) studied how fire protection expenditures affect the choice of optimal rotation age. In this nonhomogeneous setup, fire protection, undertaken at each point in time during the rotation, affects the fire arrival rate but not the salvage possibilities once fire arrives. Reed and Apaloo (1991) studied the timing of commercial thinning under forest fire risk, examining cases where some types of thinning (such as spacing juvenile trees) may increase fire arrival rates by increasing fuel in a stand by opening the canopy.

The general idea of this work is to allow the landowner an additional decision that mitigates the impact of risk on his returns. This implies that the landowner then has two decision variables, rotation age and protection effort. Often protection effort is modeled by considering both its level and timing during a rotation. Some authors assume that protection effort is applied regularly, while others assume that it is applied sporadically during the rotation.

The simplest illustration of this is to consider an action x that the landowner can undertake to manipulate a stand in a way that reduces the probability that the natural hazard will arrive. This action may or may not affect forest growth; we will suppose here that the landowner trades off some growth possibilities when x is employed. Assume also that this action is undertaken at the beginning of each rotation; an example is a prescribed burn to reduce the fuels present when each new stand is started. This implies that the cost of x is covered by the terms c_1 and c_2 in our earlier setup. Once the action is taken, assume there is a decrease in the probability of the natural hazard arriving, so that the Poisson parameter and distribution assumptions become

$$\lambda = \lambda(x), \quad \text{with } \lambda'(x) < 0 \tag{10.29a}$$

$$F(T,x) \quad \text{with } F_T > 0 \text{ and } F_x < 0. \tag{10.29b}$$

Under these assumptions the expected value of the land with total destruction of the stand is

$$J = \frac{[r + \lambda(x)][pF(T,x) - c_1]e^{-[r+\lambda(x)]T}}{r(1 - e^{-[r+\lambda(x)]T})} - \frac{\lambda(x)}{r}c_2.$$

(10.30)

The following first-order conditions for rotation age and protection effort are obtained:

$$J_T = pF_T - \frac{(r + \lambda)[pF(T,x) - c_1]}{(1 - e^{-(r+\lambda)T})} = 0$$

(10.31a)

$$J_x = (r + \lambda)pF_x + \lambda'(x)\left([pF(T,x) - c_1] - \frac{(r + \lambda)[pF(T,x) - c_1]}{(1 - e^{-[r+\lambda(x)]T})} - \frac{c_2}{r}\right).$$

(10.31b)

While the condition for the rotation age here is qualitatively similar to the earlier one, it is now affected by the choice of protection effort. The first term in (10.31b) is harvest revenue lost owing to protection, and the other terms define the return that is due to a lowered risk of the natural hazard. The optimum requires that the marginal cost of protection equal the marginal increase in the expected net return that is due to a lower risk of natural hazards. Owing to this reduced risk, the rotation age is now longer than in the absence of protection effort. Moreover, the equation system (10.31a) and (10.31b) reveals that all exogenous variables, including site preparation and regeneration costs after an event, matter for decisions on optimal rotation age and protection efforts. This feature has been an integral part of empirical work on forest fires.

Some newer studies have expanded the research beyond this point and examined in models with amenities how the risk of wildfire affects a landowner's decision to undertake preburning activities on a regular basis to reduce fire loss if the stand burns. In this case, the protection effort does not affect the probability of a fire occurring, but rather it increases the salvage possibilities k (see equation 10.22). Necessarily, these precautionary actions are assumed to reduce damage should an unplanned wildfire arrive by effectively reducing losses. In some studies, preburning of the stand is also assumed to reset the fire arrival rate to its lowest value. Yoder (2004) examines this problem and allows for an exogenous off-site stand that can be damaged by fire originating from the stand in question. In his model, a potential externality exists that is similar to those discussed in chapter 3 and that depends on a given landowner's preburning decisions. Along with the fact that rotations can be shorter under risk for a single landowner, he finds that the time between preburnings for precautionary measures will be much longer than is socially optimal in the presence of the externality.

Amacher et al. (2005) take a slightly different approach. In a nonhomogeneous Poisson model with amenities, they assume that the fire occurence probability does not depend on actions that the landowner can take, such as thinning, but instead assume that these actions increase the salvage function (reduce damage) should fire arrive. Thus the salvage function is endogenous as in (10.30). Their rationale is that fire arrives in fronts of multiple fires or through uncontrollable events such as lightning strikes, but the damage once it arrives depends on fuel loadings of the stand that are functions of forest management decisions. As our simple model here suggests, in both Yoder and Amacher et al. the rotation age is not always shorter if protection is afforded by improvement actions, and there is also an important value of information concerning fire arrival and the efficacy of fuel treatments. Crowley et al. (2007) go further and examine a problem of multiple stands similar to the interdependent stand problem discussed in chapter 3, where each stand may be held by a different landowner with different information. This work is based on a nonhomogeneous Poisson process for fire arrival, but in addition, decisions by adjacent landowners may affect both the fire arrival rate and the salvage function. In addition to the findings in the other work, a new basic finding is that free riding among landowners relative to the first-best (single-owner) outcome is likely to occur, and that the extent of free riding depends on the differences in information concerning fire arrival and salvage that landowners possess.

There are numerous other studies as well where the model of this chapter has found interesting applications. Haight et al. (1995) studied the effects of hurricane risk on loblolly pine plantations, using both an age-dependent salvage function to reflect the fact that tree ages matter in terms of both survivability and value if they make it to harvest time after a storm, and an age-dependent probability (i.e., nonhomogeneous process) to reflect the fact that any storm arriving will most likely damage older stands while younger stands will escape its effects. Stainback and Alavalapati (2004) also allowed a salvage function and considered the effect of hurricanes and fires on the carbon sequestration potential of southern pine; risk affects rotation ages and causes subsequent changes in carbon loads. An interesting extension of this storm-related work is by Meilby et al. (2001), who go beyond the typical single-stand problem to consider the windthrow risk in stands as a dynamic programming problem. In a two-stand example, they postulate that the arrangement of stands on a landscape, their characteristics, and their position relative to wind direction all factor importantly in stand-specific rates for windthrow. These "shelter" effects lead to what we found in our simple model of protection effort: For any stand, a new opportunity cost of delaying harvest arises if that stand provides protective benefits to adjoining stands. This again leads to a longer rotation period under many circumstances because the effect of risk on decision making is lessened.

Finally, Reed and Errico (1987) use the Poisson process to model catastrophic risks, but they include risk both from (age independent) fire arrival and (age dependent) arrival of pests such as insects. In both cases, harvesting is required after the event to salvage the dead (zero growth) wood stock. A forest-level analysis is also conducted, based on an aggregate of stands similar to that developed in Reed and Errico (1986) for fire risk only.

10.5 Summary

The model in this chapter illustrates the conditions for which landowner decisions under risk of catastrophic loss can be examined using a relatively simple extension of conventional stand-level approaches. The key to modeling the risk of natural hazards is to use an independent or time-dependent Poisson process. This "jump" process describes many threats to forests whose timing reflects variability in natural conditions that is not known to landowners when they make forest planting and harvesting decisions under a Faustmann or Hartman approach.

The basic approach was described under the assumption that a stand is totally destroyed by a natural hazard. In both this and the partial destruction case, risk-neutral landowners will respond by harvesting sooner. Also, the relationship between the Faustmann and Hartman rotation ages remains identical to what we found under certainty. That is, when amenity benefits from old stands are valued, the expected Hartman age is longer than the expected Faustmann age.

This basic problem can be modified for the case where there is either partial destruction once a natural hazard arrives, or the landowner can employ protective effort prior to its arrival. Reed (1984) demonstrated that random partial destruction increases rotation age relative to the case of total destruction. Furthermore, the impact of risk on rotation age is reduced if landowners can undertake actions, such as timber stand improvements, that either protect against loss should a catastrophic event arrive before the end of the rotation period, or reduce the probability of its arrival. While most analyses in this case are empirical in order to achieve the required specificity, our simple model in this chapter demonstrates how stand protection effort is chosen at the margin. Finally, we would like to stress that the model could and should be expanded to cover the case of multiple stands. As Prestemon and Holmes (2000) and others have pointed out, one possible way to hedge against catastrophic risk is to simply accumulate more forest land, since the exposure to rent loss is lower. This brings to mind several financial portfolio models that could be applied to catastrophic risk problems (e.g., as in Caulfield 1988 and Zinkhan et al. 1992).

The Poisson process used in this chapter certainly explains many nature-based threats to forests and is a logical first step in learning how risk has been introduced in forest economics problems. In closing, we should point out that the risk-induced interest rate adjustment of these models is valid only if all other economic parameters and forest growth possibilities are constant over all future rotations, and if uncertain events arrive independently over time according to a Poisson process. In the next chapter we introduce uncertainty associated with economic variables that follow more complicated stochastic processes.

11   Stochastic Rotation Models

In chapter 10 we considered catastrophic risk realized through discrete jumps in potential harvesting returns during any rotation. In its purest form, such risk is realized through an increase in the effective discount rate, and the landowner is always worse off when a catastrophe arrives before the rotation ends. For natural events such as fire or ice storms, this downward jump in returns makes sense. However, when other uncertain parameters are modeled, this generally cannot be assumed. Consider that stumpage prices tend to evolve over time (in an unknown way) without large changes in their values in any one period. There is risk faced by a landowner because a low price could exist around the time of harvesting, so that waiting to harvest renders the landowner worse off than if he had harvested sooner. On the other hand, the landowner could be better off if a high price exists at harvesting.

These types of risks are best handled using general stochastic processes, first proposed in the finance and investment literature, which provide a means for specifying both drift and volatility in the path of unknown variables over time.[1] Stochastic process models also provide a richer way to describe uncertainty than the mean-variance approaches discussed in chapter 9 for two-period life-cycle models. When one considers that the dynamic programming problems arising from stochastic processes can be represented by a two-period analogue called a "Bellman equation," the adaptation of these processes for modeling uncertainty in forest economics is clear.

In this chapter we survey the theory of stochastic processes as applied to forest economics problems. We start in section 11.1 by first describing what is meant by a stochastic process and review a critical method needed to use these processes called "Ito's lemma." We pay particular attention to the way that uncertain prices

---

1. Like Poisson-based models, the stochastic process models here retain the assumption of the continuous-time Markov property governing uncertainty. Discrete-time Markov processes will also be discussed when we take up stochastic dynamic programming in chapter 12.

and other parameters have been modeled because there has been considerable difference across studies in this area of forest economics. In section 11.2 we show how these processes can be embedded into general decision models called "optimal stopping problems" to compute rotation age solutions under a variety of risk assumptions. We show in section 11.3 how the rotation age decision can be solved in an optimal stopping problem for both single-rotation and ongoing-rotation problems. Some other modifications in the literature are discussed in section 11.4, and finally in 11.5 we offer our summary with some thoughts about future work. Throughout this chapter we will continue to assume that the landowner is risk neutral, as does nearly all of the literature surrounding these problems.[2]

While recent literature is discussed in more detail later, before proceeding it is worth noting some milestones that have shaped the field. Nordström (1975) is often mentioned as a pioneer study in the influence of stochastic timber prices on optimal rotation age. Another important general contribution with a specific application to forestry is that of Brock et al. (1988). Brazee and Mendelsohn (1988), while not making use specifically of the stochastic processes discussed in this chapter, formulated an optimal stopping problem with uncertainty present because the landowner must draw a price from a normal distribution of prices in the next period (their model is discussed in chapter 12). Haight and Holmes (1991) further continue this idea and show how various empirical models of price movement change rotation age solutions. The first applications of real options with underlying stochastic processes to forest decision making were made by Thomson (1992) and Morck et al. (1989), who were soon followed by Plantinga (1998). Clarke and Reed (1989) and Reed and Clarke (1990) developed continuous-time optimal stopping rules for rotation age when stand value is at risk, with risk modeled as stochastic processes for either price or biological growth of a forest over time. Alvarez and Koskela (2003) introduced stochastic interest rates using continuous-time optimal stopping methods to define harvesting thresholds and analyze the nature of this threshold.

11.1 Preliminaries—Stochastic Processes and Ito's Lemma

We first briefly describe some key properties of stochastic processes that forest economists should know and discuss a famous result known as Ito's lemma. This lemma is an important tool in working with stochastic processes in forest risk problems. Not only is it needed to mathematically define a change in a stochastic parameter over time, but it is also the basic building block of an optimal stopping

---

2. Motoh (2004) is one of the few exceptions. He considers a risk-averse owner of an in situ natural resource (not necessarily a forest) where the value of the natural resource stock is stochastic and follows Brownian motion.

problem with underlying stochastic processes governing prices or forest value. More detailed treatments can be found in Dixit (1993), Dixit and Pindyck (1994, chapter 3), Mikosch (1998), and F. Chang (2004, chapter 3).

11.1.1   Stochastic Processes

A Wiener process is the foundation of many stochastic process models. Any variable, such as stumpage price, that is assumed to be stochastic is so because its realization over time is some function of a Wiener process. Wiener processes are white-noise continuous-time stochastic processes that are Markov in nature, which we know from chapter 10 means that the probability distribution of the random variable at the next point in time only depends on the realization of the variable in the current point in time. Furthermore, a Wiener process has the desirable property that the change in the random variable during a given time interval is normally distributed. This conveniently means that functions of the Wiener process are also normally distributed.

Let $z(t)$ represent a Wiener process. A change in this process, which itself is stochastic, over a time interval of Δt can be denoted by Δz. The Wiener process implies that $\Delta z = \varepsilon_t \sqrt{\Delta t}$, where ε_t is a normally distributed mean zero random variable with a standard deviation equal to one. This random variable is uncorrelated over time: $E(\varepsilon_t \varepsilon_s) = 0$ for $t \neq s$. Thus the values of Δz in any two different time intervals are independent. Reducing the time increment Δt to an infinitesimally small number, we then write the change in z as dz, and by definition, in continuous time we have

$$dz = \varepsilon_t \sqrt{dt}, \tag{11.1}$$

where $E(dz) = 0$ and $\text{var}(dz) = E[(dz)^2] = dt$ because $E(\varepsilon_t) = 0$. As will be clear later, by the properties of a Wiener process, the following multiplication table is useful (see Kamien and Schwartz 1991, section 21):

	dz	dt
dz	dt	0
dt	0	0

The difficulty with Wiener processes in stochastic forestry problems is the fact that their time derivative does not exist. To see this, consider the time derivative of (11.1) written in discrete form as $\Delta z / \Delta t = \varepsilon_t / (\Delta t)^{1/2}$. If we let Δt approach zero as we would with a nonstochastic derivative, we see that $\Delta z / \Delta t$ approaches infinity (see Dixit and Pindyck 1994). Because of this problem, we must use Ito's lemma (section 11.2) to obtain differentials of stochastic processes that have underlying Wiener processes as their basis.

There are many types of stochastic processes based on Wiener processes that are relevant for forest economics problems. The most common form in early articles was to write a random process of either stand value or stumpage price, loosely defined as $x(t)$, using a Brownian motion with drift:

$$dx = \alpha\, dt + \sigma\, dz, \tag{11.2}$$

where dz is a continuous-time increment of a Wiener process, α is a "drift" parameter, and σ is a "diffusion" or "volatility" parameter, with both α and σ being non-random constants. In (11.2), the drift parameter measures the change in the mean (or trend) of the variable x over time, while the volatility parameter measures how the variance of the variable changes over time around this mean. In a graph of a stochastic variable over time, the trend in the variable over time would be given by α, while the spread of its distribution around the trend would be measured by σ. In forestry jargon, if x were forest stand value, then α would measure the growth rate in value and σ would measure fluctuations in this value in a way similar to that of a mean preserving spread. Brownian motions are similar to random walks in econometrics, explained only through pure diffusion and volatility. For this process, the mean and variance are $E(\Delta x) = \alpha\, dt$ and $\mathrm{var}(\Delta x) = \sigma^2\, dt$, respectively.

Another version of a stochastic process important to forest modeling is to assume that the drift and diffusion of the process are directly proportional to the size of the random variable x at each point in time:

$$dx = \alpha x\, dt + \sigma x\, dz. \tag{11.3}$$

This is known as geometric Brownian motion. In forestry, the interpretation of (11.3) is that the higher is stand value, the larger is the volatility and drift in stand value.

Brownian and geometric Brownian motions have been used to analyze forest harvesting under stochastic timber prices and forest stock in a number of articles. Forest or resource stock values have been modeled using Brownian motion in Motoh (2004). Stumpage prices have been assumed to follow geometric Brownian motion in Thomson (1992), Yin and Newman (1995, 1996a, 1997), Miller and Voltaire (1980, 1983), Mallaris and Brock (1982), Morck et al. (1989), Brock et al. (1989), Clarke and Reed (1989, 1990b), Yoshimoto and Shoji (1998), and Insley (2002) (who also compares results with mean reversion in a single-rotation stopping problem). Other examples are Reed and Clarke (1990), who model both stochastic price and biological growth; Reed (1993) and Bulte et al. (2002), who model amenity services of an old-growth forest (Reed included stochastic stand value as well); and Chladna (2007), who models forest carbon storage prices along with mean reverting stumpage prices.

A more general specification than (11.3) is to write the stochastic process for x as

$$dx = a(x,t)\,dt + b(x,t)\,dz, \tag{11.4}$$

where $a(x,t)$ and $b(x,t)$ are known and nonrandom functions. Equation (11.4) is called an Ito (or diffusion) process. Ito processes have the following properties: $E(dz) = 0$, $E(dx) = a(x,t)\,dt$ because $E(dz) = 0$, and $\text{var}(dx) = b^2(x,t)\,dt$.

Brownian motion has been used to model both uncertain forest stocks and stand values. Another type of process is especially relevant to the evolution of stumpage prices or interest rates over time. Brownian motion is not ideal for these variables because in (11.2)–(11.3) volatility and drift grow without bound over time. This may not be the case with prices since market forces would dampen such expansion through corresponding changes in demand and supply. Instead, unknown prices and interest rates have often been modeled using a stochastic process called "mean reversion." Mean reversion covers cases where a variable, once perturbed, eventually returns to some long-run mean value. The simplest case is reversion to a long-run linear mean. This type of process is written

$$dx = \theta(\bar{x} - x)\,dt + \sigma\,dz. \tag{11.5}$$

In (11.5), the nonrandom parameter θ measures the speed with which x reverts back to its long-run mean level $\bar{x}$. As derived in Dixit and Pindyck (1994), a percentage rate of change in x over the interval dt comes from the following expectation: $(1/dt)E(dx/x) = \theta(\bar{x} - x)$. Thus, if x is above $\bar{x}$, than this implies $E(dx) < 0$ and we expect a decrease in x over the next interval of time as x adjusts toward the long-run mean. Conversely, observing x below $\bar{x}$ implies that $E(dx) > 0$. The particular process in (11.5) is also called an "Ornstein-Uhlenbeck" process. As Gjolberg and Guttormsen (2002) point out, mean reversion tends to reduce the long-term price uncertainty perceived by a forest landowner, leading to greater valuations of the asset compared with one where prices follow Brownian motion. It is interesting, however, that they show how option value is reduced, given that there is less long-term price uncertainty with mean reversion.

Mean reversion can be explained by first-order autoregressive error processes in econometrics because the value of x in the next period depends on the value of x in this period, meaning there is some memory in the stochastic process. Articles that have used mean reversion to study stochastic forest economics problems include those by Plantinga (1998), Haight and Holmes (1991), Insley (2002), and Insley and Rollins (2005) for stumpage prices.[3] Some evidence in the literature

---

3. Most of the forestry work using mean reversion has been simulation based. To actually solve for closed-form solutions with these processes requires methods that go beyond the scope of this book. Interested readers are referred to Borodin and Salminen (2002), and Dixit and Pindyck (1994, pp. 161–167).

supports the autoregressive assumption, at least for stumpage prices. These include Brazee et al. (1999), Hultkrantz (1995), Yin and Newman (1996b), Reeves and Haight (2000), and Gjolberg and Guttormsen (2002). Gjolberg and Guttormsen further point to several studies with other commodities where mean reversion has been demonstrated to exist in data, and they provide heuristic evidence of mean reversion in coniferous timber prices reported by the UN Food and Agriculture Organization. Saphores et al. (2002) further find evidence of autoregressive trends and jumps in softwood stumpage price data for the U.S. Pacific Northwest. However, contrary to this, Prestemon (2003) uses data from U.S. southern pine stumpage prices and cannot reject the presence of a unit root using Dickey-Fuller tests, suggesting a stochastic process that is more consistent with Brownian motion and not a mean reversion. In another interesting twist concerning this debate, Yoshimoto and Shoji (2002) show how to test and directly compare Brownian motion, geometric Brownian motion, and many forms of mean reversion using simulations based on thirteen Asian timber species. They find that neither simple mean reversion nor geometric Brownian motion may be sufficient to explain the price dynamics for any of these species.

11.1.2 Ito's Lemma

Ito's lemma is a basic building block for models incorporating stochastic processes. Here we provide some intuition behind this result. To see why Ito's lemma is important, define a general stochastic process $F(x, t)$ as a function of x, the random variable, and time, and suppose that this function is twice differentiable in x and once differentiable in t. Because x is a stochastic process with the normality properties discussed earlier, $F(x, t)$ is also a stochastic process with similar normality properties. Suppose further that x is governed by Brownian motion in (11.3), and suppose we want to find the differential for the function $F(x, t)$ with respect to x and t in order to describe its total movement over time.[4]

The correct differential of the stochastic function $F(x, t)$ has been shown with Ito's lemma to equal

---

4. The conventional calculus approach discussed in the mathematical review at the end of this book would be to obtain the total derivative

$$dF = \frac{\partial F}{\partial x} dx + \frac{\partial F}{\partial t} dt + \frac{1}{2} \frac{\partial^2 F}{\partial x^2} (dx)^2 + \frac{1}{6} \frac{\partial^3 F}{\partial x^3} (dx)^3 + \cdots.$$

A basis of this derivative and result is the fact that higher-ordered terms approach zero in the limit and can be ignored, implying that the total differential we actually use is $dF = (\partial F/\partial x)\, dx + (\partial F/\partial t)\, dt$. From what we know here, though, this total derivative cannot be used if x, and thus $F(x, t)$, are based on an underlying Wiener process.

$$dF = \frac{\partial F}{\partial t} dt + \frac{\partial F}{\partial x} dx + \frac{1}{2} \frac{\partial^2 F}{\partial x^2} (dx)^2. \tag{11.6}$$

This is nothing other than a stochastic version of the chain rule of differentiation for a function of a stochastic process given by $F(x, t)$, and it looks like a Taylor series expansion in higher-ordered terms. The formal proof of (11.6) can be found in Kamien and Schwartz (1991) and Dixit and Pindyck (1994). In some problems, as we discuss in the next section, the function may only depend on x and not on time explicitly, i.e., $F(x)$. In this case, Ito's lemma is simpler to apply, given that there is no specific time derivative:

$$dF = \frac{\partial F}{\partial x} dx + \frac{1}{2} \frac{\partial^2 F}{\partial x^2} (dx)^2.$$

In both cases, a complexity in the stochastic differential is the term containing $(dx)^2$. This must be explicitly derived using whatever assumption is imposed for the stochastic process. For example, under Brownian motion, using (11.2) for dx we obtain $(dx)^2 = \alpha^2 dt^2 + \sigma^2 dz^2 + 2\alpha dt\sigma dz$, which simplifies using the Wiener process multiplication table to $(dx)^2 = \sigma^2 dt$. Under an Ito diffusion process in (11.4), we would have $(dx)^2 = b(x, t)^2 dt$. Given these definitions, for x following Brownian motion and an Ito process, respectively, Ito's lemma applied to the function $F(x, t)$ implies

$$dF(x, t) = F_t(x, t) dt + F_x(x, t)[\alpha dt + \sigma dz] + \tfrac{1}{2} F_{xx}(x, t)\sigma^2 dt, \tag{11.7a}$$

and

$$dF(x, t) = [F_t(x, t) + a(x, t)F_x(x, t) + \tfrac{1}{2} b(x, t)^2 F_{xx}(x, t)] dt + b(x, t)F_x dz. \tag{11.7b}$$

Finally, as noted earlier, many forest economics models have used the geometric Brownian motion given in (11.3). In nearly every case where this process is used, a specific parametric structure is placed on the stochastic process by writing $F(x, t) = \log x$, where $dx = \alpha x \, dt + \sigma x \, dz$. When we discuss stopping problems in section 11.4, we will need an expression for the differential $dF(x, t)$. To obtain this using Ito's lemma, we first compute the requisite derivatives, $\partial F/\partial t = 0$, $\partial F/\partial x = 1/x$, and $\partial^2 F/\partial x^2 = -(1/x^2)$. Then, using the process in (11.3) we obtain

$$dF = \frac{1}{x} dx - \frac{1}{2x^2} (dx)^2 = \frac{\alpha x \, dt + \sigma x \, dz}{x} - \frac{(\alpha x \, dt + \sigma x \, dz)^2}{2x^2}$$

$$= \alpha \, dt + \sigma \, dz - \frac{(\alpha^2 \, dt^2 + \sigma^2 \, dz^2 + 2\alpha\sigma \, dt \, dz)}{2}.$$

Now, using the multiplication table we have $dt\,dt = dt\,dz = 0$ and $dz\,dz = dt$. Thus Ito's lemma implies that $dF = \alpha\,dt + \sigma\,dz - (\sigma^2\,dt/2) = [\alpha - (\sigma^2/2)]\,dt + \sigma\,dz$. Over any finite time interval $dt = T$, the change in $F(x, t) = \log x$ is therefore normally distributed, as we knew already because it is a function of an underlying Wiener process, with a mean equal to $E(dF) = [\alpha - (1/2)\sigma^2]T$ and variance equal to $\sigma^2 T$. The mean is smaller than the parameter α because of the concavity of $F(x, t)$. This means that under uncertainty the expected value of $F(x, t)$ changes by less than the logarithm of the expected value of x. This is none other than a stochastic version of Jensen's inequality.[5]

11.2   Continuous-Time Stochastic Optimal Stopping

We are now ready to discuss the basic problems in forest economics that have been based on stochastic processes and Ito's lemma. The prototypical problem has been to examine a landowner's decision at each point in time regarding whether to harvest and plant a new stand (or maybe sell the land), or whether to wait and retain the option of harvesting later. The harvesting decision is irreversible in that the landowner must wait a considerable amount of time after harvesting before a new stand becomes merchantable. In this problem, for any instant in time, prior reforestation costs are sunk and do not matter, and the presence of uncertainty means there is always some "option value" associated with waiting. Option value measures the value of receiving additional information about unknown parameters, such as prices or the greater rent to be captured from additional growth or high prices in the next period. Stochastic process-based stopping problems are sometimes called "option value" or "real options" problems for this reason.[6]

Once a stochastic process is chosen to explain an economic parameter, the goal is to embed this into a general discrete-choice dynamic programming problem in continuous time, where there may be a transition equation explaining the state of the system, such as forest growth. If only price is stochastic, we need an equation explaining the change in price over time, dp, such as geometric Brownian motion or mean reversion, and a transition equation explaining the nonrandom change over time in the forest stock such as $ds = g[s(t)]\,dt$, where $g[s(t)]$ is a known

---

5. In deterministic terms, Jensen's inequality is defined by the following relationship: If $g(x)$ is a convex function and x is a random variable, then $Eg(x) \geq g(Ex)$ provided the expectation exists.
6. The irreversible harvesting decision and the option value of waiting implies that the forest harvesting problem has characteristics that are similar to problems studied in the finance literature, for example, by Björk (1998), Dixit and Pindyck (1994), and Ingersoll and Ross (1992). In natural resource economics, the real option approach has its origin in seminal papers by Arrow and Fisher (1974) and Henry (1974), who formalized irreversible land development decisions when the future value of development is uncertain.

growth function and $s(t)$ is known forest stock at time t. If price is deterministic but forest growth is stochastic, the deterministic transition equation for the forest stock is replaced by a stochastic process explaining the change in the stock, ds. Finally, many researchers simply refer to stand value as $x = p(t)s(t)$ and use a stochastic process for dx without specifically defining price or forest stock by their own stochastic process. In this case, the forest stock transition is conveniently wrapped into the stochastic process assumption, which somewhat simplifies the analysis.

Whatever is assumed, we solve these problems by finding the time instant at which it is optimal to stop and harvest rather than continuing a rotation. It is often easier to find the size of the forest stock or stand value where it is optimal to stop. In this section, we focus on these continuous-time decision problems that are appropriately termed optimal stopping problems. In chapter 12, we will return to discrete-time versions of these problems under both certainty and uncertainty assumptions to focus more on the dynamics of the forest.

We will work through this section by assuming that stand value x is stochastic, although we will point to some articles where price and forest stock uncertainty have been separated. Also, we will consider a single rotation period for the moment. Let stand value x follow an Ito diffusion process in (11.4), so that $dx = a(x, t)\,dt + b(x, t)\,dz$, where dz is an increment of a Wiener process; recall that the current stand value is known but future stand values are not. Let $B(x)$ denote a flow of rents, such as amenities, from letting a forest stand grow an additional period, and let $H(x)$ denote the termination payoff from harvesting. Since $H(x)$ in theory includes harvesting costs and possibly a land sale at the end of the rotation, it is some function of stand value that need not be made explicit. Notice that $B(x)$ and $H(x)$ are themselves stochastic, and since x has an underlying Wiener process, both of these functions have normal distribution properties. The resulting continuous-time harvesting stopping problem is therefore Markov in nature. Also, in both of these functions we can think of any x as representing any point in time during a rotation without loss of generality (in this literature, often x_t or x is used interchangeably, given the definition of dx). In addition, sometimes one introduces additional notation about the landowner's choice set, i.e., by defining $u(t)$ as a set of possible actions undertaken at time t, where $u(t) = 1$ implies that the landowner harvests at t, and $u(t) = 0$ implies that the landowner waits and does not harvest at t. When $u(t) = 1$, the landowner receives $H(x)$ and the time horizon ends.

In both numerical-based stochastic forestry models and those in which closed-form solutions are possible, a basic theory is followed in framing the stopping problem. The basic stopping problem is a discrete-choice dynamic programming problem given by

$$\max_{x^*} Ee^{-rT^*} H(x^*), \tag{11.8}$$

subject to $dx = a(x,t)\,dt + b(x,t)\,dz$ and an initial condition on stand value, $x(0)$. This problem is sometimes called an "American call option" from finance because the decision to harvest can be made at any point in time (see Insley and Rollins 2005 and Gjolberg and Guttormsen 2002 for a more precise definition of this term). We could put bounds on this problem by picking a maximum time at which harvesting must occur, but conceivably this could be an infinite time-horizon problem where the forest stand is preserved forever for amenities. In (11.8), T^* and x^* refer to the time and stand value at which it is optimal to stop and harvest, or where $u(T^*) = 1$. No integral exists in the maximand in (11.8) because a return is obtained only at the instant that $u(t) = 1$.[7] For the derivations in this section, let us consider (11.8) as a finite time-horizon single-rotation problem.

Since the decision to stop can be reexplained as one where the landowner at each point in time picks the greater of the harvest value or the value of continuing a rotation, the problem in (11.8) can be written as a basic dynamic programming problem in Bellman form using a value function $M(x)$ written in continuous time:

$$M(x) = \max\{H(x), B(x) + \beta E_x[M(x'|x)]\}, \tag{11.9}$$

where x' is next instant's value of x given the current value of x, and β is the discount factor. The second element on the RHS in (11.9) is the discounted continuation value from not harvesting at the instant that the landowner observes x. This continuation value includes a flow of amenities plus the expected value of additional rents from the change in x between the current and future period. With (11.9), we have managed to rewrite the stopping problem as a two-period problem since we could successively substitute for $M(x'|x)$ using (11.9) and arrive back at (11.8).

The goal at this point is to characterize the solution conditions for the stopping problem by obtaining a partial differential equation in terms of the value function that in theory can be solved for the optimal stopping stand value of x that satisfies it; i.e., the level of x where harvesting occurs (this then gives the optimal value function as well). To do this, we consider a time interval in which it is optimal to wait (i.e., an interval where the landowner resides in the "continuation" region, or

---

7. For the most part, the classic stochastic forestry problem is a stopping one because the most basic decision is one of ending the rotation by harvesting. The methods we discuss here apply to any stochastic control problem. For example, consider a natural resource depletion problem where price is stochastic but the stock is known: $\max_{u(t)} \int_0^\infty Ee^{-rt}\{H[x_t, s(t), u(t)] + B[s(t)]\}\,dt$ subject to $dx = a(x,t)\,dt + b(x,t)\,dz$, and $ds = s(0) - u$ where x is the unknown resource price, u is the resource extraction rate at each instant t, and s is the resource stock in situ at each instant. The reader should compare this with the nonstochastic optimal control problems discussed in chapter 12.

a time $t < T^*$). Then we make time explicit and write the problem in discrete time by assuming that each time interval is given by Δt.

$$M(x,t) = \max\{B(x)\Delta t + (1 + r\Delta t)^{-1} E_x[M(x_{t+\Delta t}, \Delta t + t|x)]\}, \tag{11.10}$$

where $x_{t+\Delta t}$ is the value of a stand in next time interval. We now follow well-known steps detailed in Dixit and Pindyck (1994, chapter 4), Kamien and Schwartz (1991, section 21), and Miranda and Fackler (2002, chapter 10) and multiply (11.10) by $(1 + r\Delta t)/\Delta t$ and rearrange to obtain

$$rM(x,t) = \max\left\{ B(x)(1 + r\Delta t) + \frac{E_x[M(x_{t+\Delta t}, \Delta t + t|x) - M(x,t)]}{\Delta t} \right\}. \tag{11.11}$$

Now we take Δt to zero in (11.11), which implies that Δt approaches the infinitesimally small dt increment of time. What we are left with is a continuous-time version of (11.11), which defines a Bellman equation when there is an underlying stochastic process concerning x:

$$rM(x,t) = \max\left\{ B(x) + \frac{E_x\, dM(x,t)}{dt} \right\}. \tag{11.12}$$

This equation has a clear economic interpretation. The rate of return on the forest investment (LHS) equals the amenity or rent flow from holding it for one more instant of time (the first term on the RHS) plus the rents expected in the next period from potentially harvesting one instant later (the second term on the RHS). This second term incorporates the uncertainty in the value of next period's stand from waiting to harvest indicated by the drift and volatility parameters of the stochastic process. Recall also that $dM(x,t)$ is the differential of $M(x,t)$, and that stand value x is stochastic and follows an Ito process, $dx = a(x,t)\,dt + b(x,t)\,dz$. This means, by Ito's lemma (11.6), that $M(x)$ is also stochastic and has the following differential from (11.7b): $dM(x,t) = [M_t(x,t) + a(x,t)M_x(x,t) + (1/2)b(x,t)^2 M_{xx}(x,t)]\,dt + b(x,t)M_x\,dz$. Now, to obtain the expression we need in the second RHS term of (11.12), we divide this differential by dt and take expectations, recalling from section 11.2 that $E\,dz = 0$ by definition and that $a(x,t)$ and $b(x,t)$ are nonrandom constants. Doing this, we can finally rewrite equation (11.12) in a form that many articles have derived in the stochastic forest economics literature:

$$rM(x,t) = \max_x\{B(x) + M_t(x,t) + a(x,t)M_x(x,t) + \tfrac{1}{2}b(x,t)^2 M_{xx}(x,t)\}. \tag{11.13}$$

Equation (11.13) becomes simpler by assuming an infinite time horizon. Then the value function does not explicitly depend on time, and equation (11.13) becomes $rM(x) = B(x) + a(x)M_x(x) + (1/2)b(x)^2 M_{xx}(x)$. This equation is only a function of x and is easier to solve (see Dixit and Pindyck 1994, page 107, for further details concerning this point).

Equation (11.13) is a partial differential equation that can in principle be solved for the critical level of stand value x^* and a value function $M(x^*, T^*)$ that satisfies the optimal stopping problem in (11.8), as long as $M(x, t)$ is twice differentiable in x and once differentiable in t. The solution to the partial differential equations obtained under either finite or infinite time horizons can be obtained in one of two ways. First, we can assume functional forms and solve the equation numerically, usually working backward from the maximum time period allowable to find the greatest value function at harvest time. Second, if a closed-form solution is desired and if the problem is simple enough, we can guess the form of the solution, substitute this form back into (11.13) to obtain what is known as a characteristic equation, and then find the roots of this equation to determine x^* and $M(x^*, T^*)$.

There is one more important concept here used in nearly all forest economics articles in this area, both numerical and closed-form versions. Because we derived (11.13) under the assumption of continuation, it holds for $x > x^*$ (or $t < T^*$), where x^* is the optimal stopping value for the stand. We also need two boundary conditions that must hold right at the instant of stopping, i.e., at (x^*, T^*). The first is called the "value matching" condition, and it says that at the optimal stopping time, the value function must equal the harvesting return, $M(x^*, T^*) = H(x^*, T^*)$. Taking the derivative of both sides gives us the second boundary condition, called the "smooth pasting" condition, $M_x(x^*, T^*) = H_x(x^*, T^*)$. This says that the change in the value function at the optimal stopping point must equal the change in harvesting returns, since it is optimal to stop at the (T^*, x^*) pair.[8]

Equation (11.13) is derived in most of the stochastic forest economics articles in one form or another. We briefly mention a few of these. Insley (2002) studies a single-rotation stopping problem and derives (11.13) under the assumption of stochastic timber price following an Ito process of the form, $dp = a(p, t) dt + b(p, t) dz$ and time-dependent deterministic forest growth, $ds(t) = g(t) dt$, where $s(t)$ is the known forest stock, $g(t)$ is the known growth function, and p is stumpage price. Defining $R(t) = ps(t)$ as gross returns in time t from harvesting, and K as a constant harvest cost, Insley's resulting value function corresponding to (11.10) is $M(R, t) = \max\{R(t) - K, B(t)\Delta t + (1 + r\Delta t)^{-1} E_x[M(R_{t+\Delta t}, \Delta t + t | R)]\}$. Considering only the continuation region, taking the steps outlined earlier and applying Ito's lemma, the addition of a forest growth transition equation to the problem results in a slightly different partial differential equation than (11.13): $rM(R, t) = B(t) + M_t(R, t) + [a(R, t)X + [g(t)/s(t)]R(t)]M_R(R, t) + (1/2)b(R, t)^2 X^2 M_{RR}(R, t)$. The value

---

8. For more details concerning these conditions, see Dixit (1993), Dixit and Pindyck (1994, pp. 130–132), Malliaris and Brock (1982, pp. 124–128), and F. Chang (2004, p. 302).

of continuing the rotation now separates the change in the forest stock (third term on the RHS) from changes in amenities and changes in revenues at harvest that come from the evolving stumpage price assumption. They numerically solve the problem using boundary conditions similar to those given here.

Insley and Rollins (2005) consider an ongoing-rotations stopping problem, but this time assume that the stochastic stumpage price dp follows the mean reversion in (11.4). The difference in (11.10) for the ongoing rotations problem is to modify the value obtained by stopping to include both harvest revenues and the value function defined at that point since the forest is assumed to be planted and harvested in perpetuity. In this case, following the notation given here and presenting a slightly simplified form of their equation, the Bellman equation for an ongoing infinite-rotations problem is written $M(p, t) = \max\{(p - K)S + M(p, t); B(x)\Delta t + (1 + r\Delta t)^{-1} E_x[M(p_{t+\Delta t}, \Delta t + t|x)]\}$. The difference with the single-rotation Insley (2002) version is that the value $M(p, t)$ is now received as soon as the landowner harvests and plants to begin the remaining sequence of rotations in perpetuity. Other than this, the interpretation remains the same as other problems.

Finally, in another interesting recent article, Chladna (2007) arrives at a version of (11.10) that includes two forms of stochasticity: stumpage prices that follow a mean reversion and carbon prices that follow a geometric Brownian motion. The key difference is the presence of two stochasticities, which renders the last stochastic term in the Bellman equation equal to $(1 + r\Delta t)^{-1} E_x[M(S_{t+\Delta t}, p_{t+\Delta t}, p^c_{t+\Delta t}, \Delta t + t|S_t, p_t, p^c_t)]$, where p^c_t is a stochastic carbon price. Ito's lemma now applies to the two stochastic prices together and is similar to the single-price rule (for the generalization of Ito's lemma to many stochastic processes, see Kamien and Schwartz 1991, pp. 265–267). In all of these cases, the resulting partial differential equation is solved numerically using backward recursion under the assumption of an ending point. For the interested reader, Miranda and Fackler (2002, chapter 11), provide a detailed discussion of some techniques that can be used in solving this recursion.

Before continuing, it is worth pointing out a pedagogical twist in interpretation of the stopping problem in this section. Many authors, such as Willassen (1998), Sodal (2002), and Yin and Newman (1996a), for example, refer to the stopping point as the time when an option is exercised to "invest" in forestry. This comes from a classic option value interpretation of the rotation problem where the landowner invests the harvesting cost in order to receive the value of the option by harvesting.

11.3   Harvesting Thresholds

Many of the articles cited in sections 11.1 and 11.2 address problems too complex to solve for a closed-form optimal stopping time (or threshold of stand value). In this section we will show how one constructs a closed-form solution for single and ongoing rotation models. The reader will see how even the simplest problem quickly becomes difficult to work with, making it clear why the majority of research in this literature is numerically based. A word of caution is needed first. Some of the solution concepts and results used in this section are quite complex to prove. While some will have to be taken as fact, we will provide references and a heuristic sketch of one proof in particular.

In closed-form problems, the optimal threshold value of the stand in the stopping problem of (11.8) defines an expected rotation age because the actual rotation length becomes a random variable. In this section we will examine the qualitative nature of the harvesting threshold and expected rotation age for problems using single and ongoing rotations.

11.3.1   Single-Rotation Problem

Assume a basic stopping problem as in (11.8)–(11.13). Following closely F. Chang's (2004, 2005) and Willassen's (1998) model,[9] but using a stand value interpretation denoted by a capital X, assume that X is stochastic and follows a geometric Brownian motion, $dX = \alpha X\, dt + \sigma X\, dz$, where X is the size of forest stand value at time t. Let the initial stand value, be denoted X_0, and let A denote the size of forest stand value beyond which the trees are harvested. The parameter A is known as the harvesting threshold. We will refer to the cutting time (also called a first passage time) as T. This implies that $X_T = A > X_0$. All other parameters are known and assumed constant. Let the regeneration cost of the stand equal c_2 and the harvesting cost equal c_1.

The stochastic single-rotation stopping problem for a risk-neutral landowner is now a specific version of (11.8):

$$\max_T V^s(A) = (A - c_1)E_X[e^{-rT}] - c_2. \tag{11.14}$$

There are two obvious differences in this problem with the deterministic one studied in chapter 2. First, an expectation is taken over X of the discount factor, $E_X[e^{-rT}]$, even though the interest rate is known; this must be done because T is stochastic given that X is stochastic. Second, in the literature where these models have been used, the cost c_2 is often omitted from (11.14), making the problem a "markup" version where timber price is written as a net price, and the planting cost of the first stand is assumed to be sunk at any point in time used to form the Bellman equation. In a markup problem, c_1 is reinterpreted as the lump sum har-

---

9. Reed and Clarke (1990) analyze a similar problem under conditions of stochastic biological and age-dependent growth, as does Reed (1993) for the case where forest stock and amenity services are stochastic.

vesting and planting costs of the next rotation should one be started, and thus both forest value and costs are discounted. If we consider the von Thunen-Jevons model from chapter 2 in a similar fashion, we would need to define the objective function as $V(T) = [pf(T) - c]e^{-rT}$, and the first-order condition as $pf'(T) = r[pf(T) - c]$. The markup problem for a single rotation must be compared with this modified Jevon's problem for the comparisons to be meaningful.

The difficulty in solving for an explicit solution to (11.14) is that $E_X[e^{-rT}]$ cannot be computed using conventional means. Ito's lemma needs to be applied, and as Chang suggests, it proves much easier to work with a monotonic transformation of forest stand value X, $F(X,t) = Y = \log X$. We already used Ito's lemma earlier in section 11.2 for this and showed that $dY = [\alpha - (1/2)\sigma^2]\,dt + \sigma\,dz$, given the known parameters α and σ. We also know from this derivation that over any finite time interval $T = dt$, the change in Y is normally distributed with a mean (drift) of $[\alpha - (1/2)\sigma^2]\,dt$ because $E\,dz = 0$, and the variance equals $\sigma^2\,dt$. Furthermore, given that this is a monotonic transformation, if the stand is cut at a forest stand value of size $X = A$, then it is uniquely cut at $Y = \log A$. Thus the first "passage time" can also be written formally as $T = \inf\{t \geq 0 : Y_t = \log A\}$, where the stochastic process Y_t defines the value of Y at time t, and the change in Y over time follows $dY = [\alpha - (1/2)\sigma^2]\,dt + \sigma\,dz$. Notice that it makes no difference whether we refer to Y_t or Y_T at the first passage time given the definition of T.

Using these preliminaries, the next step is to determine a parametric representation of the expected discount factor at the first passage time T. A well-known but very long and complicated proof exists for this (Harrison 1985, pp. 38–44). It involves guessing the form of a solution to a resulting partial differential equation that arises by applying Ito's lemma to $E_Y[e^{-rT}]$. In appendix 11.1 we provide a derivation of equation (11.16) and in appendix 11.2 we provide a heuristic argument for (11.15). For now, the reader will have to accept that the expected discount factor equals

$$E_Y[e^{-rT}] = e^{-\gamma(\log A - \log X_0)} = \left(\frac{X_0}{A}\right)^{\gamma}, \tag{11.15}$$

where we take expectations over Y, and γ is the positive root of the fundamental quadratic equation that arises because of the geometric Brownian motion assumption for X in the function $Y = \log X$ (see e.g., Dixit 1993, p. 11):

$$Q(\gamma) \equiv \frac{\sigma^2}{2}\gamma(\gamma - 1) + \alpha\gamma - r = 0. \tag{11.16}$$

Equation (11.15) has an intuitive interpretation since $A > X_0$ and this difference increases as time increases and the forest stand value grows at a rate of α. The further away the stopping value A is from the value of the initial stand value (or the bigger the difference $A - X_0 > 0$), the smaller is the RHS of (11.15), implying that we are discounting future returns by a larger amount. Thus (11.15) is mathematically equivalent to a conventional discount factor.

There are two roots of the second-order quadratic equation (11.16), one positive and one negative. The positive root is the only one that it makes sense to use, and we refer to it as γ in what follows:

$$\gamma = \frac{-\left(\alpha - \frac{\sigma^2}{2}\right) + \sqrt{\left(\alpha - \frac{\sigma^2}{2}\right)^2 + 2\sigma^2 r}}{\sigma^2} > 1. \tag{11.17}$$

Using (11.15), we can now rewrite the landowner's single-rotation problem in (11.14) as

$$\max_{\{A\}} V^s(A) = (A - c_1)\left(\frac{X_0}{A}\right)^\gamma. \tag{11.18}$$

Notice from the proof in appendix 11.1 and from (11.15) that the stochastic process for forest stand value is embedded in the discount factor in (11.18) through the relationship between A and X_0, and through the parameter γ. Differentiating (11.18) with respect to the optimal stopping value A gives the following condition for the optimal threshold of forest stand value at the first passage time:

$$V_A^s = A(1 - \gamma) + c_1\gamma = 0. \tag{11.19}$$

The second-order condition holds, as $V_{AA}^s = (1 - \gamma) < 0$. Denote the optimal threshold stand value as A^{*s}, where $A^{*s} = c_1\gamma/(\gamma - 1)$ and the superscript denotes the single-rotation solution. To develop a stochastic harvesting rule for this single-rotation model that is most similar to the rule under certainty, we multiply both parts of the first-order condition (11.19) by the growth rate of stand value α and rearrange to obtain

$$\alpha A^{*s} = \alpha\gamma(A^{*s} - c_1). \tag{11.20}$$

Equation (11.20) defines the cutting rule for the optimal rotation age under stochastic forest stand value following a geometric Brownian motion. The LHS represents the marginal benefit from delaying harvest one period in terms of additional forest stand value, since the growth rate is captured by the drift parameter. Because A is related to X, this term is the stochastic counterpart of a growth in value in the deterministic model. The RHS is the marginal cost of delaying harvest in terms of the opportunity cost of lost investment, because as argued earlier, the size of $(A^{*s} - c_1)$ is net rent and γ is directly related to the discount factor because $\gamma_r > 0$ from (11.17). At the optimal threshold forest stand value, these marginal costs and benefits must equate.

Armed with this necessary condition, several conclusions can be made regarding the optimal rotation in the basic stochastic model. First, from equation (11.17) the root parameter γ and thus the harvesting threshold depend on the growth rate of the forest stand value, α and volatility, σ. The root parameter also depends on the interest rate r. From (11.17), as volatility approaches zero, $\sigma \to 0$, the harvesting threshold will decrease. Higher volatility will therefore increase the harvesting threshold since we can show that $\partial A^{*s}/\partial \gamma < 0$ (see later discussion). The reason for this effect on the harvesting threshold is that higher volatility means the landowner has a higher chance of obtaining a large increase in forest stand value from waiting, and thus the harvesting threshold increases, delaying expected harvest time.

Proposition 11.1 In a single-rotation mark-up model with a stochastic forest stand value, the optimal rotation age is longer than it would be under certainty.

The next step is to compute comparative statics of the harvesting threshold defined by (11.19). In stochastic optimal stopping models, the comparative statics are slightly more involved than in deterministic models. Some variables may affect only the harvesting threshold, but others may affect the root of the fundamental quadratic equation (11.17), given that it depends on the interest rate and features of the stochastic process. Except for regeneration cost, we must obtain comparative statics effects in two steps. First we examine how the root changes and then we see how the threshold changes from (11.19). Proceeding in this way, the solution to (11.19), $A^{*s} = c_1\gamma/(\gamma - 1)$, is differentiated with respect to γ to obtain $\partial A^{*s}/\partial \gamma = (-c_1)/(\gamma - 1)^2 < 0$ given $\gamma > 1$ from (11.17). The optimal harvesting threshold A^{*s} therefore depends negatively on the root of the fundamental quadratic equation. Second, we determine how the root parameter changes with a change in the exogenous parameter. From (11.16) we have for any parameter n and using the implicit function theorem, $\partial\gamma/\partial n = -[(\partial Q(\gamma)/\partial n)/(\partial Q(\gamma)/\partial\gamma)]$. We are interested in the parameters $n = c_1, \alpha, \sigma^2, r$. We can show

$$\frac{\partial\gamma}{\partial c} = 0,$$

$$\frac{\partial\gamma}{\partial\alpha} = -\frac{\gamma}{\sigma^2\left(\gamma - \frac{1}{2}\right) + \alpha} < 0, \quad \frac{\partial\gamma}{\partial\sigma^2} = -\frac{\frac{1}{2}\gamma(\gamma - 1)}{\sigma^2\left(\gamma - \frac{1}{2}\right) + \alpha} < 0, \quad \text{and}$$

$$\frac{\partial\gamma}{\partial r} = \frac{1}{\sigma^2\left(\gamma - \frac{1}{2}\right) + \alpha} > 0.$$

Combining these derivatives with $\partial A^{*s}/\partial\gamma < 0$ establishes the following results:

$$A_c^{*s} = -\frac{\gamma}{1-\gamma} > 0; \quad A_\mu^{*s} = \frac{\partial A^{*s}}{\partial\gamma}\frac{\partial\gamma}{\partial\alpha} > 0;$$

$$A_{\sigma^2}^{*s} = \frac{\partial A^{*s}}{\partial\gamma}\frac{\partial\gamma}{\partial\sigma^2} > 0; \quad A_r^{*s} = \frac{\partial A'^s}{\partial\gamma}\frac{\partial\gamma}{\partial r} < 0.$$

We now collect our findings in lemma 11.1.

Lemma 11.1 In a single-rotation markup model with a stochastic forest stand value, the harvesting threshold for optimal stopping depends positively on the regeneration cost, the growth rate of forest stand value, and the volatility of forest stand value, but negatively on the interest rate.

The parameters that increase the harvesting threshold make sense because these contribute to a greater expected value of waiting to harvest at the margin. The opposite is true for an increase in the interest rate because this increases the expected opportunity cost of waiting through lost investment income, leading to a decrease in the harvesting threshold.

Ultimately we are interested in what happens to expected rotation age. This is an easy extension because recall that A^{*s} is positively related to the stopping time given that the stand value grows at a rate α over time. Thus we have proposition 11.2.

Proposition 11.2 In a single-rotation markup model with a stochastic forest stand value, the expected rotation age depends positively on the regeneration cost, the growth rate of forest stand value, and the volatility of forest stand value, but negatively on the interest rate.

Proposition 11.2 presents an important difference between the deterministic single-rotation markup problem and a stochastic version. Now all economic factors matter to the rotation age solution.

11.3.2 Ongoing-Rotations Problem

F. Chang (2005) derives a version of the stochastic Faustmann problem with uncertainty in the biological growth of a forest, and similar models are discussed in Sodal (2002) and Willassen (1998). We follow this work here. We retain assumptions similar to those in the single-rotation model in the previous section, and continue to assume forest stand value is stochastic and follows geometric Brownian motion. The value of the forest under ongoing rotations is a sum, evaluated at the threshold forest stand value size of A, of the expected net present value of harvesting at the first cutting time T, plus the expected discounted value of all future rotations after the first cutting time:

$$V(A) = E_X[(X_T - c)e^{-rT}] + E_X[e^{-rT}]V(A). \tag{11.21}$$

Given the definition of the first cutting time $X_T = A$, (11.21) can be solved for $V(A)$ to give:

$$V(A) = \frac{(A - c)EX_x[e^{-rT}]}{[1 - E_X(e^{-rT})]}. \tag{11.22}$$

Both the numerator and denominator of (11.22) include the expected discount factor. Once again, it is more convenient to work with $Y = \log(X)$, so that the forest is harvested uniquely at a stock level of $\log(A)$. This also means that the expected discount factor continues to be given by (11.15). Using (11.15) in (11.22) allows us to express the stochastic Faustmann formula in terms of the harvesting threshold and the root of the fundamental quadratic equation in (11.16) as follows:

$$\max_{\{A\}} V(A) = \frac{(A - c)[X_0/A]^\gamma}{1 - [X_0/A]^\gamma}, \tag{11.23}$$

where X is stochastic forest stock at time T, A is the threshold forest stock at harvest time, and $\gamma > 1$ is given by (11.17). Differentiating (11.23) with respect to A characterizes the optimal forest stock threshold value,

$$V_A = \left[1 - \left(\frac{X_0}{A}\right)^\gamma\right]\left[1 - (A - c)\frac{\gamma}{A}\right] - (A - c)\frac{\gamma}{A}\left(\frac{X_0}{A}\right)^\gamma = 0,$$

which upon simplifying gives, $V_A = (1 - \gamma)A + \gamma c - X_0^\gamma A^{1-\gamma} = 0$. This defines the critical threshold and thus implicitly defines the expected rotation age.[10] Before interpreting the first-order condition, it makes sense to derive a more intuitive harvesting rule as we did in the single-rotation problem by dividing both sides by A and solving to obtain: $A = [\gamma(A - c)]/[1 - (X_0/A)^\gamma]$. Using the notation A^* to denote the optimal harvesting threshold and multiplying both sides of this equation by the growth rate parameter α yields a Faustmann-like harvesting rule under a stochastic forest stand value:

$$\alpha A^* = \alpha\gamma(A^* - c) + \alpha\gamma(A^* - c)\left(\frac{\left(\frac{X_0}{A^*}\right)^\gamma}{\left[1 - \left(\frac{X_0}{A^*}\right)^\gamma\right]}\right). \tag{11.24}$$

The LHS of (11.24) is the marginal benefit of waiting to harvest, which equals the growth rate of forest stand value times the stand value size. The first term of

---

10. The second-order condition holds and is given by $V_{AA} = (1 - \gamma)[1 - (X/A)^\gamma] < 0$.

the RHS is the marginal cost of not harvesting in terms of forgone rents, and the second term is essentially the land rent term from not harvesting in all future rotations. Equation (11.24) can be re-expressed as:

$$\alpha A^* = \left[\frac{1}{1 - (X_0/A^*)^\gamma}\right] \alpha\gamma(A^* - c). \tag{11.25}$$

Comparing (11.20) with (11.25) reveals that the difference between the two solutions arises from the infinite chain of rotations in the stochastic Faustmann model, embodied mainly in the discount term (the first term in brackets on the RHS of 11.25). This term is greater than one, $1/[1 - (X_0/A^*)^\gamma] > 1$, given that $\gamma > 1$. In economics terms, this implies that the marginal cost of not harvesting is higher in the ongoing stochastic rotations problem than in the markup single-rotation problem, implying that expected rotation age is shorter in the ongoing-rotations problem. This mirrors what we found in the deterministic models of chapter 2. How different the two solutions are depends on the magnitude of the term $(X_0/A^*)^\gamma > 0$. The further away the cutting time is from the initial seedling stock size, or the higher the discount factor (embodied in the parameter γ), the greater is the difference between the single- and ongoing-rotations solutions under stochasticity.

Another interesting exercise is to consider how the deterministic Faustmann problem falls out of the stochastic version in (11.25). Chang (2005) makes the following argument: If we remove uncertainty, we set $\sigma = 0$ so that the Brownian motion reduces to a deterministic forest stand value growth $dX = \alpha X \, dt$, where α is the growth rate. If we then recognize from the previous discussion that $(X_0/A^*)^\gamma$ at the first passage time T^* is the stochastic equivalent of e^{-rT^*}, the Faustmann rule from chapter 2 can be written as

$$\alpha A^* = r(A^* - X_0) + \left[\frac{e^{-rT^*}}{(1 - e^{-rT^*})}\right] r(A^* - c). \tag{11.26}$$

Now for (11.26) to equal the stochastic version (11.25), one needs only to show that $\alpha\gamma$ is really a stochastic version of the interest rate r. Chang rigorously proves that $\alpha\gamma \to r$ as the volatility parameter, σ, goes to zero, or $\sigma \to 0$. Given this result, (11.26) can be rearranged into an equation that is quite similar to the Faustmann rule:

$$\alpha A^* = r(A^* - c)\left(\frac{1}{1 - e^{-rT^*}}\right). \tag{11.27}$$

Comparing the stochastic version (11.25) now with the deterministic version (11.27) reveals that their only difference is through the discount factor and the positive root $\gamma > 1$. Given that from (11.19) this root decreases in the volatility parameter σ, the difference between the two models is entirely related to the size of the volatility parameter. When this parameter is larger, the discount factor is lower and the marginal opportunity cost of continuing a rotation in the stochastic model is higher than in the deterministic model. Thus the expected rotation age increases as the threshold A^* increases. Putting all of this together gives proposition 11.3.

Proposition 11.3   In the stochastic ongoing-rotations model, the expected optimal rotation age is longer under a stochastic forest stand value than it is under certainty.

Stochastic forest stock modeled as Brownian motion results both in an increased expected net present value of harvesting in the current period, and in an increased value of waiting by increasing the expected net present value of future harvesting. Because the increased value of waiting (or option value) dominates, the optimal harvesting threshold increases and the expected rotation age increases.

The message of proposition 11.3 is not as general as one is tempted to think. By this result, one would think that uncertainty always means longer rotation ages. However, recognize that this result concerns the variance of the distribution of forest stand value, which after all is a good measure of uncertainty in this problem. An increase in volatility is also similar to a mean preserving spread. This type of result has support in other models, such as the continuous search-based stopping problems we will discuss in chapter 12. However, recall that in chapter 10 we showed that risk shortens the optimal rotation age when it is modeled as a Poisson jump process. The difference stems from how risk is important to the decisions. With risk of catastrophic loss, the rotation age is shorter because there is a corresponding increase in the marginal cost of continuing a rotation that results from an increase in the effective interest rate. This is not the case for stochastic models in this chapter.

There is also another interpretation beyond this. Catastrophic risk is always the type of risk that makes a landowner worse off. This is because it normally is assumed to cause a drastic decrease in rents when an event arrives. This is not necessarily the case with stochastic forest stock or the risk to the value of a stand modeled as it is in this chapter. For these processes, the landowner may actually be better off with greater risk (i.e., greater volatility of the process). Consider risk concerning the stumpage price. Brownian motion implies that we may observe a higher price in the future if we wait, especially if volatility is high. This chance of receiving a higher price, along with the larger return from harvesting that would come from it, makes the landowner better off in expected terms. Thus, the rotation age can be longer. As we saw in chapter 10, the only way rotation age can ever be longer in the catastrophic risk case is if landowners can undertake costly actions to protect themselves against that risk, such as thinning to remove fuels or planting in a way that reduces storm damage. Finally, it is worth noting that as Gjolberg and Guttormsen (2002) point out, mean reverting processes offer less benefit to landowners because the chance of a high price in the future is dampened by the corrective nature of these processes.

The final issue with the ongoing-rotations model is to examine the comparative statics of the optimal harvest threshold that is the solution to (11.25). As in the single-rotation stochastic model, with the exception of only the harvest and planting cost, all other parameters affect the harvesting threshold both directly and indirectly through a change in the fundamental quadratic equation root.

Differentiating (11.25) we obtain $dA^*/dc = [\gamma/(\gamma - 1)][1 - A^{*-\gamma}]^{-1} > 0$; $A^*_\gamma = (\partial A^*/\partial \gamma)(\partial \gamma/\partial \mu) > 0$; $A^*_{\sigma^2} = (\partial A^*/\partial \gamma)(\partial \gamma/\partial \sigma^2) > 0$; and $A^*_r = (\partial A^*/\partial \gamma)(\partial \gamma/\partial r) < 0$. It can be shown in this model that $\partial A^*/\partial \gamma < 0$, indicating a negative relationship between the harvesting threshold and the positive root of the quadratic equation. Now, the effects of various parameters on the root are the same as those we found in the single-rotation problem:

$$\frac{\partial \gamma}{\partial \sigma^2} = -\frac{\frac{1}{2}\gamma(\gamma - 1)}{\sigma^2\left(\gamma - \frac{1}{2}\right) + \mu} < 0, \quad \frac{\partial \gamma}{\partial \mu} = -\frac{\gamma}{\sigma^2\left(\gamma - \frac{1}{2}\right) + \mu} < 0, \quad \text{and}$$

$$\frac{\partial \gamma}{\partial r} = \frac{1}{\sigma^2\left(\gamma - \frac{1}{2}\right) + \mu} > 0.$$

Taking all of this together we have lemma 11.2.

Lemma 11.2   In the Faustmann model with stochastic forest stand value, the optimal harvesting threshold A^* depends positively on regeneration cost, the growth rate of forest stand value, and the volatility of forest stand value, but negatively on the interest rate.

So the comparative statics effects in the stochastic Faustmann model are qualitatively similar to those in the stochastic single-rotation markup model. Finally, we also have proposition 11.4.

Proposition 11.4   In the Faustmann model with a stochastic forest stand value, the expected rotation age depends positively on regeneration cost, the growth rate of forest stand value, and the volatility of forest stand value, but negatively on the interest rate.

We find the same interpretation as we had before—that higher volatility raises the optimal harvesting threshold and increases expected rotation age. The more uncertain the future value of a stand, the larger the marginal benefit required to make the irreversible decision of harvesting. This means that for a landowner, a higher volatility in stand value will increase the value of waiting to harvest at the margin.

Our analysis here was based on the assumption of a geometric Brownian motion. As we discussed in section 11.3, finding an analytical rotation age solution under mean reversion has proven to be quite difficult. Furthermore, as with the single-rotation stochastic model, the issue of jointly stochastic prices and stand growth modeled using the types of stochastic processes in this chapter remains open.

11.4 Modifications

In this section we briefly mention some other approaches to modeling stochastic forestry problems; namely, the extension of stochasticity to other parameters relevant to forest decisions, such as interest rates and uncertainty concerning amenity services.

11.4.1 Stochastic Interest Rates

Since the dawn of stochastic Faustmann rotation models, the assumption of a known interest rate has persisted. The fact that interest rates are unknown has had some interest in financial economics (see Cochrane 2001 and Björk 1998). If investment projects are liquid, or in the forestry case if land markets are perfect, then interest rate volatility does not matter for decisions. However, suppose, a landowner's investment is not liquid. Then, the long-term investment nature of forestry means that a waiting time must be endured before the forest reaches a merchantable age. Alvarez and Koskela (2003) consider a rotation model with stochastic interest rates, using a single-rotation stopping problem for a risk-neutral landowner. This basic problem is a modification of (11.8) with the following value function:

$$M(x, r) = \max_{\tau} \ E_{(x,r)} [e^{-\int_0^{\tau} r_s \, ds} x], \tag{11.28}$$

where x is the known stand value, τ is an arbitrary stopping time, and an integral is used to describe the (stochastic) time path of the interest rate. The interest rate and value of the forest stand follow mean reversion and geometric Brownian motion with only drift:

$$dr = (a - br) \, dt + c \, dz \tag{11.29a}$$

$$dx = \alpha x \, dt, \tag{11.29b}$$

where parameters a, b, c, and other variables are constants. The equation in (11.29a) is known in financial economics as a Vasiceck model, which has the empirically relevant property that the interest rate tends to revert to its mean level a/b in the long run (see e.g. Vasiceck 1977).

The main result from this model is that higher volatility in interest rates increases the optimal harvesting threshold and expected rotation age.[11] This result is similar to others in this chapter concerning the volatility of economic parameters, since the value of waiting is increased at the margin by the possibility that a high interest rate might exist at the time the landowner harvests and invests the proceeds in the alternative financial investment. The problem of ongoing rotations with stochastic interest rates remains unsolved.

11.4.2 Stochastic Amenity Services

Stochastic amenity services have been a relatively recent problem studied in stochastic process models. The first application of a stochastic Hartman problem was made by Reed (1993), who examined the conditions under which an old-growth forest should be conserved. Reed assumed that both forest stock and amenity benefits follow a geometric Brownian motion through time. His contribution was extended by Reed and Ye (1994a,b) and Alvarez and Koskela (2007b). Other applications include Conrad (1997, 2000), Forsyth (2000), and Bulte et al. (2002). The goal of much of this work has been to compute the option values of preserving forests.

Reed (1993) in his amenity model assumed a risk-neutral landowner who values both harvest revenue and amenity benefits, both of which are uncertain in the future. Reed's model also bridges the gap between the catastrophic risk problems of chapter 10 and amenity-based models in stochastic problems because a hazard function is also used to incorporate the possibility that the stand will be destroyed by a catastrophic event, leading to a (time-independent) risk-adjusted discount rate in addition to the stochasticity studied in this chapter.

At a point in time t, using our previous notation, the stochastic value of the stand is denoted by $x(t)$, and the value of amenity benefits is denoted by $B(t)$. Each follow a geometric Brownian motion:

$$dx = \alpha_1 x\, dt + \sigma_1 x\, dz_1 \tag{11.30a}$$

$$dB = \alpha_2 B\, dt + \sigma_2 B\, dz_2, \tag{11.30b}$$

where the drift parameters α_1 and α_2 and the volatility (or variance) parameters σ_1 and σ_2 are deterministic and constant, and $z_1(t)$ and $z_2(t)$ follow the standard Wie-

---

11. This has also been generalized to the case where the value of the forest stand follows a geometric Brownian motion (see Alvarez and Koskela 2007a).

ner process discussed in the first section of this chapter. The parameters here are indexed because there is a possibility that the two processes are correlated, i.e., $dz_1(t)$ and $dz_2(t)$ may be correlated, with a correlation coefficient given by $E(dz_1, dz_2) = \rho\, dt$. While correlation between stand value and amenities is hard to conceive, if we assume a constant and known stumpage price, then correlation across these two processes means that amenities are correlated over time, with fluctuations in the forest stock.[12]

With these assumptions, Reed sets up a value function that depends on amenities and harvesting, explicitly solving the optimal stopping problem. This is essentially the first treatment of a stochastic Hartman model. He finds several interesting results. First, he shows that the expected optimal rotation age is longer than the rotation age under certainty, as we also showed in the Faustmann-based model of the previous section. This finding arises because stochastic stand value and amenity benefits increase the expected present value of harvesting and the value of waiting, as well as the expected present value stream of amenities during future periods. Under risk neutrality, the value of waiting dominates the current return from harvesting, implying that the rotation age is longer than in the Hartman model of chapter 3. Second, using an analysis similar to that in section 11.4.1, but with amenities, he shows that the expected rotation age in the stochastic Hartman model depends positively on the growth rate of a stand's value, and on the volatility parameters for a stand's value and amenities, but negatively on the growth rate of amenity benefits, the correlation between stochastic stand value and amenity benefits, and the interest rate.

Naturally the expected length of the rotation period increases with the growth rate and the volatility of a stand's value. For the latter, the higher the uncertainty, the higher the marginal benefit required before making the irreversible investment of harvesting. Thus a higher volatility increases the value of waiting. Reed shows when infinite preservation of the forest is optimal. This decision depends critically on the relationship between the interest rate and the growth rates for stand value and amenity benefits. If the amenity growth rate or the stand value growth rate is high enough, and higher than the interest rate, it will never be optimal to harvest in finite time. Specifically, from (11.30a)–(11.30b) if $r - \alpha_2 < 0$, then amenity services are growing faster than the interest rate, so it is optimal to conserve the forest forever. Furthermore, if $r - \alpha_1 < 0$, then the stand value is growing faster than the interest rate, so it is also optimal to postpone harvesting for all time. We will work through an optimal control version of this result in chapter 12.

---

12. Dealing with correlated Brownian motions in the case of geometric Brownian motions requires more space than we have in this chapter, but the reader is referred to Dixit and Pindyck (1994, p. 82) for a detailed discussion.

11.4.3   Catastrophic Risks

A recurring theme has been to combine stochastic processes for prices, forest stock, or stand value with catastrophic risk and risk-adjusted discount rates, discussed in chapter 10. An example is Yin and Newman (1996a), who combine a catastrophic loss framework with stochastic timber prices. Catastrophic risk is introduced as a Poisson process, but price uncertainty is assumed to follow a geometric Brownian motion. Their homogeneous Poisson process assumes total destruction of the forest. This innovative combination of risks ends up being a relatively simple model based on the diffusion process to define stochastic profits over time, but with a risk-adjusted interest rate used to evaluate the forest investment decision. Catastrophic risk serves to decrease the value of any forest investment, as well as increasing the critical price level where investment (forest establishment) will occur. Yin and Newman find that the qualitative nature of results depends importantly on whether or not the option to invest (i.e., harvest) exists after a catastrophic event occurs. Another important recent extension along these lines is due to Motoh (2004), who has extended stochastic process models to the case of a risk-averse landowner (for the most part, all of the literature is based on a risk neutrality assumption). Motoh specifies the landowner objective function as a utility maximization problem, modeling catastrophic loss as a Poisson distribution but assuming that the value of the resource stock is stochastic and evolves according to geometric Brownian motion.

Still other interesting and relevant articles have considered catastrophic-type risks in stochastic control frameworks under risk neutrality. Thorsen and Helles (1998) consider endogenous stand risk from windthrow (or a similar weather event) with the probability of arrival being dependent on the thinning regime undertaken prior to a storm; thus both the rotation age and thinning regime are choices. The theoretical setup follows a stochastic optimal control problem of rent maximization that incorporates an evolving arrival probability at each point in time and that is subject to equations of motion specifying the forest stock, the salvageable stock after the event, and the probability of the event arriving. They find that endogenous risk matters a great deal in the effects of risk on decisions. Finally, Cairns and Lasserre (2006) study an optimal control-based model in which fire arrival over time produces greenhouse gases and changes the time patterns of carbon storage for forest stands. A Poisson probability is used to define the rate of fire arrival over time, but they do not limit themselves to the Faustmann-like constancy assumptions.

11.5   Summary

We have surveyed the theory of stochastic processes and undertaken a fairly thorough technical review of the literature in this area. Modeling uncertainty using stochastic processes is a relatively new and still developing area of forest economics. This literature has informed uncertainty work in two ways, first by showing that existing uncertainty-based results derived using other models are robust, and second by providing new extensions to our thinking about economic risk beyond simple adjustments to the discount rate used to handle catastrophic risk. This literature also provides a bridge between forests as an investment and the concept option value in the financial economics literature.

The message of this chapter, in contrast to catastrophic risk, is that general economic uncertainty leads to longer rotation ages. The stochastic approach in this chapter makes it clear that there is option value for waiting to harvest. Longer rotation ages are often observed under greater risk of the mean-preserving spread type. This confirms the robustness of results using other approaches, such as search-based dynamic programming problems with simpler distribution and uncertainty assumptions (e.g., Brazee and Mendelsohn 1988). However it also makes it clear that one should be careful about exactly what type of risk is most important to landowner decisions. If unknown economic parameters are more important, the stochastic processes in this chapter may be important to consider. Their precise use will depend on the characteristics of how these parameters are expected to evolve over time. However, if catastrophic environmental risks are more important, the landowner will tend to make decisions using a higher effective interest rate and, usually, a shorter rotation age is optimal. In many cases, and in a few examples from the literature, these risks have been combined. However, by far most authors have considered only one or the other. Thus the entire story about how risk affects forest landowner decisions in a variety of cases may not yet be fully understood.

The differences in how various risks affect forest landowners also have important policy implications. Catastrophic risk is always the type of risk that makes a landowner worse off, leading to drastic decreases in rents. This is not necessarily the case with stochastic forest stand value risk modeled as it is in this chapter. For these processes, the landowner may actually be better off with greater risk (i.e., greater volatility of future prices or forest stand values). It would be interesting to consider the government tax revenue implications of the various forms of uncertainty. In addition, very little is known about both the design of policy instruments or their effects on landowner behavior in the more general economic uncertainty models of this chapter.

We leave the reader with one other thought about an important future topic that is within stochastic control modeling of forest economics problems. Recently there has been some interest in other fields of natural resource economics concerning robust control methods. In the uncertainty modeled in this book, we have always assumed that we knew the distribution of a random variable, so what was not known was the actual realization of the random variable at a future date that comes from this known probability distribution. However, for many risks, or for correlated risks, the distribution itself may not be known. These types of problems have recently been developed in the macroeconomics literature to deal with fiscal policy shocks in economies. By and large, the results from these models show that precautionary decision making becomes the rule. In a forestry context, precautionary decisions could reverse some of the results about the option value of longer rotations derived in stochastic models.

Appendix 11.1   Derivation of Equation (11.16)

The precise derivation of equation (11.16) is as follows: Recall that $F(X)$ is a function of the stochastic process X that evolves according to $dX = \alpha X\, dt + \sigma X\, dz$. Ito's lemma gives $dF = F_X\, dX + (1/2)F_{XX}(dX)^2$, and using geometric Brownian motion this becomes $dF = F_X[\alpha X\, dt + \sigma X\, dz] + (1/2)F_{XX}[(\alpha X\, dt)^2 + (\sigma X\, dz)^2 + 2\sigma\alpha X^2\, dz\, dt]$. Using the multiplication table for a Wiener process, we have $(dt)^2 = dz\, dt = 0$ and $(dz)^2 = dt$, so that dF becomes $dF = F_X[\mu X\, dt + \sigma X\, dz] + (1/2)F_{XX}(X)[(\sigma X)^2\, dt]$. Therefore the expected value equals $E(dF) = \mu X F'(X)\, dt + (1/2)\sigma^2 X^2 F''(X)\, dt$. To consider this further, we return to the interpretation of the stochastic Bellman equation in (11.12) which implies that $rF = E(dF)/dt$. This says that the return in the current period from harvesting must equal the expected return from waiting one more period at the optimal cutting time. The term rF is the opportunity cost of waiting one more period to harvest, and the derived term $E(dF)$ is the expected benefit of waiting one more period under uncertainty that determines the forest stock in the next period. Inserting the differential dF from geometric Brownian motion derived earlier into this Bellman equation gives the following second-order differential equation:

$$\alpha X F_X(X)\, dt + \tfrac{1}{2}\sigma^2 X^2 F_{XX}(X)\, dt - rF(X) = 0. \tag{A11.1}$$

Equation (A11.1) looks quite complicated for general $F(X)$, but it is easy to characterize if one assumes a form for the general solution as follows: $F(X) = GX^\gamma$. Substituting this form of solution into (A11.1) gives the following specification: $\alpha\gamma X G X^{\gamma-1} + (1/2)\sigma^2 X^2 G\gamma(\gamma-1)X^{\gamma-2} - rGX^\gamma = 0$. This is equivalent to (11.16) if one divides all terms by GX^γ.

Appendix 11.2   A Heuristic Proof of Equation (11.15)

The exact proof for equation (11.15) in the text is quite complicated and can be found in Harrison (1985). As we have found in the text, the expected discount factor of the first passage time in the case of a logarithmic transformation of the stochastic forest stand value X is written as $F(X) = E_{\log X}[e^{-rT}]$. Assume that the current stand value level is such that we wait over the next time increment, dt; i.e., assume that $X < A$. During this next time increment, forest stand value changes by $(X + dX)$, given we are starting at X (see section 11.2). At any point in time during continuation, since the interest rate is known, and X is known but dX is uncertain, the following relationship must hold for $F(X) = E_{\log X}[e^{-rT}]$:

$$F(X) = \frac{1}{1 + r\,dt} E[F(X + dX)|X] = \frac{1}{1 + r\,dt}(F(X) + E[dF(X)]). \tag{A11.2}$$

Using the process discussed for equations (11.10)–(11.12), multiplying by $(1 + r\,dt)/r\,dt$ and taking dt to zero, we arrive at

$$rF(X) = \frac{E_X\,dF(X)}{dt}. \tag{A11.3}$$

Now, recalling that X follows a geometric Brownian motion with constant nonrandom drift α and volatility σ parameters, we can use Ito's lemma to obtain the RHS of (A11.3). We already derived this differential in section 11.2 for $F(X) = \log X$ and know that it equals $dF = F_t \mu X\,dt + F_X \sigma X\,dz + (1/2)F_{XX}\sigma^2 X^2\,dt$. Substituting this into (A11.3) and taking expectations, realizing that $E(dz) = 0$, we obtain

$$rF(X)\,dt = F_t \alpha X\,dt + \tfrac{1}{2}F_{XX}\sigma^2 X^2\,dt. \tag{A11.4}$$

Equation (A11.4) is a partial differential equation of the form $(1/2)\sigma^2 X^2 F_{XX} + \alpha X F_t - rF = 0$. This type of differential equation has a general solution of the form $F(X) = a_1 X^{\gamma_1} + a_2 X^{\gamma_2}$, where γ_1 and γ_2 are, respectively, the positive and negative roots of the standard quadratic characteristic equation for the differential equation (see also Dixit and Pindyck 1994, p. 143). Before solving this, to determine the two constants we need to use two boundary conditions. These conditions are (1) as X approaches the threshold A the discount factor is close to zero and hence $F(X)$ is close to $F(A) = 1$, and (2) when X is close to zero, T^* is most likely large and so the discount factor is close to zero and therefore $F(0) = 0$. With these results the reader can verify that the following solution works for the partial differential equation given here, $a_1 = (1/A)^{\gamma_1}$ and $a_2 = 0$. Substituting these into the partial differential equation, we have $F(X) = E_{\log X}[e^{-rT^*}] = (X/A)^{\gamma}$, where $\gamma = \gamma_1 > 1$.

12   Dynamic Models of Forest Resources

Traditional rotation analysis is static by nature. Although time elapses in the form of ongoing discrete rotations, it typically enters only as an input to a stand's yield at the rotation age. As we saw in chapters 2 and 3, this means that the optimal rotation age can be determined through a comparison of current growth and some function of the interest rate. Time may have much more profound effects in forest resource problems, however. Consider that prices, costs, and interest rates may evolve over time, and other aspects of forest land, such as amenities or carbon stocks, may change through time. In fact, the land-use decision itself may be made over time and depend importantly on these dynamic effects. For all of these cases, a more general approach to modeling forest resource decision and policy problems is needed.

Optimal control and dynamic programming are useful for problems where changes in time and space are important. Models based on dynamic programming have been used mostly for stand management problems and landowner decision making under uncertainty, while optimal control models have been used for perhaps a wider variety of nonstochastic problems. We introduce these models in this chapter and compare the results with those derived in other parts of the book, showing alternative ways that problems can be studied by abandoning the rotation model.

We begin with some preliminaries concerning dynamic optimization. Optimal control problems are then considered and four areas are studied: Faustmann-Hartman interpretations, old-growth forest interpretations, land-use interpretations, and other interpretations (such as carbon sequestration and biodiversity). In the third section, we take up dynamic programming and examine both perfect foresight and stochastic interpretations. Finally, we offer a summary, paying particular attention to the differences among models.

12.1   Dynamic Optimization

Dynamic analysis of forestry problems requires the application of dynamic optimization methods. Dynamic optimization follows a general type of approach for framing problems that is similar across specific methods. There are several comprehensive treatments of dynamic optimization in economics, including those of Kamien and Schwartz (1991), Chiang (1997), and Leonard and Van Long (2002). Clear presentations of dynamic programming can be found in Bertsekas (1987), Sargent (1987, pp. 12–56), and Adda and Cooper (2003). Miranda and Fackler (2002) is an excellent reference for solving dynamic models numerically.[1]

12.1.1   Optimal Control

The mathematical structure of any dynamic optimization problem is similar across fields of science. A generic finite time-horizon–optimal-control problem is

$$\max_u V(u) = \int_0^T F[t, y(t), u(t)]\, dt, \quad \text{s.t.} \tag{G.1}$$

$$\dot{y} = f[t, y(t), u(t)] \tag{G.2}$$

$$y(0) = y_0; \quad y(T) \text{ free/given} \tag{G.3}$$

$$u(t) \in U \quad \text{all } t \in [0, T]. \tag{G.4}$$

In the generic problem, t refers to time, $u(t)$ is the control (or choice) variable or vector, $y(t)$ is the state variable or vector that describes the system at each point in time, and F in the integral denotes an "objective functional." The integral in (G.1) defines the planning horizon; it is finite here but could be infinite in other problems. Equation (G.2) is called an equation of motion, or a transition equation. The notation $\dot{y}$ is the time derivative of $y(t)$, $\dot{y} = dy(t)/dt$. The equation of motion explains how the impact of the control variable translates to a change in the state of the system at each point in time. Equation (G.3) defines the initial state $y(0)$ and the ending value of the state at terminal time, $y(T)$. In infinite-time horizon problems, the end point of the state may be free, and then a special "transversality" condition is needed to pin down what happens at the terminal time. The objective of the problem here is to solve for a path of the control variable over time that maximizes the objective functional, with the equation of motion describing how the state of the system changes along the optimal control path.

---

1. A classic piece explaining the economic interpretations of optimal control theory is Dorfman (1969). There are also some renewable resource and forestry-specific reviews. Williams (1989) has an early yet complete discussion of dynamic modeling applied to natural resources, while Brazee (2006, 2003) shows how dynamic forest models are logical steps beyond Faustmann problems in different contexts.

The generic optimal control problem in (G.1)–(G.4) can be solved by applying the maximum principle developed by the Russian mathematician Pontryagin. The maximum principle builds on something called a Hamiltonian function and an auxiliary, or co-state (or adjoint) variable $\lambda(t)$ that in economic application defines the dynamic shadow value of the state. The Hamiltonian function for the generic problem (G.1)–(G.4) is defined by the objective function and the state as follows:

$$H = F[t, y(t), u(t)] + \lambda(t) f[t, y(t), u(t)]. \tag{G.5}$$

The maximum principle defines the first-order differential functions for the state and co-state variables. In addition, it is required that the Hamiltonian be maximized with respect to the control variable u at every point of time. Thus we can express the conditions for the solution of the generic problem (G.1)–(G.4) as follows:

$$\max_{u} \; H = F[t, y(t), u(t)] + \lambda(t) f[t, y(t), u(t)], \quad \forall \in [0, T] \tag{G.6}$$

$$\dot{y} = \frac{\partial H}{\partial \lambda} \tag{G.7}$$

$$\dot{\lambda} = -\frac{\partial H}{\partial y} \tag{G.8}$$

$$\lambda(T) = 0. \tag{G.9}$$

The sufficient second-order conditions require that the Hamiltonian be concave in $u(t)$ and $y(t)$. Depending on the specified forms of the time horizon and the objective function, the outcomes defined by (G.6)–(G.9) differ. The optimal path may also not be unique unless the objective function is linear in the control variables. In this case, the solution is called "bang-bang," and there are only two possibilities for optimal control; it either exists at a steady state, or at its maximum or minimum (for example, no harvesting). In addition to a transversality condition, some problems in forestry may include a "salvage" or "scrap" function as a requirement at the end of the planning horizon.[2] The sale of land is an example of a salvage value.

---

2. An alternative to the optimal control approach is the calculus of variations. This is applicable to cases where the objective function is linear in the control, so that the primary focus can be on the state of the system and how it changes over time. However, these restrictions mean it has found more limited applications in forestry problems than optimal control and dynamic programming have. In forestry, thinning problems have been the main area where calculus of variations has been applied. Original contributions are those of Cawrse et al. (1984), Clark and Depree (1979), Clark (1990), Betters et al. (1991), Steinkamp and Betters (1991), and Hyytiäinen et al. (2005). For recent applications of thinning in other models under uncertainty, see Brazee and Bulte (2000) and Lu and Gong (2003), and in a dynamic programming context, Teeter and Caulfield (1991).

For optimal control problems involving discounting, the necessary condition for the co-state variable (G.8) depends on whether the Hamiltonian in (G.6) is written in present-value or current-value form.[3] For a Hamiltonian function written in present-value (discounted) form, we must have $\dot{\lambda} = -(\partial H/\partial y)$. For a Hamiltltonian written in current-value (undiscounted) form, this condition is slightly modified and equals: $\dot{\lambda} - r\lambda(t) = -(\partial H/\partial y)$. We will see examples of both in this chapter.

We will not be concerned with "comparative dynamics," results showing how the optimal control path depends on exogenous parameters. Given that the optimal path is nonunique for non-bang-bang solutions, comparative dynamics becomes quite complex, especially because the equations of motion describing the state of the forest (i.e., forest stock) are nonlinear enough to render it a fairly difficult venture. A reader interested in comparative dynamics methods in optimal control models is referred to Caputo (2003, 2005) for innovative methods based on envelope results.

12.1.2 Dynamic Programming and the Bellman Equation

In dynamic programming, multiperiod discrete-time problems can be simplified into a two-period analogous decision using a Bellman equation and the related principle of optimality. The key component of a Bellman equation is to define an indirect (maximized) objective function for the next period onward, and then set up a problem in which the focus is on maximizing the value of the objective function in the current time period plus the value function for the next time period on. The value function of the next period itself incorporates all optimal decisions from that time period forward. The principle of optimality guarantees that we can select the optimal path using this equation because given an equation of motion defining the relationship between control and state variables, at a given time t the choices made from t onward need only be optimal for then on, because the value of the state at time t results from all optimal decisions made up to that point in time. Nothing that happened prior to time t is of any consequence to the problem because those decisions are irreversibly sunk once we arrive at t.

A generic dynamic programming problem can be written in many forms. A discrete-time–continuous-control form example is

$$V[y(0)] = \max_{u(t)} \sum_{t=0}^{\infty} F[t, y(t), u(t)], \quad \text{s.t.} \tag{G.10}$$

---

3. For a discussion of current- and present-value Hamiltonian functions, see Kamien and Schwartz (1981, pp. 152–153).

$y(t + 1) = f[t, y(t), u(t)]$

$y(0) = y_0; y(T).$ $\hspace{6cm}$ (G.11)

The term $V[y(0)]$ is called a "value function" and designates the optimal value of the objective function when the state is $y(0)$, i.e., at time zero. The first line of (G.11) is the equation of motion written now in discrete time. The control variable continues to be $u(t)$. The second-order conditions hold as long as $V[y(0)]$ is differentiable and concave in the control.

In order to derive an optimal path for harvesting in this problem, we must form the Bellman equation for any interval $(t, t + 1)$ (Adda and Cooper, 2003, chapter 2),

$V[y(t)] = \max_{u(t)}\{F[t, y(t), u(t)] + V[y(t + 1)]\}.$ $\hspace{3cm}$ (G.12)

This condition states that at any time t the optimal control is one that maximizes the objective function (first term) plus the maximum value of this objective function starting at time $t + 1$ onward given by the value function at time $t + 1$. The principle of optimality implies that maximizing this equation at any time t fully characterizes the optimal control path. Often, before obtaining the necessary condition, (G.12) can be simplified using the equation of motion and writing $V[y(t)] = \max_{u(t)}\{F[t, y(t), u(t)] + V[f[t, y(t), u(t)]]\}.$

Discrete dynamic programming has been the choice for forest resource problems where there is one-time harvesting, which from chapter 11 we know to be called "optimal stopping problems." As we will see, using the Bellman equation form for a solution allows this framework to be easily applied where future states of the system or future parameters are not known; then the second-period value function in (G.12) is written in expected-value terms. The main disadvantage for theoretical work is that the necessary conditions often result in partial differential equations that are difficult to solve in closed form without numerical methods. Still, the straightforward numerical methods made possible by the recursive nature of the Bellman equation have led to dynamic programming forming the basis of several rich and complex empirical and simulation-based studies in forestry.

An important distinction in dynamic programming numerical solutions is made for finite and infinite time-horizon problems. In some infinite time-horizon problems with certain stationarity features, the optimal policy function (as the solution for the control path is sometimes called) can be solved by looking for forms of a solution that converge over time as t is increased in infinity (see Spring et al. 2005a,b for recent examples, and Levhari and Mirman 1980 for a classic renewable resource open-access example). In finite-time problems, however, the path can be solved in two ways. The first is a recursive solution in which the problem is solved backward one point at a time starting from the end point, keeping track of the state and control variables at each step using the equations of motion, and picking the path back at each step that has the highest (Bellman) value of the objective function. The second is a forward solution, where the best path forward is picked as the state equations are updated according to each possible set of control paths evaluated. The recursive solution method is more common. For an interesting application of the adaptive forward solution method, see Meilby et al. (2003).

12.2 Applications of Optimal Control Theory

We consider first under what conditions Faustmann and Hartman solutions emerge from dynamic problems. We then consider the preservation of old-growth forests and dynamic land-use choices.

12.2.1 Faustmann Interpretations

An early treatment of optimal control applied to renewable resources is due to Plorde (1970). He used a model based on logistic growth describing the equation of motion for the resource stock to show that maximum sustained yield rules are nonoptimal in a steady state. The applicability of this notion for forest resources was not directly demonstrated until Anderson (1976), who showed that the Faustmann model emerges as a steady state in Plourde's problem. This formally establishes an equivalence between the Faustmann solution and the optional control solution for an even-aged forest stand.

In Anderson's model, the landowner maximizes a flow of discounted utility from consumption over time:

$$V(h) = \int_0^\infty U[h(t), c(t)]e^{-rt}\, dt, \tag{12.1}$$

where $h(t)$ is the volume of forest stock harvested at time t, $c(t)$ is the consumption of a commodity produced using harvested timber, and r is the discount rate. The equation of motion and the relationships among consumption, the forest stock, and the commodity are defined as follows:

$$\dot{V}(t) = g[V(t)] - h(t) \tag{12.2}$$

$$c(t) = f[V(t), h(t)], \tag{12.3}$$

where $f[.]$ is a concave production function for transforming wood into other goods, and $V(t)$ is growth in the forest stock. It is natural to assume that $V(t) > h(t) \geq 0$, which essentially implies that the forest stock is not harvested completely (i.e., resource use is sustainable), $h(t) \geq 0$. In (12.2)–(12.3), $h(t)$ is the only control variable and uniquely determines $c(t)$, $V(t)$ is the state of the system, and $\dot{V}(t)$ is the time derivative of the forest stock state variable.

Substituting (12.3) using (12.2), we first form the present-value Hamiltonian:

$$H(t) = U[h(t), f[V(t), h(t)]]e^{-rt} + \lambda(t)\{g[V(t)] - h(t)\} + \gamma(t)h(t), \tag{12.4}$$

where $\lambda(t)$ is the co-state variable and $\gamma(t)$ is the multiplier for the non-negativity constraint on harvests at each point in time. Applying the maximum principle defined earlier requires the following necessary conditions to hold:

$$H_h = U_h(.)e^{-rt} + U_c(.)f_h(.)e^{-rt} - \lambda(t) + \gamma(t) = 0 \tag{12.5a}$$

$$-\lambda(\dot{t}) = U_c(.)f_V(.)e^{-rt} + \lambda(t)g_V \tag{12.5b}$$

$$\dot{V}(t) = g[V(t)] - h(t) = 0, \tag{12.5c}$$

which hold along with the constraint condition on harvesting $\gamma(t)h(t) = 0$, with $\gamma(t) \geq 0$. The co-state variable $\lambda(t)$ is the dynamic shadow price of the forest stock to landowner utility in (12.2) at each point in time. Immediately, from (12.5a) we see that harvesting will not occur in any time period if $e^{-rt}[U_h(.) + U_c(.)f_h(.)] < \lambda(t)$, because $\gamma(t) > 0$. In this case the discounted marginal increase in utility from harvesting is less than the shadow price of the in situ (i.e., unharvested) forest stock at time t. The landowner would wait to harvest.

Assuming an interior solution $[\gamma(t) = 0]$, we can examine the characteristics of the optimal path for harvesting over time. To do this, we follow one standard procedure for reducing the necessary conditions by eliminating the co-state variable. First, we use equation (12.5a) to solve for the co-state variable, $e^{-rt}[U_h(.) + U_c(.)f_h(.)] = \lambda(t)$. Second, we take the time derivative of this equation to obtain $\dot{\lambda}(t) = -r[U_h(.) + U_c(.)f_h(.)]e^{-rt} + \{d[U_c(.)f_h(.) + U_h(.)]/dt\}e^{-rt}$. Finally, substitute this time derivative into (12.5b) for $\dot{\lambda}(t)$ to arrive at:[4]

$$-r[U_h(.) + U_c(.)f_h(.)]e^{-rt} + \{d[U_c(.)f_h(.) + U_h(.)]/dt\}e^{-rt}$$

$$+ U_c(.)f_h(.)e^{-rt} + \lambda(t)g_V(.) = 0. \tag{12.6}$$

We next evaluate (12.6) in a steady-state. By definition, a steady state indicates no change and requires setting the time derivatives in (12.6) equal to zero, so that the equation of motion has a growth equal to harvesting, $g[V(.)] = h$. Then we can

---

4. Alternatively, one can reduce the necessary conditions by integrating the co-state time derivative (equation 12.5b in this example) and then substituting for $\lambda(t)$ in the condition for the control (equation 12.5a in this example).

use the fact that from (12.5a), $U_h(.)e^{-rt} + U_c(.)f_h(.)e^{-rt} = \lambda(t)$, and consider the behavior of the model for $t = 0$ since we are in a steady state. Doing all of this and simplifying, we arrive at

$$-r + \frac{U_c(.)f_h(.)}{\lambda(0)} + g_V(.) = 0. \tag{12.7}$$

If we look at a pure forest harvesting problem, and rule out other commodities, then the second term in (12.7) is zero because $c(t) = 0$, and we are left with simply $r = g_V(.)$. This solution is the same as found in two-period and Von Thunen-Jevons models we discussed earlier in the book. Note also that only if $r = 0$, will we harvest in the steady state at the maximum sustained yield point of chapter 2.

Now consider whether the Faustmann harvesting rule emerges from a long-run steady state of the model defined in (12.7). To do this, assume $f_h(.) = 0$, so that the steady state has $r = g_v(.)$. Now, taking the time derivative of the equation of motion gives $\ddot{V}(t) = g_V(.)\dot{V}(t) - \dot{h}(t)$. This becomes $g_V(.) = \ddot{V}(t)/\dot{V}(t)$ given that $\dot{h}(t) = 0$ by definition because harvests are constant in the steady state. Thus the optimal harvest rule from (12.7) can be written alternatively as $r = \ddot{V}(t)/\dot{V}(t)$. Given that the model is expressed in biomass terms, the steady state can hold for even-aged forest management provided that the forest stock consists of an age class distribution of evenaged stands, with the oldest age class removed by harvesting during each interval dt, and new growth replacing the oldest forest age class that is removed (see chapter 8).[5]

Next, define the biomass of the oldest stand by $v(t)$. This is constant for every instant in the steady state since the oldest stand is harvested at each instant of time, and growth in the entire forest biomass equals harvesting. Mathematically, this also means that: $\dot{V}(t) = v(t)$. Now, the growth in biomass is given by a typical biometric forest yield function, $\dot{v}(t) = \psi[v(t)]$, so that by integrating both sides, growth in all stands from youngest to oldest is $\dot{V}(t) = \int_0^{t^*} \psi[v(s)]\,ds$, where t^* is the stationary rotation age for the oldest stand. Also, equation (12.7) can now be re-expressed as $\dot{v}(t)/v(t) = g_V = r$; that is, the oldest stand is harvested in the steady state so that its instantaneous growth in biomass equals the interest rate. This is the dynamic equivalent of the Jevon's rule we derived in chapter 2.

The rule just derived omits the land rent component discussed early in this book. To see how land rent falls out of Anderson's model, we follow his derivations again and first write the Faustmann rule in a steady state for the single (oldest) stand, $\lambda(0)\dot{v}(t) = a + r\lambda(0)v(t)$, where a is the carrying cost (land rent), and the

---

5. Heaps (1984) provides a formal proof that the oldest age class will always be harvested in a typical forestry optimal-control–rent-maximizing problem where the change in forest stock (or volume in the oldest age class) is described by an equation of motion similar to (12.2). He calls this the forestry maximum principle or generalized forest management problem.

steady-state shadow value $\lambda(0)$ is the stumpage price in the steady state.[6] Going back to the definition of the harvesting rule in terms of the whole forest stock (all ages), and using the relationship between biomass and the volume of the oldest age class in the steady state, we can write the Faustmann rule using the notation and relationships here as $\lambda(*)\ddot{V}(t) = a + r\lambda(*)\dot{V}(t)$, or $-r - a/[\lambda(*)\dot{V}(t)] + g_V = 0$.[7] It is interesting that we now can see that this rule is equivalent to (12.7) only if the land rent term a is defined as $a = -U_c(.)f_V(.)\dot{V} = -U_c(.)f_V(.)v(t)$. This has a nice economic intuition. The land rent that emerges from a dynamic control model based on utility maximization is simply the lost utility associated with holding the forest stock longer (which also means harvesting the oldest stand later).

Anderson's article is truly a classic and one of the earliest, albeit somewhat undiscovered, applications of optimal control to a utility maximization problem with forest capital. This article predated many other applications in the optimal control framework that were based on maximizing utility over time. A more general application is the problem presented in Tahvonen and Salo (1999) and Tahvonen et al. (2001), in which a forest landowner seeks to find an optimal path of consumption to maximize utility over time. Unlike Anderson (1976), these authors introduce equations of motion for both the forest stock (as a function of harvesting and growth) and the level of financial assets. Financial assets depend negatively on consumption and positively on exogenous income and income from harvesting. In periods without harvesting, the asset equation of motion is $\dot{A}(t) = rA(t) - c(t) + m(t)$, where $A(t)$ is the level of assets held by the landowner at time t, r is the real interest rate, $c(t)$ is the consumption rate, and m is exogenous income. At the harvest time for any rotation period in an infinite sequence of rotations, assets increase (jump) discontinuously, owing to rents collected from harvesting net of logging costs. The controls for this problem are rotation ages for a series of rotations through time, and the path of consumption over time. The assumption of constant rotation ages need not hold in this general formulation and depends on the nature of borrowing constraints faced by the landowner.

12.2.2 Hartman Interpretations

Snyder and Bhattacharyya (1990) examine what class of optimal control problems led to the standard Hartman model of amenity production we discussed in

---

6. If planting costs are added to the model, the Faustmann rule can again be written in the notation here, assuming that the oldest stand is harvested, as $[1/(1 - e^{-rt})][\lambda(*)v(t)e^{-rt} - Z(0)]$. This is a standard result for the harvesting rule in chapter 2 for the land rent term defined as $a = r[1 - Z(0)\lambda^{-1}(*)v^{-1}(t)]/[(1 - e^{-rt})]$.

7. Using, for example, $\dot{V}(t) = v(t)$, the fact that $\dot{V}(t) = \int_0^{t^*} \psi[v(s)]\,ds$, and that $\ddot{V}(t)/\dot{V}(t) = g_V$ in the steady state.

chapter 3. They describe a landowner who pays maintenance and provision costs to maintain a flow of recreational services of a given quality over time. To develop this model, let $G(T)$ equal the value of stumpage at harvesting for a rotation age of length T, where the sign of $G'(T)$ measures growth in value and hence can be positive or negative. Define $F[R(t), t]$ as a concave function describing the value of recreational services as a function of the level of recreational services at time t, denoted $R(t)$, and the age t of the stand. These recreational services are driven to zero when the forest is cut. The quality of recreational services also decays in each period by a proportional rate β, but these services can be enhanced at a constant rate α by incurring a maintenance cost equal to $C(t)$. The rent captured from recreational services therefore equals $F(R(t), t) - C(t)$.

Letting C_0 denote stand regeneration costs of the first rotation, the discounted net rent function for a single rotation is

$$R_1 = \int_0^T \{F[R(t), t] - C(t)\} e^{-rt}\, dt - C_0 + G(T)e^{-rT}. \tag{12.8}$$

The equation of motion describing the state $R(t)$ is $\dot{R}(t) = \alpha C(t) - \beta R(t)$ where α and β are constants. In the steady state, all rotation ages have identical rents equal to $R = R_1(1 - e^{-rT})^{-1}$. Thus Snyder and Bhattacharyya can write the landowner's problem as an infinite time horizon–control problem as follows:

$$\max_{C(t), T} N = \left\{ A(T) \left[\int_0^T (F[R(t), t] - C(t)) e^{-rt}\, dt - C_0 + G(T)e^{-rT} \right] \right\}, \tag{12.9}$$

where $A(T) = (1 - e^{-rT})^{-1}$. Boundary conditions describe the initial stock of recreational services $R(0) = R_0$, and the end point is assumed free, $R(T) \geq 0$, because rotation age can be infinite if it is optimal to preserve the forest forever. The present-value Hamiltonian for the problem in (12.9) is

$$H = A(T)\{(F[R(t), t] - C(t)) e^{-rt} - C_0 + G(T)e^{-rT}\}$$

$$+ \lambda(t)[\alpha C(t) - \beta R(t)] + \gamma(t)C(t). \tag{12.10}$$

Applying the maximum principle, the necessary conditions for the optimal solution are

$$H_C = -A(T)e^{-rt} + \lambda(t)\alpha + \gamma(t) = 0 \tag{12.10a}$$

$$-\dot{\lambda}(t) = H_R = A(T)F_R[R(t), t] - \lambda(t)\beta R(t), \tag{12.10b}$$

along with the equation of motion and the non-negativity constraint condition $\gamma(t)C(t) = 0$.

We immediately see that no maintenance of amenities is undertaken at t if $A(T)e^{-rt}/\alpha > \lambda(t)$, i.e., the shadow price of additional amenities is too low (this condition is made more likely if the marginal benefit of amenity maintenance to the recreational service stock, α, is small). To examine the optimal path of these costs $C(t)$ paid over time at an interior solution, we take the time derivative of H_C and then use this along with the condition H_C in H_R. By simplifying, we get $F_R(.) = -(re^{-rt}/\alpha) + (e^{-rt}/\alpha)\beta R(t)$. Hence, throughout time, the marginal benefit of applying $C(t)$ is the LHS, which is the marginal benefit of amenities over the next instant in time, while the RHS is the marginal loss in amenities because of decay over time net of interest costs.

The rotation age interpretation comes from the transversality condition that must hold at time T given this is a free endpoint problem. The salvage value is independent of the state variable, and thus the endpoint condition requires that the shadow value of amenity services must equal zero at the instant the stand is harvested, or $\lambda(T) = 0$ (see condition ii, p. 147 in Kamien and Schwartz, 1981), and most important, at terminal time T, $\{F[R(T), T] - C(T)\} + G'(T) - rG(T) = rN$, where rN is the opportunity cost term for the land rent, given by the lost value of not starting a new sequence of infinite rotations from (12.9) (see condition v on p. 148, in Kamien and Schwartz, 1981). This condition determines the rotation age in this stationary control problem. Rewriting it, we have

$$F[R(T), T] - C(T) + G'(T) = rG(T) + rN. \tag{12.11}$$

From (12.11), the rotation age equates the marginal benefit from delaying harvesting (LHS) to the marginal cost of delaying harvesting for another year (RHS). This condition also forms the basis of determining when an infinite rotation exists and when (12.11) equals the Hartman rule derived in chapter 3. The stand is never harvested if from (12.11), the discounted net returns from amenity services accumulated forever (LHS) are greater than the interest lost by not harvesting the stand (RHS). Equation (12.11) differs from the basic Hartman model only due to the presence of maintanence costs for recreational services $R(T)$. Note that the Faustmann solution arises immediately if we assume $F[R(T), T] = C(T) = 0$.

Tahvonen and Salo (1999) have also introduced amenities as a concave function of the age of each unharvested stand in a sequence of rotations through time. They make use of a more complicated version of the general optimal control model described earlier, again assuming that the oldest stands are harvested. Among other results, they find that nonconstant rotation ages are possible unless the representative forest owner's subjective rate of time preference equals the market interest rate. However, the comparative statics of the rotation period differ in this model from the results we found in the Faustmann and Hartman models in chapters 2 and 3 because now substitution and income effects from the utility function complicate the results, as was shown in the simple life-cycle model in chapter 2.

12.2.3   Old-Growth Forests: Mining, Amenity Benefits, and Deforestation

Old-growth forests provide many special aspects for analysis using dynamic optimization methods. In these forests, the volume of dying and decaying wood is replaced by new growth, so that overall stand growth is approximately equal to zero (in terms of renewable resources, the stock of an old-growth forest reflects the carrying capacity of the land). The question becomes how to harvest these stands over time. A tempting answer is to simply deplete them in a manner similar to that of nonrenewable resources following Hotelling's rule. According to this rule, an exhaustible resource should be extracted so that the rate of change over time in the royalty or user cost (net price) equals the rate of interest, which under costless extraction implies that $\dot{p}(t)/p(t) = r$, where $p(t)$ denotes the royalty.

This simple solution may not translate to old-growth forests because these forests have high amenity values in situ (i.e., as standing stock). Furthermore, harvesting native forests is irreversible; high amenity values are lost by harvesting irrespective of whether the land is kept in forestry or allocated to other uses. Thus amenities seem to be a compelling reason to preserve at least some old-growth stock for future generations. It is not surprising that amenities and old growth have been a fiercely fought area between forest industry and environmental groups worldwide, especially in the Amazon and the vast tropical forests of Asia. The remaining old-growth public forests in the western United States have also seen numerous confrontations and changes in forest policy as a result of these amenities, such as spotted owl habitat.

We now set out to examine what optimal control approaches can tell us about these problems. The initial focus will be on forests where there is no plantation establishment, such as tropical forests, but we will comment on how the presence of plantation forests changes the results and require modifications. Consider an old-growth forest whose stock is depleted by harvesting and there is zero growth, so that $\dot{V}(t) = -h(t)$, where $V(t)$ is the volume of forest stock at time t, and $h(t)$ is the harvest rate at t. Also suppose, for example, that this overmature forest is harvested by pulse harvesting (i.e., harvested when net rents are positive), so it might describe a situation where property rights to native forests are poorly defined and there is no incentive to start plantation forestry. Suppose amenities are increasing and concave in forest stock and in time, or $A[V(t), t]$, with $A_V(.) > 0$, $A_{VV}(.) < 0$, $A_t(.) > 0$, and $A_{tt}(.) < 0$. Let the timber price net of harvest costs at time t equal $p(t)$. The problem facing society is to choose the optimal path of harvesting (depletion) of the old-growth forest over time. Using a finite time horizon, for example, society is assumed to choose a path of harvesting rates, $h(t)$, to maximize

$$R(h) = \int_{t0}^{t1} \{p(t)h(t) + A[V(t), t]\}e^{-rt} dt, \quad \text{s.t.} \tag{12.12a}$$

$$\dot{V}(t) = -h(t), \tag{12.12b}$$

$$V(t_0) = x_0; \quad V(t_1) = x_1, \tag{12.12c}$$

$$h(t) \geq 0 \quad \text{and} \quad V(t) \geq 0. \tag{12.12d}$$

Equation (12.12b) defines the equation of motion; (12.12c) defines the boundary conditions (exhaustion of the native forest stock therefore means that $x_1 - x_0 = 0$); and (12.12d) gives the non-negativity conditions for harvesting and forest stock variables, the latter condition telling us at what point we reach a condition where the stock of old growth is exhausted. The present-value Hamiltonian is

$$p(t)h(t)e^{-rt} + A[V(t)]e^{-rt} - \lambda(t)h(t) + \gamma(t)h(t) + \eta(t)V(t). \tag{12.13}$$

Applying the maximum principle yields as the necessary conditions,

$$H_h = p(t)e^{-rt} - \lambda(t) + \gamma(t) = 0 \tag{12.14a}$$

$$-\dot{\lambda}(t) = H_V = A_V(V(t), t)e^{-rt} + \eta(t), \tag{12.14b}$$

which hold along with the equation of motion, the boundary conditions, and the non-negativity constraint conditions, $\gamma(t)h(t) = 0$ and $\eta(t)V(t) = 0$, where for the multipliers $\gamma(t), \eta(t) \geq 0$.

From (12.14a), harvesting becomes zero if and only if the marginal benefit is less than the shadow value of the old-growth resource left unharvested, or $p(t)e^{-rt} < \lambda(t)$; that is, the stumpage price is too low (for instance, the forest may be too far from markets or the land quality too poor). Assuming interior solutions, and reducing the necessary conditions using the procedure discussed in section 12.2.1, we arrive at a singular path condition for optimal harvesting of old growth equal to $-\dot{p}(t)e^{-rt} + rp(t)e^{-rt} = A_V(.)e^{-rt}$. Simplifying, this becomes

$$r = \frac{\dot{p}(t)}{p(t)} + \frac{A_V(.)}{p(t)}. \tag{12.15}$$

This is a modified Hotelling's rule for old-growth forests. If amenities are absent, (12.15) reduces to simply Hotelling's rule. But the presence of amenities associated with old growth means that the instantaneous increase in the price of timber (first term on the RHS) is lower than it would be in the absence of amenities. In other words, old growth is depleted slower when these amenities are important. Another useful form of (12.15) is to write $rp(t) - A_V(.) = \dot{p}(t)$. The RHS of this equation is the marginal benefit of delaying harvest, given by the increase in the stumpage price for the next instant, while the LHS is the marginal cost of delaying harvest, given by the lost interest income for one instant (first term); this term is reduced by amenities that accrue to unharvested forest stock. Amenities therefore decrease the marginal cost of delaying harvest. Although we do not have space to discuss this solution further, we should point out that an explicit solution to the problem of depleting an old-growth forest is sometimes possible under certain linearity assumptions. Interested readers are referred to Brazee and Southgate (1992), Dasgupta and Heal (1980, pp. 153–156), and Farzin (1984).

In this model, the old-growth forest stock did not grow or decline over time. This does not need to be true, however. Suppose that the forest contains many tree species with differing ecological roles or stages within forest succession. The growth of any given species could be negative while it is positive for others, even though overall the forest stock does not change through time. This aspect is addressed in Conrad et al. (2005), who consider a multiple-species tropical forestry harvesting problem where the growth in various species can differ according to their biological relationships on the site.[8] Their focus is on a harvester who decides the type of species and the rate of harvesting for a tropical forest concession consisting of native forest. In particular, they examine incentives for selective harvesting, such as high-grading, by solving for the depletion rates of each species class and then determining how parameters and especially taxes affect the incentives to selectively log certain species over others. Their model is an application of dynamic Lagrange multiplier methods.

For our purposes, it is useful to convert the Lagrangian problem of Conrad et al. into a continuous-time–optimal-control problem for two species, and then examine the nature of the Hotelling depletion path. Let the two tree species be labeled as $(1, 2)$. The stock (volume) of forest of species i at time t is $V_i(t)$ $(i = 1, 2)$. Let the initial endowment of that species volume on a unit of land be $V_i(0)$. For any time $t > 0$, the growth function for species i depends on the stock of both species types, and is givens by $G_i[V_1(t), V_2(t)]$, where the derivative $G_{V_i}[V_1(t), V_2(t)]$ can be positive, zero, or negative. Thus, harvesting any species causes a change in growth for both species (this is similar to an uneven-aged two age class setup we discussed in chapter 8). The equation of motion defines the stock of each species at time t as a function of species-specific harvesting control and growth, or $\dot{V}_i(t) = G_i(.) - h_i(t)$. Harvesting costs also depend on all species; $C_i[V_1(t), V_2(t)]$ and $p_i(t)$

---

8. Amacher et al. (2001) allow for high- and low-valued species and examine dynamic royalty design in an optimal control model of forest concessions. Potts and Vincent (2008) in turn examine an optimal control-based species harvesting problem and find conditions under which rent-driven harvesting of a single species causes other species to become extinct because of changes in ecosystems caused by the distortions of selective harvesting. Finally, Maestad (2001) considers an optimal control model where a concessions harvester can harvest from two areas that differ in species quality (value) and harvesting costs within a fixed concessions period.

is the species-specific stumpage price. The harvester's problem is to maximize the discounted net rents over time:

$$R(h) = \int_0^{t_1} \sum_i [p_i(t)h_i(t) - C_i(.)]e^{-rt}\,dt, \tag{12.16}$$

subject to the equation of motion and the initial endowment for each species. The problem in (12.16) is linear in the two controls, $h_1(t)$ and $h_2(t)$, and there are two state variables, $V_1(t)$ and $V_2(t)$. This problem is bang-bang and produces a singular path. The end point is not critical because we are interested here only in the species-specific optimal harvesting paths. We therefore omit any salvage function at time t_1, implicitly defining the residual value of the forest after harvesting stops as zero (this has no bearing on the optimal path since it only affects the stock at the terminal time).

The Hamiltonian associated with (12.16) is $H = \sum_i \{[p(t)h_i(t) - C_i(.)]e^{-rt} + \lambda_i(t)[G_i(.) - h_i(t)]\}$. The necessary conditions for the two-control–two-state problem are similar to what we had earlier ($i = 1, 2$ species):

$$H_{h_i} = p_i(t)e^{-rt} - \lambda_i(t) \tag{12.17a}$$

$$-\dot{\lambda}_i(t) = H_{Vi} = -C_{Vi}(.)e^{-rt} + \lambda_i(t)G_{Vi}(.). \tag{12.17b}$$

There are two ways to gain intuition from the first-order conditions. First, we can substitute from (12.17a) into (12.17b) and obtain an equation explaining the path of the co-state variable over time, $\dot{\lambda}_i(t)/\lambda_i(t) = [C_{Vi}(.)/p(t)] - G_{Vi}(.)$. This shows how the shadow price of the forest stock for each species changes over time. It is easier to interpret the price path, however. Reducing (12.17a) and (12.17b) in the usual manner by solving (12.17a) for the co-state variable and taking its time derivative before substituting this derivative into (12.17b), we arrive at the following modified Hotelling rule for each species class i:

$$\frac{\dot{p}_i(t)}{p_i(t)} = r - \frac{C_{Vi}(.)}{p_i(t)} + G_{Vi}(.). \tag{12.18}$$

This has an interpretation similar to problems where there is a varying quality of the stock embodied in the cost function (e.g., see Slade, 1982). The forest case is modified for species-specific growth. Notice that the rate of depletion (or rate of price increase) on the LHS is higher for higher-valued species with lower harvesting costs. This lends support to the argument we asserted (but did not prove) in chapter 6 that higher-valued trees are indeed depleted from the tropical landscape faster. Faster harvesting of the highest-valued species is a rent-maximizing strategy for pulse harvesting of native forests.

Brazee and Mendelsohn (1990) propose another type of extension to the harvesting problem of old-growth forests. They mix old-growth and plantation stumpage in a problem in which private ownership of forests is a well-defined property rights regime. Plantation stumpage fills in behind a sequence of harvests where the oldest age class of trees is harvested in each period as trees grow into it. Brazee and Mendelsohn impose a specific forestland holding cost term on the harvesting problem in addition to planting and harvesting costs. This land rent is the opportunity cost of holding acres of unharvested forest. Plantation establishment happens over time, depending on the price and rents of land.

The control variables in their model are the area of the oldest trees to harvest and the acres that are planted. For tractability, the planting decision must be treated as separate from harvesting of the oldest age class each period by assuming that land either comes into or out of forest plantations according to a rent function, where land rent is constant and land supply is perfectly elastic; therefore land is always available at positive net rents. The following Hotelling result emerges from the optimal control problem of harvesting the oldest stumpage over time, $\dot{p}(t)V[a(t), t] + p(t)\dot{V}[t, a(t)](1 + \dot{a}) = rp(t)V[t, a(t)] + R$, where $a(t)$ is the age of the oldest harvestable stand at time t, $V(.)$ is volume harvested as a function of time and age of the oldest trees, and R is the land rent-based cost of holding unharvested forest stock. In this model, at each time t the oldest stand of age $a(t)$ is harvested, so the result here is similar to those derived earlier. The LHS is the marginal benefit from delaying harvest in terms of additional growth in price and volume, the latter depending on time and how the age structure of the forest changes. The RHS is the marginal cost of delaying harvest and equals lost interest from not harvesting (first term) plus an additional cost of land rent.

Alternatively, this equation can be rearranged to obtain a Hotelling-type result:

$$\frac{\dot{p}(t)}{p(t)} + \frac{\dot{V}[t, a(t)](1 + \dot{a})}{V[a(t), t]} = r + \frac{R}{p(t)}.$$

This shows that along the optimal harvesting path, harvesting occurs so that the instantaneous rate of change in the price of the oldest age class equals the interest rate plus the land rent term (weighted by the price of harvesting), less the instantaneous rate of change in volume of the trees in the oldest age class.

An important feature of modeling is evident in comparing Brazee and Mendelsohn's model with that of Conrad et al. (2005); these are similar if we interpret different species as age classes. In the latter tropical deforestation depletion problem, land rent is less meaningful because property rights could be so poorly defined that protection of plantations is too costly (i.e., site investment costs are too high). We showed this as an important feature of land use in chapter 6. In the purest open-access deforestation problem, represented easily by (12.12), harvesting of old-growth native forests is simply a short-term resource depletion problem of pulse harvesting and thus land rent is not viewed as a cost to the forest user. However, a land rent-based cost would clearly be important in any dynamic model where plantations are established and protected through enforced laws and well-defined property rights, or any tropical forest depletion problem where site investment costs are small enough to render sustainable forestry rents positive.

12.2.4 Land-Use Interpretations

We previously visited land-use problems using rotation models in chapters 2–3 and 6. There, land use was assumed to be chosen along a quality continuum in the first instant of time, and those land uses implicitly were assumed to persist unless some policy or parameter changed and instantly perturbed the market allocation through distortions in relative rents. One weakness of these static approaches is that they do not allow an examination of the transition to various states of land use that dynamic path results can reveal. It is therefore not surprising to find optimal control models concerning many land use-related topics.

One especially fruitful thrust of dynamic forest models has been to study how conversion of land from forests to alternative uses, or vice versa, happens over time. A special case of this problem is to solve for a path of deforestation using an equation of motion that explains the rate of land conversion over time when land exists in different uses (e.g. see Wirl, 1999, Hartwick et al. 2001, Kant 2000, and Barbier et al. 2005).[9] Another topic receiving increasing interest in dynamic models is carbon sequestration, where land or forest stocks set aside depend on emerging markets for carbon storage (e.g., see Cairns and Lasserre 2006, Stavins and Jaffee 1990, Stavins 1999, Sohngen and Mendelsohn 2003).

In this section we discuss an innovative model by Hartwick et al. (2001) that shows how land conversion occurs when some land is reforested as second-growth forests through time when prices and costs make this more profitable than crop production. An interesting steady state can result where a cycle of timber harvesting and allowing regrowth of converted areas occurs over some period, so that timbering and regrowth becomes a land use that is preferred to conversion of forests to agriculture.[10]

---

9. Kant (2000) considers the optimal choice of a resource management regime over time in village-based economies. Barbier et al. (2005) examine land-clearing incentives of a government and how they depend on political contributions from interested agents.

10. Before proceeding, we should mention that Wirl (1999) considers a problem similar to Hartwick et al. (2001) discussed below. Wirl models the dynamics of land use that can be divided among agricultural or forest production area. He develops an optimal control model that establishes the optimality (or nonoptimality) of the typical cycles between deforestation and eventual reforestation observed in many countries. The objective functional maximized includes welfare from forest harvesting, rents

We will consider a slightly simpler version of Hartwick et al.'s partial-equilibrium–optimal-control problem. We first consider a case where land clearing is irreversible; i.e., once converted to agriculture, land units are assumed to remain in that use forever, and so no regrowth of forests occurs on the cleared land. We normalize total land endowment to equal one. Let $R_f(L)$ and $R_a(1 - L)$ and $p_c R_c(t)$ be the rents for forest harvesting practiced on L acres, agricultural production practiced on $1 - L$ acres, and the revenues from conversion of land area $R_c(t)$ from forests to agriculture, respectively. The net return from conversion comes from selling trees for $p_c R_c(t)$ less clearing costs defined as $C[R_c(t)]$, where p_c is the (constant over time) stumpage price and $R_c(t)$ is interpreted as the forest-land clearing rate measured in tree volume. The change in available land for forest use is defined by an equation of motion, $\dot{L}(t) = -R_c(t)$. To ensure that land clearing will eventually stop in the steady state, we assume as in chapter 6 that the marginal rent functions for forest and agriculture cross only once.

The control problem is to choose a path for the land-clearing rate that maximizes the following objective functional defined as discounted net rents for forest and agriculture uses plus net conversion benefits:

$$V(R_c) = \int_0^\infty \{R_f(L) + R_a(1 - L) + p_c R_c(t) - C[R_c(t)]\}e^{-rt}\,dt, \qquad (12.19)$$

subject to an initial condition of the amount of forest land within the continuum, $L(0) = L_0 \in [0, 1]$ (the state), and that conversion is either zero or positive, $R_c(t) \geq 0$.

This time it proves useful to define the current-value Hamiltonian (that is, we do not include the discount factor of the objective function 12.19 in the Hamiltonian):

$$H^c = \{R_f(L) + R_a(1 - L) + p_c R_c(t) - C[R_c(t)]\} - \lambda(t)R_c(t). \qquad (12.20)$$

The necessary conditions for this Hamiltonian differ only for the state variable. Defining the co-state variable as $\lambda(t)$, these conditions are

$$H^c_{R_c} = p_c - C'[R_c(t)] - \lambda(t) = 0 \qquad (12.21a)$$

$$\dot{\lambda} - r\lambda = -H^c_L = -[R'_f(L) - R'_a(1 - L)], \qquad (12.21b)$$

---

from standing forests, and crop production net of land conversion costs and subject to equations of motion for forest stock and land devoted to developed-use production; he then evaluates the optimality of multiple steady states by checking the stability of these states.

along with transversality conditions ensuring that land conversion is either in transition and changing, or is equal to zero in a steady state, $\lim_{t\to\infty} e^{-rt}\lambda(t) \geq 0$ and $\lim_{t\to\infty} e^{-rt}\lambda(t)L(t) \geq 0$.

The optimal path is found from the necessary conditions by substituting the equation of motion $\dot{L} = -R_c(t)$ into (12.21a) for $R_c(t)$, and taking the time derivative of the resulting equation. Then substitute this time derivative for $\dot{\lambda}$ in (12.21b), which, after rearranging, gives the following path of land conversion over time to the steady state where conversion has stopped:

$$C''(-\dot{L})\ddot{L} = rp_c - rC'(-\dot{L}) - R_f'(L) - R_a'(1 - L). \tag{12.22}$$

Equation (12.22) essentially implies that the rate of change in land use over time depends on the marginal net rents of conversion, the marginal rents to forests on unconverted land, and the marginal rents to agriculture on converted land. The steady-state amount of land in forestry, $L(\infty) = L_\infty$, is the land present when conversion ceases (when $\dot{L} = 0$). This steady state is obtained by setting $\dot{\lambda} = 0$ and substituting (12.21a) into (12.21b), evaluating the resulting equation at $R_c(\infty) = 0$ because there is no conversion, and $L(\infty) = L_\infty$, to obtain $r[p_c - C'(0)] = R_f'(L_\infty) - R_a'(1 - L_\infty) = 0$. The LHS of this equation is the marginal cost in terms of interest lost on conversion profits by not converting land. The RHS is the marginal net benefit from not converting land measured in terms of the relative difference in rents between forestry and agriculture.

These equations allow one to detail three phases, or transitions, that conversion follows on the way to any steady state. Recalling that L changes as time proceeds with conversion from the equation of motion, these phases happen along a continuum as L increases from L_0 (no conversion of forest to agriculture) to L_∞. First there is a period where the marginal return from conversion, $p_c - C'[R_c(t)]$, becomes positive, so that it becomes profitable to convert forests to agricultural land in and of itself regardless of the subsequent agricultural returns. When this marginal conversion return is negative, conversion to agriculture is driven only by agricultural returns, so that money is lost by converting forests to agriculture, but the incentive of high agricultural rents still encourages conversion. The second phase of conversion happens when agricultural and sustainable forest marginal rents are equal, $R_f'(L) = R_a'(1 - L)$ and $\lambda(t) = p_c - C'[R_c(t)] > 0$, and remain positive. Now timber from conversion can be sold at a profit, while agricultural land is established on profitable land for crops, so that there are greater incentives to clear land than in the first phase. Finally, the third phase is consistent with sustainable harvesting of timber being a higher rent venture than agriculture on cleared land. Driven by high enough relative timber rents, clearing eventually stops and a steady state is reached that has some land along the continuum in agriculture and some sustainable forestry in perpetuity.

There is an interesting feature in land conversion transition and steady-state comparisons that can explain why at some point it could be more profitable to let timber regrow after clearing land, therefore beginning a cycle of regrowing and harvesting for some period of time, rather than pursuing agriculture on the cleared land unit forever. Hartwick et al. call this new land use "cyclical timbering." It could emerge where timber is harvested as if the land is being cleared, but then regrowth is allowed until the next harvest point in time, when the regrown timber is cleared again. This process could be repeated for a land unit over an interval of time until conversion to agriculture is started again. The cyclical timbering use may be a higher-rent land use than converting to agriculture. Whether it is depends on the relative scarcity of land used for timber production.

Hartwick et al. extend their model to consider as additional choices the area of forest regrowth and the length of time that conversion proceeds after the cyclical timbering period ends and conversion commences again. Once conversion starts again, there can still be some area in forest from the regrowth period. The problem is now defined for a period $N + M$ in place of T, where N is the regrowth period during which cyclical timbering occurs on some area of land (which is itself a choice), and M is the time interval of conversion once conversion begins after the period of length N ends. Thus the cyclical timbering and following conversion periods repeat each $N + M$ period in perpetuity, and this interval length does not change as long as the parameters of the problem remain stationary.

Two key equations emerge from the necessary conditions of the new Hamiltonian for this problem. The first is a conversion path for land that is augmented by cyclical timbering after conversion begins again at the end of N periods. This is similar to the optimal path shown in (12.22), except that now it depends on the point on the land continuum when cyclical timbering commences and the area on which this occurs. The most interesting equation is the choice of area for the regrowth and timbering phase of length N. This is given by

$$\{p_c - C'[R_c(N)]\}e^{-rN} = \frac{(1 - e^{-rN})}{r} R_a'(1 - L_C - \Delta^*),$$

where L_C is a critical value of the land area on the continuum where the period of cyclical timbering of time length N begins, and Δ^* is the optimal area of land for which cyclical timbering is practiced. It is interesting that this equation shows that the area in cyclical timbering is chosen to equate the marginal benefit of conversion (LHS) to the marginal loss in agricultural rents over the regrowth period, written in perpetuity to account for each future conversion period.

12.2.5 A Note on the Stability of Steady States in Dynamic Models

As we have seen in this section, infinite continuous time control theoretic problems of land conversion often produce steady states that are as meaningful to consider as the path that conversion takes to these steady states. When forest land is considered, along with a nonlinear forest growth function that determines forest stock growth on this land, any steady state solved by setting all time derivatives equal to zero in the necessary conditions of the control problem need not be unique. Thus, the stability of different steady states that satisfy the necessary conditions must be formally checked to ensure that the steady state is either stable or a saddle point. The usual way to evaluate steady state stability is familiar from other literature such as fisheries economics. First, the necessary conditions are reduced typically into a set of equations comprised of the equations of motion and any co-state variables. These equations are created by solving for the optimal control functions in the necessary conditions (as implicit functions of state and co-state variables) and then substituting these functions into the state and co-state equations. Given the assumption of forest growth, these resulting equations are nonlinear. Now, to evaluate the stability of the steady state, the nonlinear nature of the equations requires that we first evaluate the stability of the reduced system once it has been linearized by taking first-order Taylor series expansions of these equations evaluated at the steady state. Once this new linear system has been constructed, the eigenvalues of the resulting Jacobian matrix are determined. If the eigenvalues are alternating in sign and real, then the steady state is a stable saddle point. If both eigenvalues are real and nonpositive, then the steady state is stable. If the eigenvalues are complex numbers but their real parts of negative, then the steady state is also stable but characterized by a cyclical approach. For a formal introduction into checking the stability of steady states, the reader is referred to Kamien and Schwartz (1991, chapter 9).

12.3 Applications of Dynamic Programming

Dynamic programming allows an analysis of exceedingly complex problems that may include varying species and age classes, as well as uncertainty in multiple parameters. In this section we illustrate this approach by returning to the deforestation model of Conrad et al. (2005) by rewriting and examining the problem in discrete time. We then discuss uneven-aged management models and briefly consider uncertainty.

12.3.1 Perfect Foresight Interpretations

We begin by solving our basic old-growth forest and amenity problem from section 12.2, but modify it for growth in the forest stock following Conrad et al. (2005). This problem in discrete-time–continuous-control form is

$$S[V(0)] = \max_{h(t)} \left\{ \sum_{t=0}^{\infty} \beta^t [\hat{p}(t)h(t) + A[V(t)]] \right\} \tag{12.23}$$

$$\hat{p}(t) = p(t) - c, \quad \text{s.t.}$$

$$V(t+1) = V(t) + G[V(t)] - h(t) \tag{12.24}$$

$$V(0) = V_0,$$

where $\beta = 1/(1+r)$ is now the discount factor. The second equation is the old equation of motion $\dot{V}(t) = -h(t)$ but written now in discrete time, and c is a constant harvesting cost. The term $A[V(t)]$ continues to denote amenity benefits that accrue according to the unharvested forest stock in each time period. The value function $S[V(0)]$ designates the optimal value of discounted net benefits at time zero when the beginning state is $V(0)$. The control variable continues to be volume harvested at time t, and the state variable continues to be volume of forest stock present in situ at time t. The second-order conditions hold as long as $S[V(0)]$ is differentiable and concave in the control.

The Bellman equation for the problem in (12.23)–(12.24) is

$$S[V(t)] = \max_{h(t)} \{ \hat{p}(t)h(t) + A[V(t)] + \beta S[V(t+1)] \}. \tag{12.25}$$

This condition states that at any time t, the optimal harvest strategy is one that maximizes current net rents (first two terms) plus discounted maximum net rents given by (12.23) starting at time $t+1$ onward and given by the value function at time $t+1$, $S[V(t+1)]$. The principle of optimality implies that maximizing this equation at any time t fully characterizes the optimal control path. This optimal path, also called the optimal "policy function" consistent with (12.25) and can be solved numerically using either recursive (backward) or forward solution methods.

The necessary condition for (12.25), applying envelope results to the third term, is written as $\hat{p}(t) + \beta S_{V(t+1)}(.)[\partial V(t+1)/\partial h(t)] = (1+r)\hat{p}(t) + S_{V(t+1)}(.)(-1)$, where the equation of motion was used to evaluate the derivative of the forest stock in the next period. This suggests that harvesting is optimal at time t if it equates the marginal cost of delaying harvesting, in terms of lost interest, to the marginal benefit from delaying harvest, expressed as the value of the objective function from $(t+1)$ onward. Included in this is the path of amenities. A better way of interpreting this is to do some algebra using the equation of motion and the Bellman equation. First, we rewrite (12.25):

$$S[V(t)] = \max_{h(t)} \{ \hat{p}(t)h(t) + A[V(t)] + \beta[\hat{p}(t+1)h(t+1) + A[V(t+1)]] \}. \tag{12.26}$$

Now, from the equation of motion one period ahead, $h(t+1) = V(t+1) + G[V(t+1)] - V(t+2)$, and $V(t+1) = V(t) + G[V(t)] - h(t)$. Substituting these expressions into the third term in brackets on the RHS for $h(t+1)$ and $V(t+1)$ yields a modified Bellman equation where all terms are written as a function of the control in the current period t, $h(t)$. Taking derivatives of this modified expression with respect to $h(t)$ gives the necessary condition for the optimal control path over the two time periods:

$$\frac{S[V(t)]}{\partial h(t)} = \hat{p}(t) + \beta\hat{p}(t+1)[-1 - G'(.)] - \beta A'(.) = 0. \tag{12.27}$$

Rearranging (12.27) gives a better interpretation of the optimal harvesting rule, $\hat{p}(t)(1+r) = \hat{p}(t+1)[1 + G'(.)] + A'(.)$. The LHS of this rule is the marginal cost of delaying harvesting in terms of lost interest. The RHS is the marginal benefit of delaying harvest in terms of additional growth in the value of the forest stock if it is harvested plus additional amenities that accrue from not harvesting for one more period. We can rearrange this further to obtain a Hotelling's rule-type result:

$$\frac{\hat{p}(t+1)[1 + G'(.)] - \hat{p}(t)}{\hat{p}(t)} = r - \frac{A'(.)}{\hat{p}(t)}. \tag{12.28}$$

Equation (12.28) is interesting. Harvesting proceeds so that the growth in timber value (LHS) equals the interest rate modified for the rate of increase in in situ amenity values (adjusted by the current timber price). This is simply a discrete version of (12.15) with a constant cost that was solved using optimal control. In fact, if we assume that old-growth forests do not grow, so that $G'(.) = 0$, then we have a discrete-time version identical to (12.15), or[11]

$$\frac{\hat{p}(t+1) - \hat{p}(t)}{\hat{p}(t)} = r - \frac{A'(.)}{\hat{p}(t)}.$$

The basic problem discussed here has shown up in the literature in more complicated forms, usually by assuming there is more than one age class, such as in Lyon (1981) and Haight (1985). Lyon (1981) solves a dynamic programming problem using the method of Lagrange multipliers for a social planner where there is a collection of even-aged forest stands of different ages in the economy. Harvesting

---

11. Another way of solving this problem is to rewrite the Bellman equation in terms of the state variable in the next period, $V(t+1)$, instead of the period t control variable $h(t)$. This comes from first writing (12.26) in the following form: $S[V(t)] = \max_{h(t)}\{\hat{p}(t)[V(t) + G[V(t)] - V(t+1)] + A[V(t)] + \beta S[V(t+1)]\}$. This is sometimes called the functional form of the Bellman equation. Substitution again using the equation of motion and then deriving the necessary condition with respect to $V(t+1)$ allows us to solve for an optimal path of the state variable over time. This solution mirrors the one in (12.28).

depletes these stands whereas regeneration adds to them in a sequence of harvesting over time. His problem is similar to (12.23)–(12.24) but adds equations of motion for regeneration and harvesting for each age class stand. Instead of linear harvest revenue, the term $\hat{p}(t)h(t)$ is replaced by a consumer surplus obtained from each period's timber demand function. The control variable is the acreage of various-aged trees that are harvested from an initial starting endowment. Assuming as others that the oldest stand is harvested each time period, he shows in a mining subcase, as he calls it, where growth is zero, that optimal harvesting satisfies: $\{d[1, Q(1)] - d[0, Q(0)]\}/d[0, Q(0)] = r + R$. The numerator is the instantaneous rate of price increase, written in terms of the inverse demand function as a function of the quantity of timber consumed in periods 0 and 1, given by $Q(j)$. This is identical to the LHS of (12.28) with $G'(.) = 0$. The RHS includes the interest rate, as (12.28), but now there is a new term equal to $R = (V_{22} - V_{21})/d[0, Q(0)]q^*$ that is specific to the multiple-stand acreage problem he studies. In R, V_{ij} is the value function at time j for trees of age i, and so the numerator is the difference in the value functions from harvesting in the current period and regenerating the stand, minus the value of continuing and delaying harvesting now, thereby postponing regeneration one period. The denominator is the revenue from harvesting now. This extra term appears in his optimal path because he allows for different age classes of even-aged tree stands not present in (12.26); these stands are mined as the oldest is cut and regenerated each harvest time.

As Lyon puts it, if society forgoes harvesting mature old-growth forest in any period, it must be compensated with higher-priced growth, because waiting to harvest means giving up higher current consumption but also losing the opportunity to reforest 1 year earlier. Society is compensated for forgone consumption through the interest rate term, while R measures the compensation in the price path needed because of the delayed regeneration that could take place if the mature stand were harvested. Haight (1985) considers uneven-aged stand management, solving a Bellman equation where the control in each period is how much volume of forest stock in each diameter class is left after harvesting in that class is completed. He shows that regardless of an initial condition concerning diameter classes in the starting stand, the regime solved for with dynamic programming does not converge to a steady state that is equivalent to the maximum Faustmann value of the stand. Haight (1987) goes further still to consider transitions between even-aged and uneven-aged stands and contrasts the dynamic programming approach to Faustmann stand management approaches.

Often it makes sense to write the problem so that harvesting is a discrete action undertaken once during a rotation, such as with clear-cutting. This forms an optimal stopping problem we first discussed in chapter 11. A land sale at the end of the rotation could also be assumed, effectively creating an infinite time-horizon problem given that this sale could be valued as the discounted benefits from forestry practiced in perpetuity from the harvesting time onward if land markets are perfect. For this problem, revenue from harvesting is rewritten as a function of the binary control variable $h(t) = 0, 1$ in the following manner:

$$R(t) = \begin{cases} 0 & \text{if } h(t) = 0 \\ \hat{p}(t)h(t) & \text{if } h(t) = 1. \end{cases}$$

Now, $h(t) = 1$ implies that harvesting is undertaken at time t. Appending this to the finite time-horizon problem in (12.23) and including the land sale at time N, the objective function is $S^d[V(0)] = \max_{h(t)}\{\beta^t[R(t) + A[V(t)]] + \beta^N L(N)\}$. This is then subject to a slightly different equation of motion and boundary condition compared with (12.24); in the discrete-choice problem, we write $V(t+1) = V(t) + G[V(t)]$ if $h(t) = 0$ and $V(t+1) = 0$ if $h(t) = 1$, assuming that harvesting is conducted only once before the land is sold. The initial forest stock remains as $V(0) = V_0$.

The Bellman equation for this new problem is different than (12.26) to reflect the discrete decision of whether to harvest or continue a rotation at any two-point interval of time $(t, t+1)$. To see how, define the stopping period (harvest time) as $\tilde{T}$, and assume that harvesting and the land sale are both undertaken at the beginning of the period. Then the condition describing the optimal harvest choice is simply $h(\tilde{T}) = 1$ if $\{\hat{p}(\tilde{T})h(\tilde{T}) + L(\tilde{T}) > S^d[V(\tilde{T}+1)]\}$. In this problem, there is no path to solve for the control; instead, we solve for either a stopping time $\tilde{T}$, or an optimal ending stock $V(\tilde{T})$ where revenues from harvesting plus the land sale are greater than the discounted benefit from continuing the rotation and making decisions optimally from the next time period onward. The value of continuing is simply the value function evaluated at $\tilde{T}+1$, or $S^d[V(\tilde{T}+1)]$. Thus the new Bellman equation emerges from the discrete-choice problem equal to

$$S^d[V(t)] = \max\{\hat{p}(t)h(t) + L(t), 0 + S^d[V(t+1)]\}. \tag{12.29}$$

In most cases that incorporate realistic assumptions concerning forest stocking and markets, numerical analysis must be used to solve them.

Other "mixed" dynamic programming problems also exist, with a once-per-rotation discrete decision made at a stopping time, but another decision made during each period. In this case, we solve jointly for an optimal control path and a stopping time. An example is for a landowner to choose a final discrete harvest, but also to undertake stand improvements at regular intervals from planting to harvest time. Spring et al. (2005a,b), for example, consider dynamic programming models of harvesting decisions and annual reductions in wood fuel that reduce the risk of fire (the harvesting decision can be zero or positive at any point in time).

12.3.2 A Note on Stochastic Interpretations

Dynamic programming is commonly used to model uncertainty because it allows straightforward analysis of complex stochastic processes. Stochastic dynamic programming problems are typically set up so that all information is known with certainty in the current period, but then the next period's value function is written in expected value form. Closed-loop decisions are possible, in that the decision maker can update the path of the control (i.e., reoptimize) as new information is received.

Brazee and Mendelsohn (1988) used this procedure to determine the optimal reservation price for harvesting an even-aged stand using a stopping model (Lohmander 1987 also developed a similar idea as well). They assumed a landowner must harvest before some maximum tree age, defining a terminal time. If harvesting occurs, the landowner sells the land at its bare land value, which is really the net present value of continuing from that point onward, as we discussed earlier. In each period, the landowner knows the current price and observes a distribution of prices in the next period. He harvests if a price drawn from this distribution is greater than a reservation price that's solved endogenously to the problem. Continuing a rotation has costs associated with interest lost on the harvest and land sale, while the benefits of waiting include drawing a potentially high price from the next period's price distribution. Price draws are independent over time, leading to a Markov stochastic dynamic programming stopping problem with a one-time discrete harvest choice. Since the landowner defines a reservation price in each period, a path of reservation prices is solved for over time up to the stopping point.

The Bellman equation that is solved recursively for this path is made simple by the fact that there is a known end-point where the landowner must harvest if that point is reached. At the assumed very last two periods up to the end point T, the Bellman equation equals $p^r(T)V(T-1) + E(L) = \beta[Ep(T)V(T) + E(L)]$, where E is expectations taken over the next period's price distribution and $p^r(T)$ is the reservation price. In this equation, the reservation price $p^r(t)$ equates the marginal benefit of harvesting in the current period (LHS) with the marginal benefit of waiting one period and facing the unknown distribution (RHS). For time periods before $T-1$, the Bellman equation is more complex, but the Markov nature of the problem allows the RHS to be written as the expected value of harvesting in all future periods beyond time t up to the end point period, T. The reservation price defines when in future periods harvesting occurs or does not occur. The Bellman equation is solved for the reservation price path and the land sale term L simultaneously because L should be the discounted expected value of continuing the price search process from whatever time period harvesting occurs onward in perpetuity. Simulations of the reservation price path and the land sale have shown that this dynamic programming harvesting solution outperforms the Faustmann rotation age.

Several generalizations of the dynamic programming problem above have appeared in the literature, such as those of Gong (1998, 1999), who apply adaptive planning and consider reservation price strategies under autocorrelated prices as well as harvester risk aversion, Forosbeh et al. (1996), who consider multiple products at harvest, and Gong et al. (2005), who consider amenities. Furthermore, Gong and Lofgren (2007) show how switching between a Faustmann rule and a reservation price strategy affects timber supplies and market prices, while Gong and Lofgren (2003a) develop a model to show that landowners adopting reservation price strategies tend to increase the short-run price elasticity of timber supply and dampen the effects of any demand shocks on price variation. Finally, another application of stochastic dynamic programming that links to the land-use models we discussed in the last section concerns the spatial modeling of irreversible development, such as the conversion of tropical forests to agriculture as discussed in chapter 6. For example, Albers (1996) studies a problem where the decision maker is uncertain about the relative values of future land uses. An "inflexible" open loop dynamic programming problem is constructed where all expectations of future unknown land values are taken using only the information available in the first period. Land units are linked together over space according to site units that can be developed or preserved. This type of problem is contrasted with a closed loop dynamic programming problem where future information is used to update decisions over time. In these models, there is a quasi option value of preservation given that this land use allows long-run flexibility. Finally, Costello and Kaffine (2008) examine concessions design in a dynamic programming framework. They solve for optimal harvest paths, where long-term continuation of the contract depends on harvesting methods, finding that first-best harvesting can be achieved if there is uncertainty about contract renewal.

12.4 Summary

The traditional rotation framework allows a simple solution to rotation age problems and makes the study of policies and uncertainty easy. For many problems, however, the static nature of rotation models is not enough. Dynamic analysis is called for in a variety of cases, such as transitional land-use problems, multiple forest age classes, and management of uneven-aged stands; here, dynamic optimization can provide interesting and new insights over the static approach.

The idea of harvesting once per rotation can be introduced into optimal control models under stationarity assumptions similar to constant parameter assumptions in rotation models, or in dynamic programming problems (without such constant parameter assumptions) using optimal stopping theories, albeit with added complexity. The interest in dynamic optimization is usually to solve for the optimal path of continuous harvesting over time that satisfies some target of the decision maker or to observe various steady states that might energe. Dynamic programming stopping models are closer in spirit to the idea of rotation models, since one solves for a discrete expected rotation age representing discontinuous changes to the forest stock (such changes are still possible with optimal control). The dimension of dynamic models becomes less tractable, however, as uncertainty and additional constraints are included in the problem, rendering numerical analyses important to solving them.

The types of dynamic models, the structure of decision making, and the resulting solutions differ considerably. Dynamic programming solutions often are described with partial differential equations that may not have analytical closed-form solutions. Optimal control problems are generally more straightforward because their necessary conditions usually reduce to ordinary differential equations. However, as we have seen in chapter 11, uncertainty is difficult to handle in a continuous time control framework and requires Ito calculus.

Another issue in comparing models is how the nature of decision making varies. Rotation models are invariably "open loop" problems in that all decisions are made at the beginning of the planning horizon. Constancy and known parameters make these solutions globally optimal, and even the catastrophic loss problems in chapter 10 have this distinction. In contrast, for many types of uncertainty problems, catastrophic loss included, dynamic programming allows a closed-loop approach as information is updated over time. Strictly speaking, the Faustmann and Hartman problems could be reoptimized at each point in time as new information is received. This is trivial under certainty and constant parameters, but would indeed become especially arduous if done under uncertainty. Dynamic programming, on the other hand, allows a slick solution to a problem like this, owing to the principle of optimality and the Bellman equation.

Appendix: Mathematics Review

In this appendix we briefly review the key mathematical concepts and methods used in this book. Most of these appear first in chapters 2–5. For each section of this appendix, we have indicated in parentheses where in the book each concept is first needed. We begin with a discussion of unconstrained and constrained optimization problems, including methods of comparative statics. These are followed by a discussion of the envelope theorem, the derivation of Slutsky equations, and corner solutions. We assume that readers are familiar with partial differentiation, exponential and logarithmic differentiation, the product rule, chain rule, and the quotient rule from multivariable calculus. Readers wishing to delve into more details about basic mathematics concepts for economics study are referred to Silberberg and Suen (2001, chapters 2–3) and Simon and Blume (1994, chapters 2–5).

A1 Unconstrained Optimum: Single Decision Variable (Chapter 2)

Suppose that we have the following objective function for a single (endogenous) choice variable x, and an exogenous (nondecision) variable, θ:

$$\max_x V(x; \theta). \tag{A1.1}$$

Restricting attention to interior solutions, the first-order condition is obtained by differentiating (A1.1) with respect to the decision variable x and setting the resulting derivative equal to zero,

$$V_x(x; \theta) = 0. \tag{A1.2}$$

Suppose that $V_x(x; \theta) = 0$ is an implicit function and does not yield a closed-form explicit expression for x as a function of θ. Assume that the second-order condition holds, i.e., $V_{xx}(x; \theta) < 0$, and define a unique optimal solution to (A1.2) as $x \equiv x(\theta)$ [this is unique if the function $V(x; \theta)$ is strictly concave in x]. To determine how the choice x depends on θ, we substitute $x \equiv x(\theta)$ into the first-order condition, which makes (A1.2) an identity and allows the use of the implicit function theorem. Total differentiation of this identity then yields:

$$V_{xx}(x; \theta)\, dx + V_{x\theta}(x; \theta)\, d\theta = 0. \tag{A1.3}$$

Notice that we can now solve (A1.3) to arrive at the following expression for the effect of θ on the decision variable x:

$$\frac{dx}{d\theta} = -\frac{V_{x\theta}(x; \theta)}{V_{xx}(x; \theta)}. \tag{A1.4}$$

Because the denominator is negative owing to the second-order condition, the sign of the comparative statics result depends on the cross-partial derivative of the objective function $V(.)$. This type of procedure is used throughout this book when there is the only one choice variable for the landowner, such as rotation age.

A2   Unconstrained Optimum: Two Decision Variables (Chapter 2)

In many cases, the landowner makes more than one choice, such as rotation age and silvicultural effort. Suppose we have the following unconstrained maximization problem with two choice variables, $V = \max_{x,y} V(x, y; \theta)$, where x and y are the endogenous decision variables and θ continues to denote an exogenous variable. Assuming interior solutions for x and y, the first-order conditions provide an equation system of the following first partial derivatives of V:

$$V_x = \frac{\partial V(.)}{\partial x} = V_x(x, y; \theta) = 0$$

and   (A2.1)

$$V_y = \frac{\partial V(.)}{\partial y} = V_y(x, y; \theta) = 0.$$

The second-order conditions for this problem require the matrix of second derivatives of the system of first-order conditions in (A2.1) to be negative semidefinite. This requires that $V_{xx} < 0$, $V_{yy} < 0$, and

$$j = \begin{vmatrix} V_{xx} & V_{xy} \\ V_{yx} & V_{yy} \end{vmatrix} = V_{xx}V_{yy} - V_{xy}V_{yx} = V_{xx}V_{yy} - V_{xy}^2 > 0. \tag{A2.2}$$

The matrix of second partial derivatives in (A2.2) is known as the Jacobian matrix, j. To obtain the comparative statics of the choices x and y, we now assume that the second-order conditions hold. Substituting the optimal levels of x and y into the first-order conditions makes the conditions identities, and once again we can apply the implicit function theorem as we did earlier to investigate the effects of exogenous variables on the optimal choices. Now, though, we must use a matrix form. Totally differentiating each first-order condition in (A2.1) yields

$$V_{xx}(x, y; \theta)\, dx + V_{xy}(x, y; \theta)\, dy + V_{x\theta}(x, y; \theta)\, d\theta = 0$$

and   (A2.3)

$$V_{yx}(x, y; \theta)\, dx + V_{yy}(x, y; \theta)\, dy + V_{y\theta}(x, y; \theta)\, d\theta = 0.$$

Writing (A2.3) in matrix form gives

$$\begin{bmatrix} V_{xx} & V_{xy} \\ V_{yx} & V_{yy} \end{bmatrix} \begin{pmatrix} dx \\ dy \end{pmatrix} + \begin{pmatrix} V_{x\theta} \\ V_{y\theta} \end{pmatrix} d\theta = 0.$$ (A2.4)

As long as the determinant of second derivatives j is nonzero, we can solve (A2.4) for the comparative statics of the system using Cramer's rule, and using the definition of j from (A2.2),

$$\frac{dx}{d\theta} = -\frac{\begin{vmatrix} V_{x\theta} & V_{xy} \\ V_{y\theta} & V_{yy} \end{vmatrix}}{j} = -\frac{V_{x\theta}V_{yy} - V_{y\theta}V_{xy}}{j}$$ (A2.5a)

$$\frac{dx}{d\theta} = -\frac{\begin{vmatrix} V_{xx} & V_{x\theta} \\ V_{yx} & V_{y\theta} \end{vmatrix}}{j} = -\frac{V_{xx}V_{y\theta} - V_{yx}V_{x\theta}}{j}.$$ (A2.5b)

This approach can be generalized for any system of n choices and m exogenous parameters. For n decision variables, the Jacobian matrix is an $n \times n$ matrix.

A3   Constrained Optimization: Two Decision Variables (Chapter 4)

The methods outlined in section A2 translate quite easily to a problem in which the decision maker faces a constraint on his decisions. We can most easily present such a case for two decision variables, using an example from consumer utility maximization. The methods we discuss here apply to more than two choice variables and more than one constraint in a similar fashion.

Suppose that an individual consumer maximizes his utility by choosing consumption for two commodities subject to his budget constraint. Assume that the consumer's utility function is given by $u(c_1, c_2)$ and his budget constraint by $\sum_{i=1}^{2} p_i c_i = M$, where c_i is the amount of good $i = 1, 2$ consumed (the decision variable), p_i is the price of good i, and M is the consumer's monetary income. In this problem, prices and income are the exogenous variables that determine the decisions. One of the easiest and most straightforward ways of solving this problem is to substitute the constraint into the utility function, which then becomes a function of only one choice. For example, solving for c_2 from the budget constraint gives $c_2 = -(p_1/p_2)c_1 + (M/p_2)$, and then substituting this into the utility function renders utility only a function of c_1, $u[c_1, c_2(c_1)]$. Now we take derivatives of this utility function to define the first-order condition with respect to c_1:

$$\frac{du}{dc_1} = \frac{\partial u}{\partial c_1} + \frac{\partial u}{\partial c_2}\frac{dc_2}{dc_1} = u_1 - u_2\left(\frac{p_1}{p_2}\right) = 0. \qquad \text{(A3.1)}$$

By rearranging, one obtains the familiar condition requiring that the marginal rate of substitution between goods 1 and 2 equal their price ratio, $u_1(.)/u_2(.) = p_1/p_2$. Hence (A3.1), along with the budget constraint, defines the demand functions for both goods in terms of prices and income, $c_i = c_i(p_1, p_2, M)$.

A more useful solution method, and one that must be used for many constraints, is to define a Lagrangian function as follows: $L = u(c_1, c_2) + \lambda(M - p_1c_1 - p_2c_2)$, where $\lambda > 0$ is a multiplier associated with the budget constraint: the Lagrangian function transforms the constrained maximization problem to a free optimization problem of three variables. Now, the Lagrange multiplier is endogenous and must be solved along with the two consumption decisions. The economic interpretation of the multiplier is as the marginal utility that follows from an increase in income. Differentiating the Lagrangian with respect to the two consumption choices and the multiplier yields the following first-order conditions:

$$L_1 = u_1(c_1, c_2) - \lambda p_1 = 0 \qquad \text{(A3.2)}$$

$$L_2 = u_2(c_1, c_2) - \lambda p_2 = 0 \qquad \text{(A3.3)}$$

$$L_\lambda = M - p_1c_1 - p_2c_2 = 0, \qquad \text{(A3.4)}$$

where subscripts 1 and 2 indicate derivatives of the Lagrangian with respect to c_1 and c_2, and L_λ is the derivative with respect to the multiplier for the constraint. These conditions must be simultaneously solved for the three endogenous variables (c_1, c_2, λ). The easiest way to proceed is to eliminate the multiplier by dividing both equations at an interior solution to obtain the same optimality condition as above: $u_1(c_1, c_2)/u_2(c_1, c_2) = p_1/p_2$.

The comparative statics of the choices proceed in a similar way by using Cramer's rule applied to the second derivatives of (A3.2)–(A3.4). However, for a constrained optimum, we now must define a Hessian matrix instead of the Jacobian matrix. The Hessian matrix is simply a Jacobian matrix "bordered" by derivatives of the constraint with respect to the decision variables (Silberberg and Suen 2001, ch. 6, Simon and Blume 1994, ch. 18). This Hessian is also used to define the second-order conditions for the problem. Second-order conditions require that the Hessian matrix be negative semidefinite, which holds if the Lagrangian function has negative second derivatives with respect to each decision variable, c_1 and c_2, and if the Hessian matrix has a positive determinant. These are given by

$$L_{11} = u_{11}, \quad L_{22} = u_{22} < 0 \tag{A3.5}$$

and

$$H = \begin{vmatrix} u_{11} & u_{12} & -p_1 \\ u_{21} & u_{22} & -p_2 \\ -p_1 & -p_2 & 0 \end{vmatrix} = 2p_1 p_2 u_{12} - p_1^2 u_{22} - p_2^2 u_{11} > 0, \tag{A3.6}$$

where H is the bordered Hessian matrix; this is the matrix of second derivatives bordered by first derivatives of the constraint with respect to the decision variables (given in the last row and last column of H). Provided the second-order conditions hold, the first-order conditions can be used to implicitly define the endogenous variables as a function of exogenous parameters given by prices for both goods and income. The comparative statics then proceed using Cramer's rule as shown in (A2.5a)–(A2.5b), but the numerator equivalent now is derived using H, and substituting derivatives of (A3.2)–(A3.4) into H for the appropriate decision variable. The Jacobian matrix in (A2.5a)–(A2.5b) is replaced by the determinant in (A3.6).

A4   Envelope Theorem (Chapters 4 and 5)

The envelope theorem is an important analytical tool for economic analysis and comes up many times in this book. The theorem itself helps us to define how the maximum value of an objective function, defined by substituting the optimal choices into the objective function, depends on parameters such as policy instruments (in chapter 5, this indirect function is defined as maximum landowner rent or welfare). Sticking with our earlier notation, suppose we have the following objective function:

$$\max_x V(x; \theta), \tag{A4.1}$$

where we now reinterpret x as a vector of choices and θ as a vector of exogenous parameters. The optimal solution is given by a set of implicit equations defining

$x^* \equiv x(\theta)$, where a closed-form solution may or may not exist. Assuming that the second-order conditions hold, substituting these optimal solutions into (A4.1), we define the following maximized value of the objective function:

$$V^* \equiv V[x(\theta); \theta]. \tag{A4.2}$$

Equation (A4.2) defines an indirect objective function as is often defined in economics problems. It gives the maximum value of the objective function (A4.1) as only a function of exogenous variables once the vector of choices x is set equal to its optimal value.

Consider now how the maximum objective function depends on the level of the exogenous parameter θ. Totally differentiating (A4.2), we obtain,

$$dV^* = \frac{\partial V[x(\theta); \theta]}{\partial x} \frac{dx}{d\theta} + \frac{\partial V[x(\theta); \theta]}{\partial \theta} = \frac{\partial V[x(\theta); \theta]}{\partial \theta}. \tag{A4.3}$$

Notice that the first-order condition (A1.2) implies that the derivative $\partial V[x(\theta); \theta]/\partial x$ equals zero. Thus, the effect of θ on the indirect objective function is due only to the direct effect given by the second term of (A4.3). The first-order conditions ensure that indirect effects of the parameter through changes in the decisions, already chosen optimally, are zero and can be ignored at the optimum. Thus the effect of an exogenous parameter on the maximized objective function is due only to its direct effect.

Envelope results also apply to constrained optimization problems as well. Consider the following modification of (A4.1):

$$V(x, \theta) = \max_x V(x; \theta) \quad \text{s.t. } g(x; \theta) = 0, \tag{A4.4}$$

where $g(x; \theta) = 0$ is a compact constraint set that may be an implicit function of both the decisions and the exogenous parameters. Defining a vector of multipliers associated with the constraints by λ, and assuming that the second-order conditions for an interior solution hold for all decision variables, we can write the indirect objective function as before, V^*. Totally differentiating this and again using the first-order condition for (A4.4), $[V_x(x; \theta) + \lambda g_x(x; \theta) = 0]$, we have

$$dV^* \equiv \frac{\partial V[x(\theta); \theta]}{\partial \theta} + \lambda \frac{\partial g(x; \theta)}{\partial \theta}. \tag{A4.5}$$

So again, the direct effect determines the result. Equation (A4.5) is useful in chapter 5 when studying optimal instrument design, where the government chooses a set of tax instruments θ conditional on the response of landowners (optimally choosing x) to the instrument $x(\theta)$ and subject to a revenue constraint that depends on landowner responses as well as the level of the tax, $g[x(\theta), \theta]$.

A5 Slutsky Equations (Chapter 5)

Slutsky equations are useful theoretical constructs to more formally describe comparative statics effects by decomposing the effects into income and substitution effects. They are used in this book to consider how tax instruments affect landowner decisions since any forest tax has both price and income components.

The comparative statics methods discussed in sections A1–A3 of this appendix define the total effects of an exogenous parameter on decision variables, ceteris paribus. This type of result is useful in many places of this book. However, in some chapters, particularly those dealing with the choice of forest policies and how these choices affect social welfare, it is important to decompose these total effects into their substitution and income effects. Slutsky equations must be used.

We return to our consumer theory example of constrained optimization presented in section A.3. Recall that we used the consumer's first-order condition to solve for the optimal consumption choices $c_i^*(p_1, p_2, M)$. If these optimal choices are substituted into the utility function, we arrive at an indirect utility function $V(p_1, p_2, M) = u[c_1^*(p_1, p_2, M), c_2^*(p_1, p_2, M)]$. Given that $V_M > 0$, we can invert this function to obtain the income a consumer needs to reach a given utility level using their optimal consumption decisions, u^0; i.e., $M = h(p_1, p_2, u^0)$. Substituting this for M in the indirect utility function yields a "compensated" indirect utility function, $V[p_1, p_2, h(p_1, p_2, u^0)] = u^0$, where income is used to keep the utility level of the consumer constant at u^0 as prices change. Differentiating this compensated indirect utility function with respect to p_1 gives a direct and indirect compensation effect, $V_{p_1} + h_{p_1} V_M = 0$, where the chain rule for differentiation has been used with the second term. Rearranging this differential gives

$$h_{p_1} = -\frac{V_{p_1}}{V_M} > 0. \tag{A5.1}$$

This simply implies that a higher price requires higher nominal income to keep utility constant. Substituting $M = h(p_1, p_2, u^0)$ into the uncompensated utility function and then solving for the optimal consumption decisions gives $c_1^*[p_1, p_2, h(p_1, p_2, u^0)]$.

According to the duality theorem of economics, maximizing utility and minimizing consumption expenditure for a given utility implies that the uncompensated consumption choice, $c_1^*(.)$, is equal to the compensated consumption choice, $c_1^{*s}(.)$, so that $c_1^*[p_1, p_2, h(p_1, p_2, u^0)] = c_1^{*s}(p_1, p_2, u^0)$. Finally, differentiating this with respect to p_1 gives $(\partial c_1^*/\partial p_1) + h_{p_1}(\partial c_1^*/\partial M) = \partial c_1^{*s}/\partial p_1$. Using (A5.1) for the second term on the LHS, and rearranging, we have the definition of the Slutsky equation:

$$\frac{\partial c_1^*}{\partial p_1} = \frac{\partial c_1^{*s}}{\partial p_1} - c_1^* \frac{\partial c_1^*}{\partial M}. \tag{A5.2}$$

This equation decomposes the total effect of price on the optimal consumption decision into an income effect $[-c_1^*(\partial c_1^*/\partial M)]$ and a substitution effect $\partial c_1^{*s}/\partial p_1$. This procedure applies to any utility maximization problem with n decisions (Silberberg and Suen 2001, chapter 6; Simon and Blume 1994, chapter 22; and Varian 1992, chapter 7).

A6 Corner versus Interior Solutions (Chapter 5)

Suppose we again have the following maximization problem with one choice variable, $\max_x V(x; \theta)$ s.t. $x \geq 0$, where x is the decision (endogenous) variable and θ is an exogenous parameter but now $x > 0$ and $x = 0$ are both relevant and possible solutions. The Lagrangian function for this problem is $L = \max_x V(x; \theta) + \lambda x$, $\lambda \geq 0$, and the first-order condition is $V_x(x; \theta) + \lambda x = 0$ and $\lambda x = 0$. If there is an interior solution $x > 0$, then $\lambda = 0$ and $V_x(x; \theta) = 0$. At the corner solution for x, $x = 0$, we must have $\lambda > 0$ and therefore $V_x(x; \theta) < 0$.

Now suppose there are two choices, so that the maximization problem is then $\max_{x,y} V(x, y; \theta)$ s.t. $x, y \geq 0$. Now the first-order conditions are $V_x(x, y; \theta) + \lambda = 0$ and $V_y(x, y; \theta) + \gamma = 0$, and $\lambda x, \gamma y = 0$ where λ and γ are multipliers for the nonnegativity constraints for x and y. The analysis of corner solutions proceeds in the same way. For example, $y = 0$ implies from the first-order conditions that $\gamma > 0$ and so for the first-order condition to hold, we must have $V_y(x, y; \theta) < 0$. More formal analyses of these problems, including constraint qualification, follow the Kuhn-Tucker theorem.

In many places of the book we consider the following question: Suppose that $y > 0$ is set at its optimal level, y^*; i.e., one that satisfies its first-order condition, $V_y(x, y; \theta) = 0$. We can ask whether x is optimally positive or zero. From the corner solution discussion here, the answer is written as the following condition: $x = 0$ iff $V_x(x, y^*; \theta) < 0$, but if $x > 0$ is such that $V_x(x, y^*; \theta) = 0$, then this implies $x^* > 0$.

References

Abel, A. 1987. Operative Gift and Bequest Motives. *American Economic Review* 77, 1037–1047.

Acconcia, A., M. D'Amato, and R. Martina. 2003. Corruption and Tax Evasion with Competitive Bribes. Centre for Studies in Economics and Finance (CSEF), University of Salerno, Italy, working paper 112, 29 pp.

Adams, D. 1976. A Note on the Interdependence of Stand Structure and Best Stocking in a Selection Forest. *Forest Science* 22, 180–184.

Adams, D., and A. Ek. 1974. Optimizing the Management of Uneven-aged Forest Stands. *Canadian Journal of Forest Research* 4, 274–287.

Adams, D., and R. Haynes. 1980. Softwood Timber Assessment Market Model. Structure, Projections, and Policy Simulations. *Forest Science* Monograph 22, 64 pp.

Adda, J., and R. Cooper. 2003. *Dynamic Economics*. Cambridge, MA: MIT Press, 279 pp.

Adit, T. 2003. Economic Analysis of Corruption: A Survey. *Economic Journal* F632–652.

Albers, H. 1996. Modeling Ecological Constraints on Tropical Forest Management: Spatial Interdependence, Irreversibility and Uncertainty. *Journal of Environmental Economics and Management* 30, 73–94.

Albers, H., A. Fisher, and M. Hanemann. 1996. Valuation and Management of Tropical Forests. Implications of Uncertainty and Irreversibility. *Environmental and Resource Economics* 8, 39–61.

Alix-Garcia, J. 2007. A Spatial Analysis of Common Property Deforestation. *Journal of Environmental Economics and Management* 53, 1241–157.

Alston, L. J., G. Libecap, and B. Mueller. 1999. Titles, Conflict and Land Use: The Development of Property Rights and Land Reform on the Brazilian Amazon Frontier. Ann Arbor: University of Michigan Press.

Alston, L. J., G. D. Libecap, and B. Mueller. 2000. Land Reform Policies, the Sources of Violent Conflict, and Implications of Deforestation in the Brazilian Amazon. *Journal of Environmental Economics and Management* 39, 162–188.

Alvarez, L., and E. Koskela. 2003. On Forest Rotation under Interest Rate Variability. *Interrnational Tax and Public Finance* 10, 489–503.

Alvarez, L. H. R., and E. Koskela. 2006. Does Risk Aversion Accelerate Optimal Forest Rotation Under Uncertainty? *Journal of Forest Economics* 12, 171–184.

Alvarez, L. H. R., and E. Koskela. 2007a. Optimal Harvesting under Resource Stock and Price Uncertainty. *Journal of Economic Dynamics and Control* 31, 2461–2484.

Alvarez, L. H. R., and E. Koskela. 2007b. The Forest Rotation Problem with Stochastic Harvest and Amenity Value. *Natural Resource Modeling* 20, 477–509.

Amacher, G. 1997. The Design of Forest Taxation: A Synthesis with New Directions. *Silva Fennica* 31, 101–119.

Amacher, G. 1999. Government Preferences and Public Harvesting: A Second-Best Approach. *American Journal of Agricultural Economics* 81, 14–28.

Amacher, G. 2002. Forest Policies and Many Governments. *Forest Science* 48, 146–157.

Amacher, G. and R. J. Brazee. 1997a. Designing Forest Taxes with Varying Government Preferences and Budget Targets. *Journal of Environmental Economics and Management* 32, 323–340.

Amacher, G. and R. Brazee. 1997b. Forest Quality, government revenue constraints, and the second best efficiency of regressive forest taxes, *Canadian Journal of Forest Research* 27, 1503–1508.

Amacher, G., R. Brazee, and T. Thomson. 1991. The Effect of Forest Productivity Taxes on Timber Stand Investment and Rotation Length. *Forest Science* 36, 1099–1118.

Amacher, G., R. Brazee, E. Koskela, and M. Ollikainen. 1999. Bequests, Taxation, and Short- and Long-Run Timber Supplies: An Overlapping Generations Problem. *Environmental and Resource Economics* 13, 269–288.

Amacher, G., R. Brazee, and M. Witvliet. 2001. Royal Systems, Government Revenues, and Forest Conditions: An Application from Malaysia. *Land Economics* 76, 300–313.

Amacher, G., E. Koskela, and M. Ollikainen. 2002. Optimal Forest Taxation in an Overlapping Generations Model with Timber Bequests. *Journal of Environmental Economics and Management* 44, 346–369.

Amacher, G., C. Conway, and J. Sullivan. 2003a. Econometric Analysis of Forest Landowners: Is There Anything Left to Do? *Journal of Forest Economics* 9, 137–164.

Amacher, G., E. Koskela, M. Ollikainen, and C. Conway. 2003. Bequest Intentions and Forest Landowners: Theory and Empirical Evidence. *American Journal of Agricultural Economics* 84, 1103–1114.

Amacher, G., E. Koskela, and M. Ollikainen. 2004. Forest Rotations and Stand Interdependency: Ownership Structure and Timing of Decisions. *Natural Resource Modeling* 17, 1–43.

Amacher, G., A. Malik, and R. Haight. 2005. Not Getting Burned: The Importance of Fire Prevention in Forest Management. *Land Economics* 81, 284–302.

Amacher, G., E. Koskela, and M. Ollikainen. 2007. Royalty Reform and Illegal Reporting of Harvest Volumes under Alternative Penalty Schemes. *Environmental and Resource Economics* 38, 189–211.

Amacher, G., M. Ollikainen, and E. Koskela. 2008. Deforestation and Land Use Under Insecure Property Rights. *Environment and Development Economics* (in press).

Anderson, F. 1976. Control Theory and the Optimal Timber Rotation. *Forest Science* 22, 242–246.

Anderson, W., R. Guldin, and J. Vasievich. 1987. Assessing the Risk of Insect Attack in Plantation Investment. *Journal of Forestry* 85, 46–47.

Ando A., J. Camm, S. Polasky, and A. Solow. 1998. Species Distribution, Land Values and Efficient Conservation. *Science* 279, 2126–2128.

Angelson, A. 1999. Agricultural Expansion and Deforestation: Modeling the Impact of Population, Market Forces, and Property Rights. *Journal of Development Economics* 58, 85–218.

Angelson, A., and D. Kaimowitz. 1999. Rethinking the Causes of Deforestation: Lessons from Economic Models. *World Bank Research Observer* 14, 73–98.

Armsberg, J. 1998. Economic Parameters of Deforestation. *World Bank Economic Review* 12, 133–153.

Armsworth, P., B. Kendall, and F. Davis. 2004. An Introduction to Biodiversity Concepts for Environmental Economists. *Resource and Energy Economics* 26, 115–136.

Arnold, M., G. Köhlin, and R. Persson. 2006. Woodfuels, Livelihoods and Policy Interventions: Changing Perspectives. *World Development* 34/3, 596–611.

Arnott, J., and W. Beese. 1997. Alternatives to Clearcutting in BC Coastal Montane Forests. *Forestry Chronicle* 73, 670–678.

Aronsson, T. 1993. Nonlinear Taxes and Intertemporal Resource Management: The Case of Timber. *Scandinavian Journal of Economics* 2, 195–207.

Aronsson, T. 1990. The Incidence of Forest Taxation—A Study of the Swedish Roundwood Market. *Scandinavian Journal of Economics* 92, 65–79.

Arrow, K. J. 1965. Yrjö Jahnsson Lecture Notes. Reprinted in K. J. Arrow, *Essays in the Theory of Risk Bearing*. Amsterdam: North Holland.

Arrow, K. J., and A. C. Fisher. 1974. Environmental Preservation, Uncertainty and Irreversibility. *Quarterly Journal of Economics* 88, 312–319.

Atkinson, A. B., and J. E. Stiglitz. 1980. *Lectures on Public Economics*. New York: McGraw-Hill.

Azariadis, C. 1993. *Intertemporal Macroeconomics*. Cambridge, MA: Blackwell Press.

Balmford, A., K. J. Gaston, A. S. L. Rodrigues, and A. James. 2000. Integrating Costs of Conservation into International Priority Setting. *Conservation Biology* 14, 597–605.

Barbier, E., and J. Burgess. 1997. The Economics of Tropical Forest Land Use Options. *Land Economics* 73, 174–195.

Barbier, E., and J. Burgess. 2001a. The Dynamics of Tropical Deforestation. *Journal of Economic Surveys* 15, 413–433.

Barbier, E., and J. Burgess. 2001b. Tropical Deforestation, Tenure Insecurity and Unsustainability. *Forest Science* 47, 497–509.

Barbier, E., and J. Shogren. 2004. Growth with Endogenous Risk of Biological Invasion. *Economic Inquiry* 42, 587–601.

Barbier, E., N. Bockstael, J. Burgess, and I. Strand. 1995. The Linkages Between Timber Trade and Tropical Deforestation. *World Economy* 18, 411–442.

Barbier, E., R. Damania, and D. Leonard. 2005. Corruption, Trade and Resource Conversion. *Journal of Environmental Economics and Management* 50, 276–299.

Bardhan, P. 1997. Corruption and Development: A Review of Issues. *Journal of Economic Literature* 35, 1320–1346.

Barr, C. 2001. Timber Concession Reform: Questioning the Sustainable Logging Paradigm. Chapter 4 in *Banking on Sustainability: Structural Adjustment and Policy Reform in Post Suharto Indonesia*. Center for International Forestry Research (CIFOR), Bogor, Indonesia.

Barreto, P., P. Amaral, E. Vidal, and C. Uhl. 1998. Costs and Benefits of Forest Management for Timber Production in Eastern Amazonia. *Forest Ecology and Management* 108, 9–26.

Barro, R. 1974. Are Government Bonds Net Wealth? *Journal of Political Economy* 82, 1095–1117.

Barro, R. J. 1989. The Ricardian Approach to Budget Deficits. *Journal of Economic Perspectives* 3, 37–54.

Basu, K., S. Bhattacharya, and A. Mishra. 1992. A Note on Bribery and the Control of Corruption. *Journal of Public Economics* 48, 349–359.

Baumol, W. J., and W. E. Oates. 1988. *The Theory of Economic Policy*. 2nd ed. Cambridge: Cambridge University Press.

Bentick, B. L. 1980. Capitalized Property Taxes and the Viability of Rural Enterprise Subject to Urban Pressure. *Land Economics* 56, 451–456.

Bentley, W., and D. E. Teeguarden. 1965. Financial Maturity. A Theoretical Review. *Forest Science* 11, 76–87.

Bergeron, Y., A. Leduc, R. Harvey, and S. Gauthier. 2002. Natural Fire Regime: A Guide for Sustainable Management of the Canadian Boreal Forest. *Silva Fennica* 36(1), 81–95.

Bergstrom, T. C. 1999. Systems of Benevolent Utility Functions. *Journal of Public Economic Theory* 1(1), 71–100.

Bertsekas, D. 1987. Dynamic Programming: Deterministic and Stochastic Models. Englewood Cliffs, NJ: Prentice Hall, 376 pp.

Betters, D., E. Steinkamp, and M. Turner. 1991. Singular Path Solutions and Optimal Rates for Thinning Even-Aged Forest Stands. *Forest Science* 37, 1632–1640.

Bevers, M., and J. Hof. 1999. Spatially Optimizing Wildlife Habitat Edge Effects in Forest Management Linear and Mixed-Integer Programs. *Forest Science* 45, 249–258.

Bevers, M., J. Hof, B. Kent, and M. Raphael. 1995. Sustainable Forest Management for Optimizing Multispecies Wildlife Habitat: A Coastal Douglas-Fir Example. *Natural Resource Modeling* 1, 1–24.

Binkley, C. 1981. Timber Supply from Nonindustrial Forests. Bulletin No. 92, Yale University, School of Forestry and Environmental Studies, New Haven, CT, 97 pp.

Binkley, C. 1993. Long-run Timber Supply: Price Elasticity, Inventory Elasticity and the Use of Capital in Timber Production. *Natural Resource Modeling* 7, 163–181.

Binkley, C. 1987a. When Is the Optimal Economic Rotation Longer than the Rotation of Maximum Sustained Yield? *Journal of Environmental Economics and Management* 14, 152–158.

Binkley, C. 1987b. Economic Models of Timber Supply. In *Global Forest Sector*, M. Kallio, D. P. Dykstra, and C. Binkley, eds., pp. 109–133. New York: Wiley.

Björk, T. 1998. *Arbitrage Theory in Continuous Time*. Oxford: Oxford University Press.

Blanchard, O., and S. Fischer. 1990. *Lectures in Macroeconomics*. Cambridge, MA: MIT Press.

Blaser, J., and J. Douglas. 2000. A Future for Forests? Issues and Implications for the Emerging Forest Policy and Strategy of the World Bank. *ITTO Tropical Forest Update* 10, 9–14.

Bohn, H., and R. Deacon. 2000. Ownership Risk, Investment, and the Use of Natural Resources. *American Economic Review* 90, 526–549.

Boscolo, M., and J. R. Vincent. 2000. Promoting Better Logging Practices in Tropical Forests: A Simulation Analysis of Alternative Regulations. *Land Economics* 76(1), 1–14.

Boscolo, M., and J. R. Vincent. 2003. Nonconvexities in the Production of Timber, Biodiversity, and Carbon Sequestration. *Journal of Environmental Economics and Management* 46, 251–268.

Bowes, M., and J. Krutilla. 1985. Multiple-Use Management of Public Forestlands. In *Handbook of Natural Resource and Energy Economics*. Vol. II. A. V. Kneese and J. L. Sweeney, eds. Amsterdam: North Holland.

Bowes, M., and J. Krutilla. 1989. *Multiple-Use Management: The Economics of Public Forests*. Washington DC: Resources for the Future, 353 pp.

Bowes, M., J. Krutilla, and P. Sherman. 1984. Forest Management for Increased Timber and Water Yields. *Water Resources Research* 20, 655–663.

Boyd, R., and W. Hyde. 1989. *Forestry Sector Intervention: The Impacts of Public Regulation on Social Welfare*. Ames: Iowa State University Press.

Brännlund, R., P.-O. Johansson, and K.-G. Löfgren. 1985. An Econometric Analysis of Aggregate Sawtimber and Pulpwood Supply in Sweden. *Forest Science* 31, 595–606.

Brazee, R. J. 2003. The Volvo Theorem: From Myth to Behavior Model. In *Recent Accomplishments in Applied Forest Economics Research*, F. Helles, N. Strange, and L. Wickmann, eds., pp. 39–48. Dordrecht: Kluwer Academic Publishers.

Brazee, R. 2006. The Faustmann Face of Optimal Forest Harvesting. In *The International Yearbook of Environmental and Resource Economics 2006/2007: A Survey of Current Issues*, T. Tietenberg and H. Folmer, eds., pp. 255–288. Cheltenham, U.K. and Northampton, MA: Elgar.

Brazee, R., and G. Amacher. 2000. Duality and Faustmann: Implications for the Evaluation of Landowner Behavior. *Forest Science* 46, 132–138.

Brazee, R., and G. Amacher. 2002. Forestry at the Margin. Working Paper, Virginia Polytechnic Institute and State University, Blacksburg, VA.

Brazee, R. J., and E. Bulte. 2000. Optimal Harvesting and Thinning with Stochastic Prices. *Forest Science* 46(1), 23–31.

Brazee, R., and R. Mendelsohn. 1988. Timber Harvesting with Fluctuating Stumpage Prices. *Forest Science* 34(2), 359–372.

Brazee, R., and R. Mendelsohn. 1990. A Dynamic Model of Timber Markets. *Forest Science* 36(2), 255–264.

Brazee, R. J., and D. Southgate. 1992. Development of Ethnobiologically Diverse Tropical Forests. *Land Economics* 68(4), 454–466.

Brazee, R., Amacher, G., and C. Conway. 1999. Optimal Harvesting with Autocorrelated Stumpage Prices. *Journal of Forest Economics* 5(2), 201–216.

Brock, W., and A. Xepapadeas. 2003. Valuing Biodiversity from an Economic Perspective: A Unified Economic, Ecological and Genetic Approach. *American Economic Review* 93, 1597–1614.

Brock, W. A., M. Rothschild, and J. E. Stiglitz. 1989. Stochastic Capital Theory. In *Joan Robinson and Modern Economic Theory*. G. Feiwel, ed., pp. 591–622. New York: New York University Press.

Bulte, E., D. P. van Soest, G. C. van Kooten, and R. A. Schipper. 2002. Forest Conservation in Costa Rica when Non-use Benefits Are Uncertain but Rising. *American Journal of Agricultural Economics* 84(1), 150–160.

Buongiorno, J. 1996. Uneven-Aged Forest Management in Europe and North America: New Methods for Old Concepts. S.J. Hall Lectureship in Industrial Forestry, University of Wisconsin.

Buongiorno J. and B. Michie. 1980. A Matrix of Unven-Aged Forest Management. *Forest Science* 26, 609–625.

Buongiorno J., J. Peyron, F. Houllier, and M. Bruciamacchie. 1995. Growth and Management Decisions of Mixed Species, Uneven-aged Forests in the French Jura: Implications for Economic Returns and Tree Diversity. *Forest Science* 41, 397–429.

Buongiorno, J., and K. Gilless. 2003. *Decision Methods for Forest Resource Management*. London: Elsevier Science.

Cahuc, P., and A. Zylberberg. 2004. *Labor Economics*. Cambrige, MA: MIT Press.

Cairns, R., and P. Lasserre. 2006. Implementing Carbon Credits for Forests Based on Green Accounting. *Ecological Economics* 56, 610–621.

Calish, S., R. Fight, and D. Teeguarden. 1978. How Do Nontimber Values Affect Douglas-Fir Rotations? *Journal of Forestry* 76, 217–221.

Calkin, D., C. Montgomery, N. Schumaker, S. Polasky, J. Arthur, and D. Nalle. 2002. Developing a Production Possibility Set of Wildlife Species Persistence and Timber Harvest Value. *Canadian Journal of Forest Research* 32, 1329–1342.

Caputo, M. 2003. The Comparative Dynamics of Closed-Loop Controls for Discounted Infinite Horizon Optimal Control Problems. *Journal of Economic Dynamics and Control* 27, 1335–1365.

Caputo, M. 2005. *Foundations of Dynamic Economic Analysis: Optimal Control Theory and Applications*. Cambridge: Cambridge University Press, 579 pp.

Carmichael, J. 1983. On Barro's Theorem of Debt Neutrality: The Irrelevance of Net Wealth. *American Economic Review* 72, 202–213.

Carroll, C., and M. Kimball. 1996. On the Concavity of the Consumption Function, *Econometrica* 64, 981–992.

Caulfield, J. 1988. A Stochastic Efficiency Approach for Determining the Economic Rotation of a Forest Stand. *Forest Science* 34(2), 441–457.

Cawrse, D., D. Betters, and B. Kent. 1984. A Variational Solution Technique for Determining Optimal Thinning and Rotational Schedules. *Forest Science* 37, 793–802.

Chander, P., and L. Wilde. 1992. Corruption in Tax Administration. *Journal of Public Economics* 49, 333–349.

Chang, F. 2004. *Stochastic Optimization in Continuous Time*. Cambridge: Cambridge University Press.

Chang, F. 2005. On the Elasticities of Harvesting Rules. *Journal of Economic Dynamics and Control* 29, 469–485.

Chang, S. 1982. An Economic Analysis of Forest Taxation's Impact on Optimal Rotation Age. *Land Economics* 58, 310–323.

Chang, S. 1983. Rotation Age, Management Intensity, and the Economic Factors of Timber Production: Do Changes in Stumpage Price, Interest Rate, Regeneration Cost, and Forest Taxation Matter? *Forest Science* 29, 267–277.

Chang, S. 1984. The Determination of the Optimal Rotation Age. A Theoretical Analysis. *Forest Ecological Management* 8, 137–147.

Chiang, A. C. 1997. *Elements of Dynamic Optimization*. New York: McGraw-Hill.

Chipman, J. 1977. An Empirical Implication of Auspitz-Lieben-Edgeworth-Pareto Complementarity. *Journal of Economic Theory* 14, 228–231.

Chisholm, A. H., 1975. Income Taxes and Investment Decisions: The Long-Life Appreciating Asset Case. *Economic Inquiry* 13, 565–578.

Chladna, Z. 2007. Determination of Optimal Rotation Period under Stochastic Wood and Carbon Prices. *Forest Policy and Economics* 9, 1031–1045.

Chomitz, K., and D. Gray. 1996. Roads, Land Use, and Deforestation: A Spatial Model Applied to Belize. *World Bank Economic Review* 10, 487–512.

Christiansen, V. 1993. A Normative Analysis of Capital Income Taxes in the Presence of Aggregate Risk. *Geneva Papers on Risk and Insurance Theory* 18, 55–76.

Clark, C. 1990. *Mathematical Bioeconomics. The Optimal Management of Renewable Resources*. 2nd ed. New York: Wiley.

Clark, C., and J. DePree. 1979. A Simple Linear Model for the Optimal Exploitation of Renewable Resources. *Applied Mathematical Optimization* 5, 181–196.

Clarke, H. R. and W. J. Reed. 1988. A Stochastic Analysis of Land Development Timing and Property Valuation. *Regional Science and Urban Economics* 18, 357–381.

Clarke, H. R., and W. J. Reed. 1989. The Tree-Cutting Problem in a Stochastic Environment. The Case of Age-Dependent Growth. *Journal of Economic Dynamics and Control* 13, 569–595.

Clarke, H. R., and W. J. Reed. 1990a. Applications of Optimal Stopping in Resource Economics. *Economic Record* 254–265.

Clarke, H., and W. Reed. 1990b. Harvest Decisions and Asset Valuation for Biological Resources Exhibiting Size-Dependent Stochastic Growth. *International Economic Review* 31, 147–169.

Clarke, H. R., W. J. Reed, and R. M. Shestra. 1993. Optimal Enforcement of Property Rights on Developing Country Forests Subject to Illegal Logging. *Resource and Energy Economics* 15, 271–293.

Cochrane, J. H. 2001. *Asset Pricing*. Princeton, NJ: Princeton University Press.

Conrad, J. 1997. On the Option Value of Old-Growth Forest. *Ecological Economics* 22, 97–102.

Conrad, J. 1999. *Resource Economics*. Cambridge: Cambridge University Press.

Conrad, J. 2000. Wilderness: Options to Preserve, Extract or Develop. *Resource and Energy Economics* 22, 205–219.

Conrad, J., and C. Clark. 1987. *Natural Resource Economics Notes and Problems*. Cambridge: Cambridge University Press.

Conrad, J., and D. Ludwig. 1994. Forest Land Policy: The Optimal Stock of Old-Growth Forest. *Natural Resource Modeling* 8, 27–45.

Conrad, C., M. Gillis, and E. Mercer. 2005. Tropical Forest Harvesting and Taxation: A Dynamic Model of Harvesting Behavior Under Selective Extraction Systems. *Environment and Development Economics* 10(5), 689–709.

Contreras-Hermosilla, A. 2002. *The Underlying Causes of Forest Decline*. Occasional Paper No. 30, Center for International Forestry Research (CIFOR), Bogor, Indonesia.

Conway, C., G. Amacher, J. Sullivan, and D. Wear. 2003. Decisions Nonindustrial Forest Owners Make. *Journal of Forest Economics* 9, 181–203.

Costello, C. and D. Kaffine. 2008. Natural Resource Use with Limited Tenure Property Rights. *Journal of Environmental Economics and Management* 55, 20–36.

Crowley, C., A. Malik, and G. Amacher. 2007. Adjacency Externalities and Forest Fire Prevention. *Land Economics* (in press).

Csuti, B., S. Polasky, P. Williams, R. Pressey, J. Camm, M. Kershaw, R. Kiester, B. Hamilton, M. Huso, and K. Sahr. 1997. A Comparison of Reserve Selection Algorithms Using Data on Terrestrial Vertebrates in Oregon. *Biological Conservation* 80, 83–97.

Cubbage, F. W., J. O'Laughin, and C. S. Bullock. 1993. *Forest Resource Policy*. New York: Wiley.

Damania, R., and E. Barbier. 2003. Lobbying, Trade, and Renewable Resource Harvesting in Developing Countries. Discussion Paper, Center for Economic Studies, University of Adelaide, Australia No. SA5005.

Dasgupta, P. S., and G. Heal. 1980. *Economic Theory and Exhaustible Resources*. Cambridge: Cambridge University Press.

Davis, L. S., K. Johnson, P. S. Bettinger, and T. E. Howard. 2001. CHIO, Classical Approaches to Forest Management Planning. *Forest Management*. 4th ed. New York: McGraw-Hill.

DeCoster, L. A. 2000. Summary of the Forest Fragmentation 2000 Conference: How forests are being Nibbled to death and what to do about it. In DeCoster (Ed.), Proceedings of the Forest Fragmentation 2000 conference, pp. 2–12, Alexandria VA: Sampson Group, Inc.

Delacote, P. 2005. Forestry Sector Concentration, Corruption, and Forest Over-exploitation. Economics Department, EUI Florence, Working Paper.

De la Croix, D., and P. Michelle. 2002. *A Theory of Economic Growth. Dynamics and Policy in Overlapping Generations*. Cambridge: Cambridge University Press.

Deacon, R. 1994. Deforestation and the Rule of Law in a Cross Section of Countries. *Land Economics* 70, 414–430.

Deiniger, K., and G. Feder. 2001. Land Institutions and Land Markets. In *Handbook of Agricultural Economics*. Vol. 1A: *Agricultural Production*. B. Gardner and G. Rausser, eds. Amsterdam: North Holland.

Di Tella, R., and E. Schargrodsky. 2003. The Role of Wages and Auditing During a Crackdown on Corruption in the City of Buenos Aires. *Journal of Law and Economics* XLVI, 269–292.

Dixit, A. 1993. *The Art of Smooth Pasting*. London and New York: Routledge.

Dixit, A. K., and R. S. Pindyck. 1994. *Investment Under Uncertainty*. Princeton, NJ: Princeton University Press.

Dorfman, R. 1969. An Economic Interpretation of Optimal Control Theory. *American Economic Review* 59, 817–831.

Dudewicz, E., and S. Mishra. 1988. *Modern Mathematical Statistics*. Singapore: Wiley.

Ehnström, B. 2001. Leaving Dead Wood for Insects in Boreal Forests—Suggestions for the Future. *Scandinavian Journal of Forest Research*. Supplement 3, 91–98.

Englin, J., and J. Callaway. 1993. Global Climate Change and Optimal Forest Management. *Natural Resource Modeling* 7, 191–202.

Englin, J. E., and M. K. Klan. 1990. Optimal Taxation: Timber and Externalities. *Journal of Environmental Economics and Management* 18, 263–275.

Englin, J., P. Boxall, and G. Hauer. 2000. An Empirical Examination of Optimal Rotations in a Multiple-Use Forest in the Presence of Fire Risk. *Journal of Agricultural and Resource Economics* 25, 14–27.

Epstein, L. G., and S. E. Zin. 1989. Substitution, Risk Aversion, and the Temporal Behavior of Consumption and Asset Returns: A Theoretical Framework. *Econometrica* 57(4), 937–9769.

Faith, D., and P. Walker. 1996. How do Indicator Groups Provide Information About the Relative Biodiversity of Different Sets of Areas?: On Hotspots, Complementarity and Pattern-based Approaches. *Biodiversity Letters* 3, 18–25.

FAO (Food and Agricultural Organization). 1996. *State of the World's Forests 2001*. Rome: Food and Agricultural Organization.

FAO (Food and Agricultural Organization). 2001. *State of the World's Forests 2001*. Rome: Food and Agricultural Organization.

FAO (Food and Agricultural Organization). 2005. *State of the World's Forests 2001*. Rome: Food and Agricultural Organization.

Fairchild, F. R. 1909. The Taxation of Timber Lands in the United States. In *Proceedings of the 2nd Annual Conference on Taxation*, pp. 69–82. International Tax Association, Columbus, Ohio.

Fairchild, F. R. 1935. Forest Taxation in the United States. *US Department of Agriculture, Miscellaneous Publications*, No. 218.

Farzin, Y. H. 1984. The Effect of the Discount Rate on Depletion of Exhaustible Resources. *Journal of Political Economy* 92, 841–851.

Faustmann, M. 1849. Calculation of the Value which Forest Land and Immature Stands Possess. Reprinted in *Journal of Forest Economics* 1, 89–114.

Fernow, B. 1911. *A Brief History of Forestry in Europe, the United States and Other Countries*. Revised and enlarged ed. Toronto University Press and *Forestry Quarterly*, Cambridge, MA.

Fina, M., G. Amacher, and J. Sullivan. 2001. Uncertainty, Debt, and Forest Harvesting: Faustmann Revisited. *Forest Science* 47(2), 188–196.

Finnish Statistical Yearbook of Forestry. 2002. Finnish Forest Research Institute. Gummerus, Jyväskylä. 352 pp.

Finnoff, D., C. Settle, J. Shogren, and J. Tschirhart. 2006. Invasive Species and the Depth of Bioeconomic Integration. *Choices* 21, 147–151.

Forman, R. 1995. Some General Principles of Landscape and Regional Ecology. *Landscape Ecology* 10, 133–142.

Forosbeh, P., R. Brazee, and J. Pichens. 1996. A Strategy for Multiproduct Stand Management with Uncertain Future Prices. *Forest Science* 42, 58–66.

Forsyth, M. 2000. On Estimating the Option Value of Preserving the Wilderness Area. *Canadian Journal of Economics* 33, 413–434.

Franklin, J. F. 1993. Preserving Biodiversity: Species, Ecosystems, or Landscapes. *Ecological Applications* 3, 202–205.

Franklin, J., and R. Forman. 1987. Creating Landscape Patterns by Forest Cutting: Ecological Consequences and Principles. *Landscape Ecology* 1, 5–18.

Franklin, J., D. Berg, D. Thornburgh, and J. Tappeiner. 1996. Alternative Silvicultural Approaches to Timber Harvesting: Variable Retention Harvest Systems. In *Creating a Forestry for the 21st Century: The Science of Ecosystem Management*, K. Kohm and J. Franklin, eds., pp. 111–139. Washington DC: Island Press.

FSC Finland. 2002. *Man and the Forest. The Finnish FSC Standard* (in Finnish).

Fullerton, A. 1982. On the Possibility of the Inverse Relationship Between Tax Rates and Government Revenues. *Journal of Public Economics* 19, 3–22.

Gaffney, M. 1957. Concepts of Financial Maturity and Other Essays. Agricultural Economics Information, Series 62. North Carolina State College, Raleigh.

Gamponia, V., and R. Mendelsohn. 1987. The Economic Efficiency of Forest Taxes. *Forest Science* 33, 367–378.

Gardiner, B., and C. Quine. 2000. Management of Forests to Reduce the Risk of Abiotic Damage—A Review with Particular Reference to the Effects of Strong Winds. *Forest Ecology and Management* 135, 261–277.

Getz, W. M. and R. G. Haight. 1989. *Population Harvesting: Demographic Models for Fish, Forest and Animal Resources*, Princeton, NJ: Princeton University Press.

Gjolberg, O., and A. G. Guttorsem. 2002. Real Options in the Forest: What If Prices Are Mean-Reverting? *Forest Policy and Economics* 4, 13–20.

Gollier, C. 2001. *The Economics of Risk and Time*. Cambridge, MA: MIT Press.

Gong, P. 1992. Multiobjective Dynamic Programming for Forest Resource Management. *Forest Ecology and Management* 48, 43–54.

Gong, P. 1998. Risk Preferences and Adaptive Harvest Policies for Even-Aged Stand Management. *Forest Science* 44, 496–506.

Gong, P. 1999. Optimal Harvest Policy with First-Order Autoregressive Price Process. *Journal of Forest Economics* 5, 413–439.

Gong, P., and K.-G. Löfgren. 2003a. Risk Aversion and the Short-Term Supply of Timber. *Forest Science* 49, 647–656.

Gong, P., and K. Löfgren. 2003b. Timber Supply Under Demand Uncertainty: Welfare Gains from Perfect Competition with Rational Expectations. *Natural Resource Modeling* 16, 69–97.

Gong, P., and K. Löfgren. 2007. Market and Welfare Implications of the Reservation Price Strategy for Forest Harvest Decisions. *Journal of Forest Economics* 13, 217–243.

Gong, P., M. Boman, and L. Mattsson. 2005. Non-Timber Benefits, Price Uncertainty, and Optimal Harvest of An Even-Aged Stand. *Forest Policy and Economics* 7, 283–295.

Gordon, R. G., and H. R. Varian. 1988. Intergenerational Risk Sharing. *Journal of Public Economics* 37, 185–202.

Gray, J. 2000. Forest Concession Policies and Revenue Systems: Country Experience and Policy Changes for Sustainable Tropical Forestry. World Bank Technical Paper, Forest Series, Washington DC.

Gregory, G. 1972. *Forest Resource Economics*. New York: Wiley, 548 pp.

Gregory G. R. 1987. *Resource Economics for Foresters*. New York: Wiley.

Guertin, C. 2003. Illegal Logging and Illegal Activities in the Forest Sector: Overview and Possible Issues for the UNECE Timber Committee and FAO European Forestry Commission. Presented at UNECE Timber Committee Market Discussions, October 7–8, 2003, Geneva.

Gutrich, J., and R. B. Howarth. 2007. Carbon Sequestration and the Optimal Management of New Hampshire Timber Stands. *Ecological Economics* 62(3/4), 441–450.

Haight, R. 1985. A Comparison of Dynamic and Static Economic Models of Even-Aged Stand Management. *Forest Science* 31, 957–974.

Haight, R. 1987. Evaluating The Efficiency of Even-aged and Uneven-aged Management. *Forest Science* 31, 957–974.

Haight, R. 1995. Comparing Extinction Risk and Economic Cost in Wildlife Conservation Planning. *Ecological Applications* 5, 767–775.

Haight, R., and W. Getz. 1987. Fixed and Equilibrium Endpoint Problems in Uneven-aged Stand Management. *Forest Science* 33, 908–931.

Haight, R. G., and T. P. Holmes. 1991. Stochastic Price Models and Optimal Tree Cutting: Results for Loblolly Pine. *Natural Resource Modeling* 5, 423–443.

Haight, R., and L. Travis. 1997. Wildlife Conservation Planning Using Stochastic Optimization and Importance Sampling. *Forest Science* 43, 129–139.

Haight, R., W. Smith, and T. Straka. 1995. Hurricanes and the Economics of Loblolly Pine Plantations. *Forest Science* 41, 675–688.

Hamaide, B., and C. ReVelle and S. Malcon. 2006. Biological Reserves, Rare Species and the Trade-off Between Species Abundance and Species Diversity. *Ecological Economics* 56, 570–583.

Hanski, I. 1999. *Metapopulation Ecology*. Oxford: Oxford University Press.

Hardie, I., and P. Parks. 1997. Land Use with Heterogeneous Land Quality: An Application of an Area Base Model. *American Journal of Agricultural Economics* 77, 299–310.

Hardner, J., and R. Rice. 2000. Rethinking Forest Use Contracts in Latin America. Environment Division Working Paper, Inter-American Development Bank, Washington DC.

Harrison, J. M. 1985. *Brownian Motion and Stochastic Flow Systems*. New York: Wiley.

Hartman, R. 1976. The Harvesting Decision When a Standing Forest Has Value. *Economic Inquiry* 14, 52–55.

Hartwick, J., N. Van Long, and H. Tian. 2001. Deforestation and Development in a Small Open Economy. *Journal of Environmental Economics and Management* 41, 235–251.

Hazell, P., and L. Gustafsson. 1999. Retention of Trees at Final Harvest—Evaluation of a Conservation Technique Using Epiphytic Bryophyte and Lichen Transplants. *Biological Conservation* 90, 133–142.

Heaps, T. 1984. The Forestry Maximum Principle. *Journal of Economic Dynamics and Control* 7, 131–151.

Heaps, T. 1995. Multiple Use Values and Optimal Steady-State Age Distributions. *Natural Resource Modeling* 9, 329–339.

Heaps, T., and P. Neher. 1979. The Economics of Forestry when the Rate of Harvest is Constrained. *Journal of Environmental Economics and Management* 6, 297–316.

Helles, F., and M. Linddal. 1997. Early Danish Contributions to Forest Economics. *Journal of Forest Economics* 3, 87–103.

Hellstien, M. 1988. Socially Optimal Forestry. *Journal of Environmental Economics and Management* 15, 387–394.

Henry, C. 1974. Option Values in the Economics of Irreplaceable Assets. *Review of Economic Studies* 41, 89–104.

Hetemäki, L., and J. Kuuluvainen. 1992. Incorporating Data and Theory in Roundwood Supply and Demand Estimation. *American Journal of Agricultural Economics* 74, 1010–1018.

Hey, J. D. and P. J. Lambert (eds.). 1987. *Surveys in the Economics of Uncertainty*. Oxford: Blackwell.

Hilmo, O. 2002. Growth and Morphological Response of Old-forest Lichens Transplanted into a Young and an Old *Picea abies* Forest. *Ecography* 25, 329–335.

Hirschleifer, J. 1970. *Investment, Interest and Capital*. London: Prentice Hall.

Hoen, H., and B. Solberg. 1994. Potential and Economic Efficiency of Carbon Sequestration in Forest Biomass through Silvicultural Management. *Forest Science* 40, 429–451.

Hof, J., and M. Bevers. 2002. *Spatial Optimization in Ecological Applications*. New York: Columbia University Press.

Holling, C. 2001. Understanding the Complexity of Economic, Ecological, and Social Systems. *Ecosystems* 4, 390–405.

Hotte, L. 2005. Natural-Resource Exploitation with Costly Enforcement of Property Rights. *Oxford Economic Papers* 57, 497–521.

Hultkrantz, L. 1992. Forestry and the Bequest Motive. *Journal of Environmental Economics and Management* 22, 164–177.

Hultkrantz, L. 1995. The Behavior of Timber Rents in Sweden 1909–1990. *Journal of Forest Economics* 1(2), 165–180.

Hultkranz, L., and T. Aronsson. 1989. Factors Affecting the Supply and Demand of Timber from Private Nonindustrial Lands in Sweden: An Econometric Study. *Forest Science* 35, 946–961.

Hunter, M. 1990. *Wildlife, Forests and Forestry*. Englewood Cliffs, NJ: Prentice Hall.

Hunter, M. L. (ed.) 1999. *Maintaining Biodiversity in Forest Ecosystems*. Cambridge: Cambridge University Press.

Hyde, W. 1980. *Timber Supply, Land Allocation and Economic Efficiency*. Baltimore, MD: Resources for the Future.

Hyde, W., and G. Amacher (eds.) 2001. *Economics of Forestry Development in Asia: An Empirical Introduction*. Ann Arbor: University of Michigan Press.

Hyde, W., G. Amacher, and W. Magrath. 1996. Deforestation and Forest Land Use. Theory, Evidence and Policy Implications. *World Bank Research Observer* 11, 223–248.

Hyytiäinen, K. and O. Tahvonen. 2001. The effects of legal limits and recommendations on timber production: The case of Finland. *Forest Science* 47, 443–454.

Hyytiäinen, K., P. Hari, T. Kokkila, A. Mäkelä, O. Tahvonen, and J. Taipale. 2004. Connecting Process-Based Forest Growth Model to Stand-Level Economic Optimization. *Canadian Journal of Forest Research* 34, 2060–2073.

Hyytiäinen, K., O. Tahvonen, and L. Valsta. 2005. Optimum Juvenile Density, Harvesting and Stand Structure in Even-Aged Scots Pine Stands. *Forest Science* 51, 120–133.

Ihori, T. 1996. *Public Finance in an Overlapping Generations Economy.* London: Macmillan.

Ingersoll, J. E., and S. A. Ross. 1992. Waiting to Invest: Investment and Uncertainty. *Journal of Business* 65, 1–29.

Innes, R., S. Polasky, and J. Tschirhart. 1998. Takings, Compensation, and Endangered Species Protection on Private Lands. *Journal of Economic Perspectives* 12, 35–52.

Insley, M. C. 2002. A Real Options Approach to the Valuation of Forestry Investment. *Journal of Environmental Economics and Management* 44, 471–492.

Insley, M. C., and K. Rollins. 2005. On Solving the Multirotational Timber Harvesting Problem with Stochastic Prices: A Linear Complementarity Formulation. *American Journal of Agricultural Economics* 87(3), 735–755.

ITTO (International Tropical Timber Organization). 2005. Status of Tropical Forest Management: Summary Report, a Special Edition of the Tropical Forest Update. *ITTO Tropical Forest Update* 16, 2006.

Jacobson, M. 2002. Ecosystem Management in the Southeast United States: Interest of Forestlandowners in Joint Management Across Ownerships. *Small-Scale Forest Economics Management and Policy* 1, 57–70.

Jalonen, J., and I. Vanha-Majamaa. 2001. Immediate Effects of Four Different Felling Methods on Mature Boreal Spruce Forest Understory Vegetation in Southern Finland. *Forest Ecology and Management* 146, 25–34.

Johansson, P., and K.-G. Löfgren. 1985. *The Economics of Forestry and Natural Resources.* Oxford: Basil Blackwell.

Jonsell, M., K. Nittérus, and K. Stighäll. 2004. Saproxylic Beetles in Natural and Man-made Deciduous High Stumps Retained for Conservation. *Biological Conservation* 118, 163–173.

Jonsson, M., T. Ranius, H. Ekvall, G. Bosted, A. Dahlberg, B. Ehnström, B. Nordén, and J. Stokland. 2006. Cost-effectiveness of Silvicultural Measures to Increase Substrate Availability for Red-listed Wood-living Organisms in Norway Spruce Forests. *Biological Conservation* 127, 443–462.

Juutinen, A., and M. Ollikainen. 2008. Trading in Nature Values as a Means of Biodiversity Conservation in Boreal Forests: Theory and Experience from Finland. Working Paper, University of Helsinki.

Juutinen, A., E. Mäntymaa, M. Mönkkönen, and J. Salmi. 2004. Cost-effective Selection of Boreal Old Forest Reserves: A Case from Northern Fennoscandia. *Forest Science* 50, 527–539.

Juutinen, A., E. Mäntymaa, M. Mönkkönen, and R. Svento. 2008. Voluntary Agreements in Protecting Privately Owned Forests in Finland—To Buy or to Lease? *Forest Policy and Economics* 10, 230–239.

Juutinen, A., M. Mönkkönen, and M. Ollikainen. 2008. Do Environmental Diversity Approaches Lead to Improved Site Selection? A Comparison with the Multi-species Approach. *Forest Ecology and Management* 255, 3750–3757.

Junntinen, K., R. Penttilä, and P. Martikainen. 2006. Fallen Retention Aspen Trees on Clear-cuts Can Be Important Habitats for Red-listed Polypores: A Case Study in Finland. *Biodiversity and Conservation.* (In Press).

Kaimowitz, D. 1996. *Livestock and Deforestation: Central America in the 1980s and 1990s: A Policy Perspective*. Center for International Forestry Research, 95 pp.

Kaimowitz, D., and A. Angelson. 1998. Economic Models of Deforestation: A Review. Center for International Forestry Research.

Kamien, M., and N. Schwartz. 1981. *Dynamic Optimization*. Amsterdam: North Holland Press.

Kamien, M., and N. Schwartz. 1991. *Dynamic Optimization: The Calculus of Variations and Optimal Control in Economics and Management*. 2nd ed. Amsterdam: Elsevier.

Kannai, Y. 1980. The ALEP Definition of Complementarity and Least Concave Utility Functions. *Journal of Economic Theory* 22, 115–117.

Kant, S. 2000. A Dynamic Approach to Forest Regimes in Developing Economies. *Ecological Economics* 32(2), 287–300.

Karlin, S., and H. M. Taylor. 1975. *A First Course in Stochastic Processes*. 2nd ed., New York: Academic Press.

Kassar, I., and P. Lasserre. 2004. Species Preservation and Biodiversity Value: A Real Option Approach. *Journal of Environmental Economics and Management* 48, 857–879.

Kemp, M. C., and E. J. Moore. 1979. Biological Capital Theory. A Question and a Conjecture. *Economics Letters* 4, 141–144.

Kennedy, J., and M. Scorgie. 1996. Who Discovered the Faustmann Condition? *History of Political Economy* 28, 77–80.

Kilkki, P., and U. Väisänen. 1969. Determination of the Optimal Cutting Policy for the Forest Stand by Means of Dynamic Programming. *Acta Forestalia Fennica* 102, 100–112.

Kim, C. S., R. Lubowski, J. Lewandrowski, and M. Eiswerth. 2006. Prevention or Control: Optimal Government Policies for Invasive Species Management. *Agricultural and Resource Economics Review* 35, 29–40.

Kimball, M. 1987. Making Sense of Two-Sided Altruism. *Journal of Monetary Economics* 20, 301–326.

Klemperer, D. 1975. The Parable of the Allowable Pump Effect. *Journal of Forestry* 73, 640–641.

Klemperer, D. 1976. Impacts of Tax Alternatives on Forest Values and Investment. *Land Economics* 52, 135–157.

Knapp, K. 1981. The Supply of Timber from Private Nonindustrial Forests. Doctoral Dissertation, Yale University.

Koivula, M. 2002. Alternative Harvesting Methods and Boreal Carabid Beetles (Coleoptera, Carabidae). *Forest Ecology and Management* 167, 103–121.

Koskela, E., 1989a. Forest Taxation and Timber Supply Under Price Uncertainty: Perfect Capital Markets. *Forest Science* 35, 137–159.

Koskela, E. 1989b. Forest Taxation and Timber Supply under Price Uncertainty: Credit Rationing in Capital Markets. *Forest Science* 35, 160–172.

Koskela, E., and M. Ollikainen. 1997a. Optimal Design of Forest and Capital Income Taxation in an Economy with an Austrian Sector. *Journal of Forest Eonomics* 3, 107–132.

Koskela, E., and M. Ollikainen. 1997b. Optimal Design of Forest Taxation with Multiple-Use Characteristics of Forest Stands. *Environmental and Resource Economics* 10, 41–62.

Koskela, E., and M. Ollikainen. 1998a. Tax Incidence and Optimal Forest Taxation Under Stochastic Demand. *Forest Science* 44, 4–16.

Koskela, E., and M. Ollikainen. 1998b. A Game-theoretic Model of Timber Prices with Capital Stock: An Empirical Application to the Finnish Forest Sector. *Canadian Journal of Forest Research* 28, 1481–1493.

Koskela, E., and M. Ollikainen. 1999a. Optimal Public Harvesting under the Interdependence of Public and Private Forests. *Forest Science* 45, 259–271.

Koskela, E., and M. Ollikainen. 1999b. Timber Supply, Amenity Values and Biological Uncertainty. *Journal of Forest Economics* 5, 285–304.

Koskela, E., and M. Ollikainen. 2001a. Forest Taxation and Rotation Age under Private Amenity Valuation: New Results. *Journal of Environmental Economics and Management* 42, 374–384.

Koskela, E., and M. Ollikainen. 2001b. Optimal Private and Public Harvesting under Spatial and Temporal Interdependence. *Forest Science* 47, 484–496.

Koskela, E., and M. Ollikainen. 2003a. Optimal Forest Taxation under Private and Social Amenity Valuation. *Forest Science* 49, 596–607.

Koskela, E., and M. Ollikainen. 2003b. Behavioral and Welfare Analysis of Progressive Forest Taxation. *Canadian Journal of Forest Research* 33, 2352–2361.

Koskela, E., M. Ollikainen, and M. Puhakka. 2002. Renewable Resources in an Overlapping Generations Economy without Capital. *Journal of Environmental Economics and Management* 43, 497–517.

Koskela, E., M. Ollikainen, and T. Pukkala. 2007a. Biodiversity Policies in Commercial Boreal Forests: Optimal Design of Subsidy and Tax Combinations. *Forest Policy and Economics* 9, 982–995.

Koskela, E., M. Ollikainen, and T. Pukkala. 2007b. Biodiversity Conservation in Commercial Boreal Forests: The Optimal Rotation Age and Retention Tree Volume. *Forest Science* 53, 443–452.

Kovenock, D. 1986. Property and Income Taxation in an Economy with an Austrian Sector. *Land Economics* 62, 201–209.

Kovenock, D., and M. Rothschild. 1983. Capital Gains Taxation in an Economy with an Austrian Sector. *Journal of Public Economics* 21, 215–256.

Kreps, D., and E. Porteus. 1978. Temporal Resolution of Uncertainty and Dynamic Choice Theory. *Econometrica* 46, 185–200.

Krcmar-Nozic, E., G. van Kooten, and B. Wilson. 2000. Threat to Biodiversity: The Invasion of Exotic Species: Conserving Nature's Diversity: Insights from Biology, Ethics and Economics. In *Studies in Environmental and Natural Resource Economics*, G. van Kooten and E. Bulte, eds., pp. 68–87. Aldershot, UK: Ashgate.

Kuuluvainen, J. 1990. Virtual Price Approach to Short-term Timber Supply under Credit Rationing. *Journal of Environmental Economics and Management* 19, 109–126.

Kuuluvainen, J., H. Karppinen, and V. Ovaskainen. 1996. Landowner Objectives and Nonindustrial Private Timber Supply. *Forest Science* 42, 300–309.

Kuuluvainen, J., and J. Salo. 1991. Timber Supply and Life Cycle Harvest of Nonindustrial Private Forest Owners: An Empirical Analysis of the Finnish Case. *Forest Science* 37, 1011–1029.

Kuuluvainen, J., and O. Tahvonen. 1999. Testing the Forest Rotation Model: Evidence from Panel Data. *Forest Science* 45, 539–551.

Kuuluvainen, J. 2002. Natural Variability of Forests as a Reference for Restoring and Managing Biological Diversity in Boreal Fennoscandia. *Silva Fennica* 36(1), 97–125.

Laaksonen-Craig, S. and M. Ollikainen. 2008. The Impact of Recycled Paper on Pulp and Sawn Wood Market: Theoretical and Econometric Analysis with Finnish Data. University of Helsinki, Department of Economics and Management, Discussion Papers 27.

Lambert, P. J. 2002. *The Distribution and Redistribution of Income*. 3rd ed. Manchester, UK: Manchester University Press.

Latacz-Lohmann, U., and C. van der Hamsvoort. 1997. Auctioning Conservation Contracts: A Theoretical Analysis and an Application. *American Journal of Agricultural Economics* 79, 407–418.

Latacz-Lohman, U., and S. Schillizi. 2005. Auctions for Conservation Contracts: A Review of the Theoretical and Empirical Literature. Report to the Scottish Executive Environment and Rural Affairs Department, 84 pp.

Leonard, D., and N. van Long. 2002. *Optimal Control Theory and Static Optimization in Economics*. Cambridge: Cambridge University Press.

Leslie, P. 1945. On the Use of Matrices in Certain Population Mathematics. *Biometrica* 33, 183–212.

Levhari, D., and L. Mirman. 1980. The Great Fish War: An Example Using a Dynamic Cournot-Nash Solution. *Bell Journal of Economics* 11(1), 322–334.

Li, C., and K. Löfgren. 1998. A Dynamic Model of Biodiversity Preservation. *Environment and Development Economics* 3, 157–172.

Lichtenberg, E. 1989. Land Quality, Irrigation Development, and Cropping Patterns in the Northern High Plains. *American Journal of Agricultural Economics* 71, 187–194.

Lindenmayer, D., and J. Franklin. 2002. *Conserving Forest Biodiversity: A Comprehensive Multiscaled Approach*. Washington, DC: Island Press.

Ljungqvist, L., and T. J. Sargent. 2004. *Recursive Macroeconomic Theory*. 2nd ed. Cambridge, MA: MIT Press.

Löfgren, K.-G. 1983. The Faustmann-Ohlin Theorem: a Historical Note. *History of Political Economy* 15, 261–264.

Löfgren, K.-G. 1984. Endowments and Timber Supply. *European Review of Agricultural Economics* 11, 17–28.

Löfgren, K.-G. 1991. Another Reconciliation between Economists and Forestry Experts: OLG-Arguments. *Environmental and Resource Economics* 1, 83–95.

Löfgren, K.-G. 1999. Ohlin versus Hechscher and Wicksell on Forestry. One Win (points) and One Draw. *Umeå Economic Studies* No. 513. Umeå University, Umeå, Sweden.

Lohmander, P. 1987. The Economics of Forest Management Under Risk. Report 79, SE-901 83, Swedish University of Agricultural Sciences, Umea, Sweden.

Lohmander, P., and F. Helles. 1987. Windthrow Probability as a Function of Stand Characteristics and Shelter. *Scandinavian Journal of Forest Research* 2, 227–238.

Lu, F., and P. Gong. 2003. Optimal Stocking Level and Final Harvest Age with Stochastic Prices. *Journal of Forest Economics* 9, 119–136.

Lueck, D., and J. Michael. 2003. Preemptive Habitat Destruction under the Endangered Species Act. *Journal of Law and Economics* 46, 27–60.

Lyon, K. 1981. Mining of the Forest and the Time Path of the Price of Timber. *Journal of Environmental Economics and Management* 8, 330–334.

Lyon, K., and R. Sedjo. 1983. An Optimal Control Theory Model to Estimate the Regional Long-term Supply of Timber. *Forest Science* 29, 798–812.

Lyon, K., and R. Sedjo. 1990. Comparative Advantage in Timber Supply: Lessons from History and the Timber Supply Model. *Journal of Business Administration*, special issue, "Emerging Issues in Forest Policy," 19, 171–176.

MacArthur, R., and E. Wilson. 1967. *The Theory of Island Biogeography*. Princeton, N.J.: Princeton University Press.

Maestad, O. 2001. Timber Trade Restrictions and Tropical Deforestation: A Forest Mining Approach. *Resource and Energy Economics* 23, 111–132.

Malcomson, J. 1986. Some Analytics of the Laffer Curve. *Journal of Public Economics* 29, 293–316.

Mallaris, A. G., and W. A. Brock. 1982. *Stochastic Methods in Economics*. New York: North Holland.

Malinvaud, E. 1987. Communication. The Overlapping Generations Model in 1947. *Journal of Economic Literature* XXV, 103–105.

Manz, P. 1986. Forestry Economics in the Steady State: The Contribution of J. H. von Thunen. *History of Political Economy* 18, 281–290.

Margules, C., and R. Pressey. 2000. Systematic Conservation Planning. *Nature* 405, 243–253.

Martell, D. L. 1980. The Optimal Rotation of a Flammable Forest Stand. *Canadian Journal of Forest Research* 10, 30–34.

Max, W., and D. Lehman. 1988. A Behavioral Model of Timber Supply. *Journal of Environmental Economics and Management* 15, 71–86.

May, R. 1990. Taxonomy as Destiny. *Nature* 347, 129–130.

McAusland, C., and C. Costello. 2004. Avoiding Invasives: Trade-related Policies for Controlling Unintentional Exotic Species Introductions. *Journal of Environmental Economics and Management* 48, 954–977.

McIntyre, S., and R. Hobbs. 1999. A Framework for Conceptualizing Human Effects on Landscapes and Its Relevance to Management and Research Models. *Conservation Biology* 13, 1282–1292.

McConnell, K., J. Daberkow, and I. Hardie. 1983. Planning Timber Production with Evolving Prices and Costs. *Land Economics* 59, 292–299.

Mehmood, S., and D. Zhang. 2005. Determinants of Forest Landowner Participation in the Endangered Species Act Safe Harbor Program. *Human Dimension of Wildlife* 10, 249–225.

Mehta, S., R. Haight, F. Homans, S. Polasky, and R. Venette. 2007. Optimal Detection and Control Strategies for Invasive Species Management. *Ecological Economics* 61, 237–245.

Meilby, H., N. Strange, and B. Thorsen. 2001. Optimal Spatial Harvest Planning under Risk of Windthrow. *Forest Ecology and Management* 149, 15–31.

Meilby, H., B. Thorsen, and N. Strange. 2003. Adaptive Spatial Harvest Planning under Risk of Windthrow. In *Recent Accomplishments in Applied Forest Economics Research*, F. Helles, N. Strange, and L. Wichman eds., pp. 49–61. Amsterdam: Kluwer Academic Publishers.

Mendelsohn, R. 1993. Nonlinear Forest Taxes. A Note. *Journal of Environmental Economics and Management* 24, 296–299.

Mendelsohn, R. 1994. Property Rights and Tropical Deforestation. *Oxford Economic Papers* 46, 750–756.

Mendoza, G., and A. Setyarso. 1986. A Transition Matrix Forest Growth Model For Valuating Alternative Harvesting Schemes in Indonesia. *Forest Ecology and Management* 15, 219–228.

Merry, F., E. Lima, G. Amacher, O. Almeida, A. Alves, and M. dos Santos. 2004. *Overcoming Marginalization in the Brazilian Amazon: Case Studies of Forests.* International Institute for Environment and Development, London, 78 pp.

Merton, R. C. 1971. Optimum Consumption and Portfolio Rules in a Continuous-Time Model. *Journal of Economic Theory* 3, 373–413.

Miceli, T., H. Munneke, C. Sirmans, and G. Turnbull. 2002. Title Systems and Land Values. *Journal of Law and Economics* 45, 565–582.

Mikosch, T. 1998. *Elementary Stochastic Calculus with Finance in View.* River Edge, NJ: World Scientific Publishing.

Miller, R., and K. Voltaire. 1980. A Sequential Stochastic Tree Problem, *Economics Letters* 5, 135–140.

Miller, R., and K. Voltaire. 1983. A Stochastic Analysis of the Tree Paradigm. *Journal of Economic Dynamics and Control*, 6, 371–386.

Miranda, M., and P. Fackler. 2002. *Applied Computational Economics and Finance.* Cambridge, MA: MIT Press, 509 pp.

Mitra, T., and H. Wan. 1985. Some Theoretical Results on the Economics of Forestry. *Review of Economic Studies* LII, 263–282.

Mitra, T., and H. Wan. 1986. On the Faustmann Solution to the Forest Management Problem. *Journal of Economic Theory* 40, 229–249.

Montgomery, C. 1995. Economic Analysis of the Spatial Dimensions of Species Preservation: The Distribution of Northern Spotted Owl Habitat. *Forest Science* 41, 67–83.

Montgomery C., and D. Adams. 1995. Optimal Timber Management Policies. In *The Handbook of Environmental Economics*, D. Bromley, ed. Oxford: Basil Blackwell.

Mood, A. M., F. A. Gaybill, and D. C. Boes. 1974. *Introduction to the Theory of Statistics.* 3rd ed. New York: McGraw-Hill.

Mookherjee, D., and I. P. L. Png. 1995. Corruptible Law Enforcers: How Should They Be Compensated? *Economic Journal* 105, 145–159.

Morck, R., E. Schwartz, and S. Stangeland. 1989. The Valuation of Forestry Resources under Stochastic Prices and Inventories. *Journal of Financial and Quantitative Analysis* 24, 473–487.

Motoh, T. 2004. Optimal Natural Resources Management under Uncertainty with Catastrophic Risk. *Energy Economics* 26(3), 487–499.

Musgrave, R., and T. Thin. 1948. Income Tax Progression, 1929–1948. *Journal of Political Economy* 56, 498–514.

Myles, G. 1995. *Public Economics.* Cambridge: Cambridge University Press.

Nalle, D., A. C. Montgomery, J. Arthur, S. Polasky, and N. Schumaker. 2004. Modelling Joint Production of Wildlife and Timber. *Journal of Environmental Economics and Management* 48, 997–1017.

Näslund, B. 1969. Optimal Rotation and Thinning. *Forest Science* 15, 446–451.

Neary, J., and K. Roberts. 1980. The Theory of Household Behavior under Rationing. *European Economic Review* 13, 25–42.

Newbery, D., and J. Stiglitz. 1981. *The Theory of Commodity Price Stabilization. A Study in the Economics of Risk*. Oxford: Clarendon Press.

Newman, D. 1984. A Discussion of the Concept of the Optimal Forest Rotation and a Review of the Recent Literature. School of Forestry and Environmental Studies, Duke University, Durham, NC.

Newman, D. 1987. An Econometric Analysis of the Southern Softwood Stumpage Market: 1950–1980. *Forest Science* 33, 932–945.

Newman, D. 2002. Forestry's Golden Rule and the Development of the Optimal Forest Rotation Literature. *Journal of Forest Economics* 8, 5–27.

Newman, D., C. Gilbert, and W. Hyde. 1985. The Optimal Forest Rotation with Evolving Prices. *Land Economics* 61, 347–353.

Nguyen, D. 1979. Environmental Services and the Optimal Rotation Problem in Forest Management. *Journal of Environmental Management* 8, 127–236.

Nordström, J. 1975. A Stochastic Model for the Growth Period Decision in Forestry. *Swedish Journal of Economics* 77, 329–337.

Ohlin, B. 1921. Concerning the Question of the Rotation Period in Forestry. Reprinted in *Journal of Forest Economics* 1, 89–114.

Ollikainen, M. 1990. Forest Taxation and the Timing of Private Nonindustrial Forest Harvests under Interest Rate Uncertainty. *Canadian Journal of Forest Research* 20, 1823–1829.

Ollikainen, M. 1991. The Effect of Nontimber Taxes on the Harvest Timing—the Case of Private Nonindustrial Forest Owners—A Note. *Forest Science* 37, 356–363.

Ollikainen, M. 1993. A Mean-Variance Approach to Short-term Timber Selling and Forest Taxation under Multiple Sources of Uncertainty. *Canadian Journal of Forest Research* 23, 573–581.

Ollikainen, M. 1996a. Essays on Timber Supply and Forest Taxation. Research Report No. 33, Government Institute for Economic Research, Helsinki, Finland.

Ollikainen, M. 1996b. The Analytics of Timber Supply and Forest Taxes Under Endogenous Credit Rationing. *Journal of Forest Economics* 2, 93–130.

Ollikainen, M. 1998. Sustainable Forestry: Timber Bequests, Future Generations and Optimal Tax Policy. *Environmental and Resource Economics* 12, 255–273.

Ovaskainen, V. 1992. Forest Taxation, Timber Supply and Economic Efficiency. *Acta Forestalia Fennica* 233, 88 pp.

Palmer, C. E. 2000. The Extent and Causes of Illegal Logging: An Analysis of a Major Cause of Tropical Deforestation in Indonesia, CSERGE Working Paper, University College London, ISSN 0967-8875.

Palmer, C. E. 2005. The Nature of Corruption in Forest Management. *World Economics* 6(2), 1–10.

Parks, P. J., and I. W. Hardie. 1995. Least-Cost Forest Carbon Reserves: Cost-effective Subsidies to Convert Marginal Agricultural Land to Forests. *Land Economics* 71(1), 122–136.

Parks, P. J., E. B. Barbier, and J. C. Burgess. 1998. The Economics of Forest Land Use in Temperate and Tropical Areas. *Environmental and Resource Economics* 11 (3–4), 473–487.

Parks, P. J., D. Hall, B. Kristrom, O. Masera, R. Moulton, A. Plantinga, J. Swishes, and J. Winjum. 1997. An Economic Approach to Planting Trees for Carbon Storage. In *Economics of Carbon Sequestration in Forestry* Special issue: *Critical Reviews in Environmental Science and Technology*, R. A. Sedjo, R. Sampson, and J. Wisniewski eds., pp. 9–22. Boca Raton, FL: CRC Press, vol. 27.

Pattanayak, S., B. Murray, and R. Abt. 2002. How Joint is Joint Forest Production? An Econometric Analysis of Timber Supply Conditional on Endogenous Amenity Value. *Forest Science* 48(3), 479–491.

Pearse, P. 1967. The Optimum Forest Rotation. *Forestry Chronicle* 43, 178–195.

Peltola, J., and K. C. Knapp. 2001. Recursive Preferences in Forest Management. *Forest Science* 47, 455–465.

Perrings, C. 2005. Mitigation and adaptation strategies for the control of biological invasions. *Ecological Economics* 52, 315–325.

Pigou, A. 1932. *The Economics of Welfare*. London: Macmillan.

Pimentel, D., Z. Rodolfo, and M. Doug. 2005. Update on the Environmental and Economic Costs Associated with Alien-Invasive Species in the United States. *Ecological Economics* 52, 273–288.

Plantinga, A. J. 1998. The Optimal Timber Rotation: An Option Value Approach. *Forest Science* 44, 192–202.

Plantinga, A., and R. Birdsey. 1994. Optimal Forest Stand Management When Benefits Are Derived from Carbon. *Natural Resource Modeling* 8, 373–387.

Plourde, C. 1970. A Simple Model of Replenishable Natural Resource Exploitation. *American Economic Review* 60, 518–521.

Polasky, S., and A. Solow. 1995. On the value of a collection of species. *Journal of Environmental Economics and Management* 29(3): 298–303.

Polasky, S., J. Camm, and B. Garber-Yonts. 2001. Selecting Biological Reserves Effectively: An Application to Terrestrial Vertebrate Conservation in Oregon. *Land Economics* 77, 68–78.

Polinsky, A., and M. Shavel. 1994. A Note on Optimal Cleanup and Liability after Environmentally Harmful Discharges. *Law and Economics* 16, 17–24.

Potts, M., and J. Vincent. 2008. Harvest and Extinction in Multi-species Ecosystems. *Ecological Economics* 65, 336–347.

Pressler, M. 1850. For the Comprehension of Net Revenue Silviculture and the Management Objectives Derived Thereof. Reprinted in *Journal of Forest Economics* 1, 89–114.

Prestemon, J. P. 2003. Evaluation of U.S. Southern Pine Stumpage Market Informational Efficiency. *Canadian Journal of Forest Research* 33, 561–572.

Prestemon, J., and T. Holmes. 2000. Timber Price Dynamics Following a Natural Catastrophe. *American Journal of Agricultural Economics* 82, 145–160.

Prestemon, J., S. Zhu, J. Turner, J. Buongiorno, and R. Rhuong. 2006. Forest Product Trade Impacts of an Invasive Species: Modeling Structure and Intervention Trade-Offs. *Agricultural and Resource Economics Review* 35, 128–143.

Pukkala, T., and J. Miina. 1997. A Method for Stochastic Multi-objective Optimization of Stand Management. *Forest Ecology and Management* 98, 189–203.

Pukkala, T., S. Kellomäki, and E. Mustonen. 1988. Prediction of the Amenity of a Tree Stand. *Scandinavian Journal of Forest Research* 3, 533–544.

Pukkala, T., and T. Kolström. 1988. Simulation of the Development of Norway Spruce Stands Using a Transition Matrix. *Forest Ecology and Management* 25, 255–267.

Ramsey, F. 1927. A Contribution to the Theory of Taxation. *Economic Journal* 37, 47–61.

Ranius, T., H. Ekvall, M. Jonsson, and G. Bostedt. 2005. Cost-efficiency of Measures to Increase the Amount of Coarse Woody Debris in Managed Norway Spruce Forests. *Forest Ecology and Management* 206, 119–133.

Reed, W. 1984. The Effects of the Risk of Fire on the Optimal Rotation of a Forest. *Journal of Environmental Economics and Management* 11, 180–190.

Reed, W. J. 1986. Optimal Harvesting Models in Forest Management—A Survey. *Natural Resource Modeling* 1, 55–79.

Reed, W. J. 1987. Protecting a Forest Against Fire: Optimal Protection Patterns and Harvest Policies. *Natural Resource Modeling* 2, 23–53.

Reed, W. J. 1993. The Decision to Conserve or Harvest Old-Growth Forest. *Ecological Economics* 8, 45–69.

Reed, W., and J. Apaloo. 1991. Evaluating the Effects of Risk on the Economics of Juvenile Spacing and Commercial Thinning. *Canadian Journal of Forest Research* 21, 1390–1400.

Reed, W. J., and H. R. Clarke. 1990. Harvest Decisions and Asset Valuation for Biological Resources Exhibiting Size-Dependent Stochastic Growth. *International Economic Review* 31, 147–169.

Reed, W. and D. Errico. 1986. Optimal Harvest Scheduling at the Forest Level in the Presence of Fire Risk. *Canadian Journal of Forest Research* 16, 266–278.

Reed, W., and D. Errico. 1987. Techiques for Assessing the Effects of Pest Hazards on Long-run Timber Supply. *Canadian Journal of Forest Research* 17, 1455–1465.

Reed, W. J., and J. J. Ye. 1994a. The Role of Stochastic Monotonicity in the Decision to Conserve or Harvest Old-Growth Forest. *Natural Resource Modeling* 8, 47–79.

Reed, W. J., and J. J. Ye. 1994b. Cost-Benefit Analysis Applied to Wilderness Preservation—Option Value Uncertainty and Ditonicity. *Natural Resource Modeling* 8, 335–361.

Reeves, L. H., and R. G. Haight. 2000. Timber Harvest Scheduling with Price Uncertainty Using Markowitz Portfolio Optimization. *Annals of Operations Research* 95, 229–250.

Richards, R. 1999. Internalizing the Externalities of Tropical Forestry: Review of Financing and Incentive Mechanisms. Overseas Development Institute, London.

Ross, S. M. 1970. *Applied Probability Models with Optimization Applications.* New York: Dover.

Routledge, R. D. 1980. The Effect of Potential Catastrophic Mortality and Other Unpredictable Events on Optimal Forest Rotation Policy. *Forest Science* 26, 389–399.

Salanie, B. 2003. *The Economics of Taxation.* Cambridge, MA: MIT Press.

Salo, S., and O. Tahvonen. 2000. On Faustmann Rotation and Normal Forest Convergence, Working Paper 243, Helsinki School of Economics and Business Administration.

Salo, S., and O. Tahvonen. 2002a. On Equilibrium Cycles and Normal Forests in Optimal Harvesting of Tree Age Classes. *Journal of Environmental Economics and Management* 44(1), 1–22.

Salo, S., and O. Tahvonen. 2002b. On the Optimality of a Normal Forest with Multiple Land Classes. *Forest Science* 48(3), 530–542.

Salo, S., and O. Tahvonen. 2002c. Optimal Evolution of Forest Age-Classes and Land Allocation between Forestry and Agriculture, Working Paper 286, Helsinki School of Economics and Business Administration.

Salo, S., and O. Tahvonen. 2003. On the Economics of Forest Age Class. *Journal of Economic Dynamics and Control* 27, 1411–1435.

Salo, S., and O. Tahvonen. 2004. Renewable Resources with Endogenous Age Classes and Allocation of Land. *American Journal of Agricultural Economics* 86(2), 513–530.

Samuelson, P. A. 1958. An Exact Consumption-Loan Model of Interest with or without the Social Contrivance of Money. *Journal of Political Economy* 66, 467–482.

Samuelson, P. 1974. Complementarity. An Essay on the 40th Anniversary of the Hicks-Allen Revolution in Demand Theory. *Journal of Economic Literature* 12, 1255–1289.

Samuelson, P. 1976. Economics of Forestry in an Evolving Society. *Economic Inquiry* 14, 466–492.

Sandmo, A. 1975. Optimal Taxation in the Presence of Externalities. *Swedish Journal of Economics* 77, 86–98.

Sanyal, A., I. N. Gang, and O. Goswami. 2000. Corruption, Tax Evasion and the Laffer Curve. *Public Choice* 105, 61–78.

Saphores, J.-D., L. Khalaf, and D. Pelletier. 2002. On Jumps and ARCH Effects in Natural Resource Prices: An Application to Pacific Northwest Stumpage Prices. *American Journal of Agricultural Economics* 84(2), 387–400.

Sargent, T. 1987. *Dynamic Macroeconomic Theory*. Cambridge, MA: Harvard University Press.

Schleifer, A., and R. W. Vishny. 1993. Corruption. *Quarterly Journal of Economics* 108, 599–617.

Scorgie, M., and J. Kennedy. 2000. Who Discovered the Faustmann Condition? *History of Political Economy* 28, 77–80.

Sedjo, R. A., and K. S. Lyon. 1990. *The Long-Term Adequacy of World Timber Supply*. Washington, DC: Resources for the Future Press.

Sedjo, R., J. Wisniewski, A. Sample, and J. Kinsman. 1995. The Economics of Managing Carbon via Forestry: An Assessment of Existing Studies. *Environmental and Resource Economics* 6, 139–165.

Sedjo, R., R. Simpson, and J. Wisniewski (eds.). 1997. *Economics of Carbon Sequestration in Forestry*. New York: Lewis.

Selden, L. 1978. A New Representation of Preferences over "Certain × Uncertain" Consumption Pairs: The Ordinal "Certainty Equivalent" Hypothesis, *Econometrica* 46, 1045–1060.

Shively, G. E. 2002. Agricultural Change, Rural Labor Markets, and Forest Clearing: An Illustrative Case from the Philippines. *Land Economics* 77, 268–284.

Shively, G. E. and S. Pagiola. 2004. Agricultural Intensification, Local Labor Markets, and Deforestation in the Philippines. *Environmental and Development Economics* 9, 241–266.

Siitonen, J. 2001. Forest Management, Coarse Woody Debris and Saproxylic Organisms: Fennoscandian Boreal Forests as an Example. *Ecological Bulletin* 49, 11–41.

Silberberg, E. 1990. *The Structure of Economics. A Mathematical Analysis*. 2nd ed. New York: McGraw-Hill.

Silberberg, E., and W. Suen. 2001. *The Structure of Economics. A Mathematical Analysis*, 3rd ed. New York: McGraw-Hill.

Sills, E., and K. Abt (eds.). 2003. *Forests in a Market Economy*. London: Kluwer.

Sills, E., S. Lele, T. Holmes, and M. Haefele. 2003. Nontimber Forest Products in a Rural Household Economy. In *Forests in a Market Economy*, E. Sills and K. Abt, eds., pp. 259–282. London: Kluwer. Academic Publishers.

Similä, M., J. Kouki, and P. Martikainen. 2003. Saproxylic Beetles in Managed and Seminatural Scots Pine Forests: Quality of Dead Wood Matters. *Forest Ecology and Management* 174, 365–381.

Simon, C., and L. Blume. 1994. *Mathematics for Economists*. New York: Norton.

Slade, M. E. 1982. Trends in Natural Resource Commodity Prices: An Analysis of the Time Domain. *Journal of Environmental Economics and Management* 9, 132–137.

Smith, J., K. Obidzinski, S. Subarudi, and I. Suramenggala. 2003. Illegal Logging, Collusive Corruption and Fragmented Governments in Kalimantan, Indonesia. *International Forestry Review* 5(3), 293–302.

Smith, W. 2002. The Global Problem of Illegal Logging. *ITTO Tropical Forest Update* 12, 3–5.

Snyder, D., and R. Bhattacharyya. 1990. A More General Dynamic Economic Model of the Optimal Rotation of Multiple-Use Forests. *Journal of Environmental Economics and Management* 18, 168–175.

Sodal, S. 2002. The Stochastic Rotation Problem: A Comment. *Journal of Economic Dynamics and Control* 26, 509–515.

Sohngen, B., and R. Mendelsohn. 2003. An Optimal Control Model of Forest Carbon Sequestration. *American Journal of Agricultural Economics* 85(2), 448–457.

Sprangenberg, J. 2007. Biodiversity Pressure and the Driving Forces behind. *Ecological Economics* 61, 146–158.

Spring, D., J. Kennedy, and R. M. Nally. 2005a. Existence Value and Optimal Timber-Wildlife Management in a Flammable Multistand Forest. *Ecological Economics* 55, 365–379.

Spring, D., J. Kennedy, and R. M. Nally. 2005b. Optimal Management of a Flammable Forest Providing Timber and Carbon Sequestration Benefits. *Australian Journal of Agricultural and Resource Economics* 49(3), 303–320.

Stainback, A., and J. Alavalapati. 2004. Modeling Catastrophic Risk in Economic Analysis of Forest Carbon Sequestration. *Natural Resource Modeling* 17, 299–317.

Stavins, R. N. 1999. The Costs of Carbon Sequestration: A Revealed-Preference Approach. *American Economic Review* 89(2), 994–1009.

Stavins, R., and A. Jaffe. 1990. Unintended Impacts of Public Investments on Private Decisions: The Depletion of Forested Wetlands. *American Economic Review* 80(3), 339–352.

Steinkamp, E. A., and D. R. Betters. 1991. Optimal Control Theory Applied to Joint Production of Timber and Forage. *Natural Resource Modeling* 5, 147–160.

Stockland, J. 1997. Representativeness and Efficiency of Bird and Insect Conservation in Norwegian Boreal Forest Reserves. *Conservation Biology* 11, 101–111.

Stokey, N. L., and R. E. Lucas. 1989. *Recursive Methods in Economic Dynamics*. Cambridge, MA: Harvard University Press.

Strang, W. 1983. On the Optimal Forest Harvesting Decision. *Economic Inquiry* 21, 576–583.

Strange, N., J. Brodie, H. Meilby, and F. Helles. 1999. Optimal Control of Multiple-Use Products: The Case of Timber, Forage and Water Protection. *Natural Resource Modeling* 12, 335–354.

Sullivan, J., G. S. Amacher, and S. Chapman. 2005. Forest Banking and Forest Landowners: Forgoing Management Rights for Guaranteed Financial Returns. *Forest Policy and Economics* 7, 381–392.

Svensson, J. 2005. Eight Questions about Corruption. *Journal of Economic Perspectives* 19, 19–42.

Swallow, S., and D. Wear. 1993. Spatial Interactions in Multiple-Use Forestry and Substitution and Wealth Effects for the Single Stand. *Journal of Environmental Economics and Management* 25, 103–120.

Swallow, S., P. Parks, and D. Wear. 1990. Policy-Relevant Nonconvexities in the Production of Multiple Forest Benefits. *Journal of Environmental Economics and Management* 19, 264–280.

Swallow, S., P. Talukdar, and D. Wear. 1997. Spatial and Temporal Specialization in Forest Ecosystem Management under Sole Ownership. *American Journal of Agricultural Economics* 79, 311–326.

Tacconi, L., M. Boscolo, and D. Brack. 2003. National and International Policies to Control Illegal Forest Activities. Center for International Forest Research, Jakarta, July.

Tahvonen, O. 1998. Bequests Credit Rationing and in situ Values in the Faustmann-Pressler-Ohlin Forestry Model. *Scandinavian Journal of Economics* 100, 781–800.

Tahvonen, O. 2004a. Timber Production versus Old-Growth Preservation with Endogenous Prices and Forest Age-Classes. *Canadian Journal of forest Research* 34(6), 1296–1310.

Tahvonen, O. 2004b. Optimal Harvesting of Forest Age Classes: A Survey of Some Recent Results. *Mathematical Population Studies* 11, 205–232.

Tahvonen, O. 2007. Optimal Choice Between Even-Aged and Uneven-Aged Forest Management Systems. Working Papers of the Finnish Forest Research Institute no. 60, p. 35.

Tahvonen, O., and S. Salo. 1999. Optimal Forest Rotation with in situ Preferences. *Journal of Environmental Economics and Management* 37, 106–128.

Tahvonen, O., S. Salo, and J. Kuuluvainen. 2001. Optimal Forest Rotation and Land Values Under a Borrowing Constraint. *Journal of Economic Dynamics and Control* 25, 1595–1627.

Teeter, L. D., and J. P. Caulfield. 1991. Stand Density Management Strategies under Risk: Effects of Stochastic Prices. *Canadian Journal of Forest Research* 21(9), 1373–1379.

Thomson, T. A. 1992. Optimal Forest Rotation When Stumpage Prices Follow a Diffusion Process. *Land Economics* 68, 329–342.

Thorsen, B. J., and F. Helles. 1998. Optimal Stand Management with Endogenous Risk of Sudden Destruction. *Forest Ecology and Management*, 108, 287–299.

Thurston, H., and H. S. Burness. 2006. Promoting Sustainable Logging in Brazil's National Forests: Tax Revenue for an Indemnity Fund. *Forest Policy and Economics* 9, 50–62.

Toppinen, A. 1998. Incorporating Cointegration Relations in a Short-run Model of the Finnish Sawlog Market. *Canadian Journal of Forest Research* 28, 291–298.

Tresch, R. 2002. *Public Finance, A Normative Theory*. London: Academic Press, 950 pp.

Usher, M. 1966. A Matrix Approach to the Management of Renewable Resources with Special Reference to Selection Forests—Two Extensions. *Journal of Applied Ecology* 6, 347–346.

Uusivuori, J. 2000. Neutrality of Forestry Income Taxation and Inheritable Tax Exemptions for Timber Capital. *Forest Science* 46, 219–228.

Uusivuori, J. 2002. Nonconstant Risk Attitudes and Timber Harvesting. *Forest Science* 48, 459–470.

Uusivuori, J., and J. Kuuluvainen. 2005. The Harvesting Decision When a Standing Forest with Multiple Age Classes has Value. *American Journal of Agricultural Economics* 87, 61–76.

Uusivuori, J. and J. Kuuluvainen. 2008. Forest Taxation in multiple stand forestry with amenity preferences. *Canadian Journal of Forest Research* 38, 806–820.

Uusivuori, J., and J. Laturi. 2007. Carbon Rentals and Silvicultural Subsidies for Private Forests as Climate Policy Instruments. *Canadian Journal of Forest Research* 37, 2541–2551.

Valkonen, S., J. Ruuska, and J. Siipilehto. 2002. Effect of Retained Trees on the Development of Young Scots Pine Stands in Southern Finland. *Forest Ecology and Management* 166(1–3), 227–243.

Vane-Wright, R., D. Humphries, and P. Williams. 1991. What to Protect? Systematics and the Agony of Choice. *Biological Conservation* 55, 235–234.

Vanha-Majamaa, I., and J. Jalonen. 2001. Green Tree Retention in Fennoscandian Forestry. *Scandinavian Journal of Forest Research*. Supplement 3, 79–90.

van Kooten, G. Rovan Kooten, and G. Brown. 1992. Modeling the Effect of Uncertainty on Timber Harvest: A Suggested Approach and Empirical Example. *Journal of Agricultural and Resource Economics* 17, 162–172.

van Kooten, G. C. Binkley, and G. Delcourt. 1995. Effect of Carbon Taxes and Subsidies on Optimal Forest Rotation Age and Supply of Carbon Services. *American Journal of Agricultural Economics* 77, 365–375.

van Kooten, G. C., R. A. Sedjo, and E. H. Bulte. 1999. Tropical Deforestation: Issues and Policies. In *The International Year Book of Environmental and Resource Economics 1999/2000*, H. Folmer, and T. Tietenberg, eds., Cheltenham, UK: Edward Elgar, pp. 199–248.

Varian, H. 1992. *Microeconomic Analysis*. 3rd ed. New York: Norton.

Vasicek, O. 1977. An Equilibrium Characterization of the Term Structure. *Journal of Financial Economics* 5, 177–188.

Viitala, E.-J. 2002. The Optimal Rotation Age of a Forest: Approaches and Their Economic Foundations (in Finnish). Metsäntutkimuslaitoksen tiedonantoja 848.

Viitala, E. J. 2006. An Early Contribution of Martin Faustmann to Natural Resource Economics. *Journal of Forest Economics* 12, 131–144.

Vincent, J. 1990. Rent Capture and the Feasibility of Tropical Forest Management. *Land Economics* 66, 212–223.

Vincent, J., and C. Binkley. 1993. Efficient Multiple-Use Forestry May Require Land-Use Specialization. *Land Economics* 69, 370–376.

Walker, R., and T. Smith 1993. Tropical Deforestation and Forest Management under the System of Concession Logging: A Decision-Theoretic Analysis. *Journal of Regional Science* 33, 387–419.

Wan, H. 1994. Revisiting the Mitra-Wan Tree Farm. *International Economic Review* 35(1), 193–198.

Wear, D., and P. Parks. 1994. The Economics of Timber Supply: An Analytical Synthesis of Modeling Approaches. *Natural Resource Modeling* 8, 199–223.

Weber, C. 2000. Two Further Empirical Implications of Auspitz-Lieben-Edgeworth-Pareto Complementarity. *Economic Letters* 67, 289–295.

Weil, P. 1987. Love Thy Children: Reflections in the Barro Debt Neutrality Theorem. *Journal of Monetary Economics* 19, 377–391.

Weil, P. 1990. Nonexpected Utility in Macroeconomics. *Quarterly Journal of Economics* 105, 209–242.

Weil, P. 1993. Precautionary Savings and the Permanent Income Hypothesis. *Review of Economic Studies* 60, 367–383.

Weitzman, M. 1992. On Diversity. *Quarterly Journal of Economics* 107, 363–405.

Weitzman, M. 1993. What to Preserve? An Application of Diversity Theory to Crane Conservation. *Quarterly Journal of Economics* 108, 157–183.

Weitzman, M. 2003. *Income, Wealth, and the Maximum Principle.* Cambridge, MA: Harvard University Press.

Wibowo, D., and N. Byron. 1999. Deforestation Mechanisms: A Survey. *International Journal of Social Economics* 26, 455–474.

Wiens, J. A. 1997. The Emerging Role of Patchiness in Conservation Biology. In *The Ecological Basis of Conservation: Heterogeneity, Ecosystems, and Biodiversity*, S. T. A. Pickett, R. S. Ostfeld, M. Shachak, and G. E. Likens, eds., pp. 93–107. New York: Chapman and Hall.

Wikström, P., and L. Eriksson. 2000. Solving the Stand Management Problem Under Biodiversity-Related Considerations. *Forest Ecology and Management* 126, 361–376.

Willassen, Y. 1998. The Stochastic Rotation Problem: A Generalization of Faustmann's Formula to Stochastic Forest Growth. *Journal of Economic Dynamics and Control* 22, 573–596.

Williams, B. 1989. Review of Dynamic Optimization Methods in Renewable Resource Management. *Natural Resource Modeling* 3(2), 137–216.

Williams. J., and J. Nautiyal. 1992. Integrating Budworm and Timber Management: Comparing Preventive versus Remedial Strategies. *Natural Resource Modeling* 6, 409–433.

Wikström, P., and L. Eriksson. 2000. Solving the Stand Management Problem under Biodiversity-related Consideration. *Forest Ecology and Management* 126, 261–376.

Winkler, N. 1997. Environmentally Sound Forest Harvesting: Testing The Applicability of the FAO Model Code in the Amazon of Brazil. Case Study 8, Rome: Food and Agriculture Organization.

Wirl, F. 1999. De- and Reforestation: Stability, Instability, and Limit Cycles. *Environmental and Resource Economics* 14, 463–479.

Yin, R., and D. H. Newman. 1995. A Note of the Tree-Cutting Problem in a Stochastic Environment. *Journal of Forest Economics* 1(2), 181–190.

Yin, R., and D. H. Newman. 1996a. The Effect of Catastrophic Risk on Forest Investment Decisions. *Journal of Environmental Economics and Management* 31, 186–197.

Yin, R., and D. H. Newman. 1996b. Are Markets for Stumpage Informationally Efficient? *Canadian Journal of Forest Research* 26, 1032–1039.

Yin, R., and D. H. Newman. 1997. When to Cut a Stand of Trees? *Natural Resource Modeling* 10(3), 251–261.

Yoder, J. 2004. Playing with Fire: Endogenous Risk in Resource Management. *American Journal of Agricultural Economics* 86, 933–948.

Yoshimoto, A., and I. Shoji. 1998. Searching for an Optimal Rotation Age for Forest Stand Management under Stochastic Log Prices. *European Journal of Operational Research*, 105, 100–112.

Yoshimoto, A., and I. Shoji. 2002. Comparative Analysis of Stochastic Models for Financial Uncertainty in Forest Management. *Forest Science* 48, 755–766.

Zhang, D. 2001. Faustmann in an Uncertain Policy Environment. *Forest Policy and Economics* 2, 203–210.

Zhang, D. 2004. Endangered Species Act and Timber Harvesting: The Case of Red-Cockaded Woodpeckers. *Economic Inquiry* 42, 150–165.

Zinkhan, F., W. Sizemore, G. Mason, and T. Ebner. 1992. *Timberland Investments*. Portland OR.: Timber Press.

Author Index

Subject Index